Symplectic Elasticity

Symplectic Elasticity

Weian Yao • Wanxie Zhong
Dalian University of Technology, P R China
Chee Wah Lim
City University of Hong Kong, Hong Kong SAR

World Scientific
NEW JERSEY • LONDON • SINGAPORE • BEIJING • SHANGHAI • HONG KONG • TAIPEI • CHENNAI

Published by

World Scientific Publishing Co. Pte. Ltd.

5 Toh Tuck Link, Singapore 596224

USA office: 27 Warren Street, Suite 401-402, Hackensack, NJ 07601

UK office: 57 Shelton Street, Covent Garden, London WC2H 9HE

British Library Cataloguing-in-Publication Data
A catalogue record for this book is available from the British Library.

SYMPLECTIC ELASTICITY

ISBN-13 978-981-277-870-3
ISBN-10 981-277-870-5

Typeset by Stallion Press
Email: enquiries@stallionpress.com

Printed in Singapore.

Contents

Preface

The use of symplectic space has been exploited in a number of fields in physics for many years particularly in quantum mechanics including the famous Yang-Mills field, relativity, gravitation, astrophysics, classical mechanics, etc. "Symplectic" is a Greek word which was first introduced in 1939 by Hermann Weyl in *The Classical Groups, Their Invariants and Representations*. In elasticity, symplectic approach was first applied in the early 1990s by one of the authors, Professor W. Zhong, to describe a new approach for solving basic problems in solid mechanics which have long been bottlenecks in the history of elasticity. It is based on Hamiltonian principle with Legendre's transformation whereby analytical solutions could be obtained by expansion of eigenfunctions. The methodology is rational and systematic with clear step-by-step derivation procedure. The advantage of symplectic approach with respective to classical approach by semi-inverse method is at least three-fold. First, the symplectic approach alters the classical practice and concept of solution methodology and hence allows the many basic problems previously unsolvable or too complicated to be solved be resolved accordingly. For instance, the conventional approach in plate and shell theories by Timoshenko has been based on the semi-inverse method with trial 1D or 2D displacement functions, such as Navier's method and the Levy's method for plates. The trial functions, however, do not always exist except in some very special cases of boundary conditions such as plates with two opposite sides simply supported. Using the symplectic approach, trial functions are no longer required. Second, it consolidates the many seemingly scattered and unrelated solutions of rigid body movement and elastic deformation by mapping with a series of zero and nonzero eigenvalues. Last but not least, the Saint-Venant problems for plain elasticity and elastic cylinders can be described in a new system of equations and solved. The difficulty of satisfying end boundary conditions in conventional problems which could only be covered using the Saint-Vanent principle can also be solved.

In this book, the authors' main objective is to introduce the major concepts and application of symplectic elasticity through discussion in some classical topics in elasticity. The rational approach of symplectic methodology has been very clearly elaborated in the selected examples. It should be emphasized that the potential of symplectic methodology for deriving analytical solutions is far more than what have been presented in this book. It has been applied by the authors and others not only in more complicated elasticity problems in thin and think plates with mixed boundary, shells, piezoelectric structures, but also in anisotropic structures, vibration and nonlinear dynamics, control theory, electromagnetism, waveguide, nanomechanics, quantum mechanics etc. The major parts of the topics will be disseminated through various scientific and technical conferences and journals and the contents will be included in a future manuscript when the theory becomes mature.

This book targets readers in engineering mechanics, nanomechanics, applied mathematics, engineering structures at postgraduate and research levels. Teachers at higher institutions may also find this book valuable for either teaching or reference. The contents are also beneficial to relevant parties in other disciplines such as physics, control, electronics, etc.

This book on symplectic elasticity was first published in Chinese entitled《辛弹性力学》by Higher Education Press in 2002. As it is a new approach with great research potential for further development and breakthrough is unavailable to many researchers in the field who have no access to Chinese language and literature, effort was initiated to translate the original manuscript into English to arouse the interest of many researchers and to promote better, wider and faster research progress using symplectic approach in elasticity.

In this English edition, a number of printing errors in the Chinese edition have been duly corrected. The authors are grateful to Professor L.H. He, Dr Ziran Li, Dr Chao-Feng Lü, Miss Ling Qiao and Professor Baisheng Wu for assisting in the translation and/or proofreading, in one way and another. Appreciation should also be extended to Higher Education Press, Beijing, for granting permission to publish this translated English edition.

Weian Yao
Wanxie Zhong
Chee Wah Lim
December 2008

Preface to the Chinese Edition

Elasticity has been one of the most complex fields in various branches of mathematical physics involving partial differential equations. The solution procedure of such problems has been a bottleneck in the development of elasticity. Considering "Theory of Elasticity" of S.P. Timoshenko as an example, the solutions of various elasticity problems using the semi-inverse method constitute a large portion of the text. The application of semi-inverse method is due to the complexity of the system of equations. The conventional method of solution is always confined to the solution of a higher-order partial differential equation with a single variable by eliminating the various unknown functions. From the viewpoint of mathematical systems, the solution of single variable systems belongs to the Lagrangian approach which inevitably results in a higher-order partial differential equation. Hence, the effective methods in mathematical physics such as variable separation and expansion of eigenfunctions become inapplicable. Consequently, the semi-inverse method has been unable to achieve major breakthrough for a long period.

In this book, the symplectic space formed by the original and dual variables is introduced into elasticity in accordance with analogy theory between structural mechanics and control optimization. As a result, it is possible to apply the direct analytical methods of variable separation and expansion of symplectic eigenfunctions. It then forms the symplectic analytical systems in elasticity and it is the breaking and unified approach that is emphasized throughout the book. The solution of symplectic analytical systems is based on a rational, systematic approach with clear step-be-step derivation procedure. It alters the classical practice in elasticity of using the semi-inverse method by presenting a new systematic and rational method of solution. The many previous problems unsolvable or too complicated to be solved using the semi-inverse approach can hence be resolved accordingly. For instance, we have presented solutions for plate bending problems with various boundary conditions, laminated composite plates and anisotropic problems. In addition, the Saint-Venant problems for plain elasticity and elastic cylinders can be described in a new system of equations and solved. The difficulty of satisfying end boundary conditions in conventional problems which could only be covered using the Saint-Vanent principle can also be solved.

Due to the difference in basic principle for elasticity problems using the classical semi-inverse method and the symplectic analytical systems, the rational approach of the latter can be completely and directly generalized to more complex problems. More problems can thus be solved analytically and larger solution domain can thus be obtained. The solution procedure for various symplectic analytical systems is identical and only the algebraic derivation becomes more complicated for more complex problems. Such algebraic complexity can be overcome by using symbolic mathematical softwares.

The analytical concept for symplectic systems and conventional approach for partial differential equation are just opposite. The conventional concept tries to eliminate unknown variables as many as possible thus increasing the order of partial differential equations. Since higher-order differential equations are not conducive to numerical solution methods such as finite element method, such elimination will cause problems in numerical analysis. On the contrary, although there are more variables in symplectic analytical systems, the order of differential equations is lower. There are numerical advantages in dealing with lower-order differential equations and an increase in the number of variables will not have significant impact on the system. In other words, the association of symplectic analytical systems with numerical methods will not only greatly highlight the excellence of the former but also enhances the usefulness of computers for solving engineering problems.

The book aims at introducing to the readers the methodology of symplectic analytical systems in elasticity in a systematic manner. From basic equations of elasticity and classical variational principle, it first describes in detail the procedure of constructing Hamiltonian mixed energy variational principle and Hamiltonian dual system of equations and subsequently the symplectic analytical systems through discussion of various basic problems in elasticity. The eigenvalue problem in the transverse direction, i.e. the symplectic eigenvalue problem, can then be derived by applying the method of separation of variables. Hence, the solution can be obtained by expanding the eigenvectors. Many solutions with specific interpretation in physics can be obtained via the eigenfunctions of specific eigenvalues and their corresponding Jordan form eigenfunctions. In general, a similar solution methodology is applied throughout this book so that the readers are able to master the solution procedure of symplectic analytical systems. Furthermore, the differences with respect to the classical semi-inverse can also be clearly observed.

Analytical method is emphasized in this book because it is a major first step in deriving new solutions for a system. The problems discussed in this book are all fundamental problems in elasticity such as Timoshenko beam, plane elasticity, laminated composite plate, plate bending, etc. It is emphasized here that the method of symplectic analytical system is absolutely applicable to three dimensional problems such as cylindrical bodies. Such complicated subjects are, however, not described in this book.

Introducing a new concept without sophisticated mathematics is a unique feature in this book. The related contents of symplectic mathematics as required in substitution of systems are discussed in detail in Chapter 1. The mathematical preliminaries for calculus and matrix algebra are at graduate level and omitting this part will not hinder the understanding and master of symplectic analytical system in elasticity. The contents in this book have been repeatedly introduced in courses for graduates and higher level undergraduates. It not only helps the students familiarize a new solution methodology but also widens very much their research vision. The effect has been remarkable.

The research outcome in this book is completed via fundings from Key Project of National Natural Science Foundation of China entitled "Hamiltonian Systems in Engineering Mechanics" and National "211" Engineering Construction Project entitled "Computational Engineering Science". Professor Haojiang Ding of Zhejiang University in China is gratefully acknowledged for painstakingly and carefully reading through the manuscript and providing many valuable suggestions. The unrelenting assistance of Dr Qiang Gao for preparing all figures in this book and Dr Yongfeng Sui for proofreading of manuscript are also gratefully acknowledged.

Although every care has been taken to ensure correctness of contents, incompleteness is bound to exist in this book. The authors thank the readers in advance for giving critical comments and pointing out mistakes.

Weian Yao
Wanxie Zhong
October 2001

Foreword to the Chinese Edition

Theoretical mechanics has been for a long time the leading discipline in the history of science. The development for centuries has resulted in multiple branches and the achievement has been remarkable. Mechanics as a fundamental subject has helped in the development of many engineering disciplines such as aeronautical and astronautical engineering, mechanical engineering, civil engineering, chemical engineering, natural resources engineering, material engineering, etc. Mean while, many theories and methodologies have been developed in applied mechanics as compelled by the requirement in various engineering applications. From the viewpoint of applied mathematics, a problem is clearly described once the basic differential equations are constructed. The remaining task is to look for a solution. Nevertheless, in many circumstances the solutions are extremely difficult although the basic equations are already constructed.

Elasticity has been one of the most complex fields in various branches of mathematical physics involving partial differential equations. The fundamental systems of equations in elasticity were established as early as the beginning of the nineteenth century. However, for over a century of development, the solutions were far from complete. Solution methodology has been a bottleneck in the development of elasticity. The difficulty for strict solutions in elasticity, in turn, drove the development of some applied branches such as structural mechanics, thin-walled structures, plate and shell theories, as well as structural dynamics, stability, soil mechanics, fluid mechanics, etc. These branches form the various systems in applied mechanics. Although these applied theories simply the governing equations, analytical solutions are still difficult to a large extent. Research collaboration between mathematicians and mechanicians not only enriched mathematical physics but also developed applied mechanics. Some representative works in this period are Methods of Mathematical Physics by R. Courant and D. Hilbert, and a set of texts by S.P. Timoshenko including Theory of Elasticity, Theory of Elastic Stability, Theory of Plates and Shells, Vibration Problems in Engineering, Mechanics of Materials, etc. This set of analytical solutions became the classical solution systems in the field. The achievement in the period was marvelous and it influenced the subsequent research development in the field.

With the advent of computing machinery and high level programming languages in the second half of the twentieth century, the surface of finite element method in applied mechanics changed the situation rapidly. Based on the theory of applied mechanics and powerful computational ability, versatile numerical methods in finite element method were developed for solving structural mechanics, solid mechanics, etc., described by linear equations. Large scale finite element systems were geared to solve sets of linear algebraic equations with tens of thousands of unknown variables. It became a powerful analytical tool for engineers and the status of computational mechanics was established. The successful application of finite element method in structural analysis swiftly extended to various aspects in computational mechanics, engineering and science with significant achievement.

The success of finite element method has not weakened the significance of analytical methods because (i) it is a kind of numerical approximation and its theory is based on analytical methods; and (ii) many problems require analytical solutions such as crack tip singularity element in fracture mechanics, element for infinite domain, etc. Besides, the application of finite element method for analyses with local effect such as boundary effects in shell theory, and the free boundaries and boundary singular points in composite materials, etc., result in stiffness problems and therefore analytical methods are still of significant importance.

Considering "Theory of Elasticity" of S.P. Timoshenko as an example, the solutions of various elasticity problems using the semi-inverse method constitute a large portion of the text. This method was introduced by Saint-Venant in 1855–1856 to obtain certain solutions for torsion and bending of elastic columns. Since then it became the classical solution methodology for elasticity and its influence extends to the present moment. It is a trial method which is valid only for a specific problem without generality. It often obtains a certain solution but it cannot ensure complete solutions. What bothers one is the way to obtain a specific trial in order to solve the problem in hand.

The application of semi-inverse method is due to the complexity of the governing system of equations. The conventional analytical method is confined to the domain of single variable, using either stress function (method of force) or displacement method (only the shallow shell theory applies the hybrid method). The various unknown functions are eliminated thus resulting in a higher-order partial differential equation with a single variable which is then solved. From the viewpoint of mathematical systems, the solution of single variable systems belongs to the Lagrangian approach which inevitably results in a higher-order partial differential equation. Hence, the effective methods in mathematical physics such as variable separation and expansion of eigenfunctions become inapplicable. Consequently, the semi-inverse method has been unable to achieve major breakthrough for a long period.

Hence, a question arises. Is it absolutely necessary to employ this classical approach of eliminating variables? In reality, the classical approach is not the only avenue and dual theory and state symplectic space is the answer.

Recalling the many years we learnt applied mechanics, we may observe some problems. Classical analytical mechanics is the most fundamental system. Lagrange equation, the principle of minimum action, Hamilton's canonical equation, canonical transformation, Hamilton-Jacoby theory, etc., are all very beautiful theoretical systems. Classical analytical mechanics is also the basis of some fundamental branches of science such as statistical mechanics, electrodynamics, quantum mechanics, etc. It is, however, insufficiently appears in courses in applied mechanics. This is because it is less relevant to courses in elasticity, structural mechanics, fluid mechanics, vibration and stability, etc. Although control theory is originated from mechanics, it is seldom introduced in courses in applied mechanics. For instance, many existing texts in elasticity do not have much relevance with analytical mechanics. Systems of theory and methodologies for these fields are independent to a certain extent.

Control theory developed into modern control theory with the impact of computational techniques. The modern control theory is not merely an extension of the classical control theory but it has undergone fundamental changes in the basic theory with major breakthrough. The state space method based on modern control theory can be traced back to the system of Hamilton's canonical equation which is in principle a system with dual variables and dual equations.

Control theory underwent changes in system representation regulated by its own rules during development. Its system of theory was thought to have deviated further from applied mechanics. However, the real situation is not. It has been proven that the mathematical problems of modern control theory and structural mechanics map one-to-one and they are mutually similar. From the viewpoint of mathematics, the similarity is based on the theory and fundamentals of dual variables and Hamiltonian systems. It indicates that mechanics is able to gain advantage from the successful experience of control theory. As a matter of fact, the teaching and research in applied mathematics develops more and more towards systems of duality. From the observation above, the systems of dual variables should be implemented in the various branches of applied mechanics in a natural and systematic manner.

The development of information techniques of contemporary science and technology has been applied to intelligent materials, intelligent structures, precision weapon, etc., and the influence of control and remote sensing via various channels has been observed. Structural control has received increasing attention from now and then. Such development trend should not be neglected in the teaching of applied mechanics. As the world now moves towards "smartness", mechanics will not be able to be "smart" if it is not linked to control theory. In the United States, the mismatch in designs by structure engineers and control engineers has not been beneficial for overall rational design and there are voices to call for a "control-structure overall design". In fact, symplectic analytical systems can be applied to many other subjects such as vibration, wave propagation, etc. The engagement of an identical system of theory will encourage and make easy the assimilation and association of various branches of science such as mechanics and modern control, etc. It is also beneficial to teaching.

The advance from Lagrangian systems to Hamiltonian systems means the advance from the conventional Euclidean geometry to symplectic geometry. It is a breakthrough of the conventional concept which causes the application of dual and mixed variables into the vast fields of mechanics. In addition, symplectic systems can also be applied to mathematical physics and further to other related disciplines. Introducing the application of this approach in elasticity as a professional fundamental course to students will unquestionably help them to achieve greater heights in research in the future.

Nomenclature

k = shear correction factor
l, m, n = direction cosines of the exterior normal vector $\boldsymbol{n}$
$\boldsymbol{n}$ = exterior normal vector of the section
$\boldsymbol{p}$ = dual variable of the original variable (generalized momentum)
$\boldsymbol{q}$ = original variable (generalized displacement)
q = density of transverse load acting on beam or plate
$\boldsymbol{u}$ = displacement vector
u = displacement component along the rectangular coordinate x-direction
u_ρ, u_φ = displacement components in polar coordinates system
v = displacement component along the rectangular coordinate y-direction
v_c = strain complementary energy density
v_ε = strain energy density
$\boldsymbol{v}$ = full state vector of symplectic space
w = displacement component along the rectangular coordinate z-direction, or deflection of beam or plate
x, y, z = rectangular coordinates
A = cross-section area
$\boldsymbol{C}$ = stiffness coefficient matrix of the material
D = flexural rigidity of plate
E = modulus of elasticity for tension (Young's modulus)
E_c = complementary energy for support displacements
E_w = external potential energy
E_p = total potential energy
E_{pc} = total complementary energy
$\boldsymbol{F}$ = body force vector per unit volume
$\boldsymbol{F}_n$ = projections of the force per unit area acting on the inclined plane with exterior normal vector $\boldsymbol{n}$

F_{nx}, F_{ny}, F_{nz} = rectangular components of projections of the force $\boldsymbol{F}_n$ per unit area
F_s, F_{sx}, F_{sy} = shear force acting on cross-section of beam or plate
F_{sx}^t, F_{sy}^t = total equivalent shear force of plate cross-section
F_x, F_y, F_z = rectangular components of body force vector $\boldsymbol{F}$ per unit volume
G = elastic shear modulus
$\mathscr{H}$ = Hamiltonian density function
$\boldsymbol{H}$ = Hamiltonian matrix
I = moment of inertia of cross-section
$\boldsymbol{I}_n, \boldsymbol{I}$ = unit matrix
$\boldsymbol{J}_{2n}, \boldsymbol{J}$ = unit symplectic matrix
$\mathscr{L}$ = Lagrange function
$\boldsymbol{M}$ = bending moment vector of the plate
M_x, M_y, M_{xy} = bending and torsional moment of beam or plate
$\boldsymbol{S}$ = flexibility coefficient matrix of the material
V = region of elastic body
V_c = strain complementary energy
V_ε = strain energy
W = exterior work
$\gamma_{xy}, \gamma_{yz}, \gamma_{zx}$ = shear strains in rectangular coordinates system
$\gamma_{\rho\varphi}$ = shear strain in polar coordinates system
δ = variational notation
$\boldsymbol{\varepsilon}$ = vector format of stain components
$\varepsilon_1, \varepsilon_2, \varepsilon_3$ = principal strains
$\varepsilon_x, \varepsilon_y, \varepsilon_z$ = linear strains in rectangular coordinates system
$\varepsilon_\rho, \varepsilon_\varphi$ = linear strains in polar coordinates system
θ = rotational angle of cross-section, or volume strain
$\boldsymbol{\kappa}$ = curvature vector of the plate
$\kappa, \kappa_x, \kappa_y, \kappa_{xy}$ = curvatures
λ = Lame' constants or eigenvalue
μ = eigenvalue of Hamiltonian matrix
ν = Poisson's ratio
ρ = density of materials, or polar radius in polar coordinates system
σ = vector format of stress components
$\sigma_1, \sigma_2, \sigma_3$ = principal stresses
$\sigma, \sigma_x, \sigma_y, \sigma_z$ = normal stresses in rectangular coordinates system
$\sigma_\rho, \sigma_\varphi$ = normal stresses in polar coordinates system

$\tau, \tau_{xy}, \tau_{yz}, \tau_{zx}$ = shear stresses in rectangular coordinates system
$\tau_{\rho\varphi}$ = shear stress in polar coordinates system
ϕ_x, ϕ_y = bending moment functions of plate bending
φ = polar angle in polar coordinates system
φ_{f} = Airy stress function
ψ = eigenfunction vector of Hamiltonian matrix
ω = circular frequency
Γ = boundary of region V
Γ_u = boundary for specified displacements
Γ_σ = boundary for specified tractions

Chapter 1

Mathematical Preliminaries

This chapter serves to improve the completeness and systemization of theory introduced in this book. First of all, the basic concepts and basic elements of mathematics relevant to this book are briefly introduced. It includes the Euclidean space, symplectic space, and Legendre transformation, etc. In addition, it also briefly reviews the Hamilton principle and the Hamilton canonical equations in analytical mechanics, as well as the reciprocal theorems which are closely related to the contents of this book. Readers who are familiar with the mathematical preliminaries may omit this chapter and go directly to Chapter 2.

1.1. Linear Space

Linear space is one of the most basic concepts in linear algebra. It has been not only extensively applied in many fields in modern mathematics, but also exists as a common mathematical structure in various models in physics. As the basic concepts and basic elements have been widely described and proved in detail in many teaching materials[1–3], only contents relevant to this book will be particularly discussed without proofs in this section.

Definition 1.1. Let a linear space V in a real number field R has n linearly independent vectors (generalized vectors) $\{\boldsymbol{\alpha}_1, \boldsymbol{\alpha}_2, \ldots, \boldsymbol{\alpha}_n\}$ and every vector $\boldsymbol{\alpha}$ in V can be expressed as a linear combination of the vectors $\{\boldsymbol{\alpha}_1, \boldsymbol{\alpha}_2, \ldots, \boldsymbol{\alpha}_n\}$ as

$$\boldsymbol{\alpha} = x_1\boldsymbol{\alpha}_1 + x_2\boldsymbol{\alpha}_2 + \cdots + x_n\boldsymbol{\alpha}_n \tag{1.1.1}$$

Then $\{\boldsymbol{\alpha}_1, \boldsymbol{\alpha}_2, \ldots, \boldsymbol{\alpha}_n\}$ is called a **basis** of V, denoted as $\{\boldsymbol{\alpha}_i\}$ in brief, and $\{x_1, x_2, \ldots, x_n\}^{\mathrm{T}}$ are the **coordinates** of $\boldsymbol{\alpha}$ referring to basis $\{\boldsymbol{\alpha}_i\}$. Here, V is regarded as a n-dimensional **linear space**.

Incidentally, the above definition indicates that a problem in an abstract n-dimensional linear space can be completely described by general n-dimensional vectors in a real number field R via a basis. In reality, many properties and operations in linear space are eventually transformed to corresponding properties and operations of general vectors and matrices by way of a basis. It is conducive to a better understanding and application if discussion is done in a general n-dimensional vector space. It will become clear through further discussion introduced as follows.

A basis in a n-dimensional linear space is not unique. Coordinate systems of a vector referring to different bases are different.

Let $\{\boldsymbol{\alpha}_i\}$ and $\{\boldsymbol{\beta}_j\}$ are two bases in V and they are related by

$$\{\boldsymbol{\beta}_1, \boldsymbol{\beta}_2, \ldots, \boldsymbol{\beta}_n\} = \{\boldsymbol{\alpha}_1, \boldsymbol{\alpha}_2, \ldots, \boldsymbol{\alpha}_n\}\boldsymbol{A} \tag{1.1.2}$$

where

$$\boldsymbol{A} = \begin{bmatrix} a_{11} & a_{12} & \cdots & a_{1n} \\ a_{21} & a_{22} & \cdots & a_{2n} \\ \cdots & \cdots & \cdots & \cdots \\ a_{n1} & a_{n2} & \cdots & a_{nn} \end{bmatrix} \tag{1.1.3}$$

is the **transformation matrix** from basis $\{\boldsymbol{\alpha}_i\}$ to basis $\{\boldsymbol{\beta}_j\}$. It must be a nonsingular matrix. If $\{x_1, x_2, \ldots, x_n\}^{\mathrm{T}}$ and $\{y_1, y_2, \ldots, y_n\}^{\mathrm{T}}$ are the coordinates of a vector $\boldsymbol{\gamma}$ referring to bases $\{\boldsymbol{\alpha}_i\}$ and $\{\boldsymbol{\beta}_j\}$, respectively, then we have

$$\begin{Bmatrix} x_1 \\ x_2 \\ \vdots \\ x_n \end{Bmatrix} = \boldsymbol{A} \begin{Bmatrix} y_1 \\ y_2 \\ \vdots \\ y_n \end{Bmatrix} \quad \text{or} \quad \begin{Bmatrix} y_1 \\ y_2 \\ \vdots \\ y_n \end{Bmatrix} = \boldsymbol{A}^{-1} \begin{Bmatrix} x_1 \\ x_2 \\ \vdots \\ x_n \end{Bmatrix} \tag{1.1.4}$$

The transformation of a basis to another basis in a general vector space is actually the transformation of coordinates.

The self-mapping of a linear space V in a linear space is usually regarded as a transformation of V. In this regard, linear transformation is the most basic and simplest transformation.

Definition 1.2. A transformation $\tilde{\boldsymbol{A}}$ in a linear space V in a real number field R is called a **linear transformation** if, for any two vectors $\boldsymbol{\xi}, \boldsymbol{\eta}$ in V and any constant k in R, we have

$$\tilde{\boldsymbol{A}}(\boldsymbol{\xi}+\boldsymbol{\eta}) = \tilde{\boldsymbol{A}}(\boldsymbol{\xi}) + \tilde{\boldsymbol{A}}(\boldsymbol{\eta}) \tag{1.1.5}$$

$$\tilde{\boldsymbol{A}}(k\boldsymbol{\xi}) = k\tilde{\boldsymbol{A}}(\boldsymbol{\xi}) \tag{1.1.6}$$

Here and subsequently, $\tilde{\boldsymbol{A}}(\boldsymbol{\xi})$ is denoted briefly as $\tilde{\boldsymbol{A}}\boldsymbol{\xi}$.

Let $\{\boldsymbol{\alpha}_i\}$ be a basis in a linear space V, then the image $\{\tilde{\boldsymbol{A}}\boldsymbol{\alpha}_i\}$ of the vector upon linear transformation $\tilde{\boldsymbol{A}}$ can be linearly represented by basis $\{\boldsymbol{\alpha}_i\}$ as

$$\left.\begin{aligned} \tilde{\boldsymbol{A}}\boldsymbol{\alpha}_1 &= a_{11}\boldsymbol{\alpha}_1 + a_{21}\boldsymbol{\alpha}_2 + \cdots + a_{n1}\boldsymbol{\alpha}_n \\ \tilde{\boldsymbol{A}}\boldsymbol{\alpha}_2 &= a_{12}\boldsymbol{\alpha}_1 + a_{22}\boldsymbol{\alpha}_2 + \cdots + a_{n2}\boldsymbol{\alpha}_n \\ &\cdots\cdots\cdots\cdots\cdots\cdots \\ \tilde{\boldsymbol{A}}\boldsymbol{\alpha}_n &= a_{1n}\boldsymbol{\alpha}_1 + a_{2n}\boldsymbol{\alpha}_2 + \cdots + a_{nn}\boldsymbol{\alpha}_n \end{aligned}\right\} \tag{1.1.7}$$

or in matrix form, it is

$$\{\tilde{\boldsymbol{A}}\boldsymbol{\alpha}_1, \tilde{\boldsymbol{A}}\boldsymbol{\alpha}_2, \ldots, \tilde{\boldsymbol{A}}\boldsymbol{\alpha}_n\} = \{\boldsymbol{\alpha}_1, \boldsymbol{\alpha}_2, \ldots, \boldsymbol{\alpha}_n\}\boldsymbol{A} \tag{1.1.8}$$

where

$$\boldsymbol{A} = \begin{bmatrix} a_{11} & a_{12} & \cdots & a_{1n} \\ a_{21} & a_{22} & \cdots & a_{2n} \\ \cdots & \cdots & \cdots & \cdots \\ a_{n1} & a_{n2} & \cdots & a_{nn} \end{bmatrix} \tag{1.1.9}$$

Matrix $\boldsymbol{A}$ is the matrix resulted from linear transformation $\tilde{\boldsymbol{A}}$ referring to basis $\{\boldsymbol{\alpha}_i\}$. Obviously, $\{\tilde{\boldsymbol{A}}\boldsymbol{\alpha}_i\}$ is another basis of the linear space and matrix $\boldsymbol{A}$ is the transformation matrix if the latter is nonsingular.

Definition 1.3. Let $\tilde{\boldsymbol{A}}$ be a linear transformation in a linear space $\boldsymbol{V}$ in a real number field R. If there exists a nonzero vector $\boldsymbol{x}$ with respect to a number μ (inclusive a complex number) such that

$$\tilde{\boldsymbol{A}}\boldsymbol{x} = \mu\boldsymbol{x} \tag{1.1.10}$$

then μ is the **eigenvalue** of the linear transformation $\tilde{\boldsymbol{A}}$ and $\boldsymbol{x}$ the **eigenvector** of the linear transformation $\tilde{\boldsymbol{A}}$ corresponding to eigenvalue μ.

Let $\boldsymbol{A}$ be the matrix of linear transformation $\tilde{\boldsymbol{A}}$ referring to basis $\{\boldsymbol{\alpha}_i\}$ in linear space V, μ be an eigenvalue of $\tilde{\boldsymbol{A}}$, and $\{x_1, x_2, \ldots, x_n\}^{\mathrm{T}}$ be the coordinates of the corresponding eigenvector $\boldsymbol{x}$ referring to basis $\{\boldsymbol{\alpha}_i\}$.

Substituting into Eq. (1.1.10) yields

$$\tilde{\boldsymbol{A}}\{\boldsymbol{\alpha}_1, \boldsymbol{\alpha}_2, \ldots, \boldsymbol{\alpha}_n\} \begin{Bmatrix} x_1 \\ x_2 \\ \vdots \\ x_n \end{Bmatrix} = \mu\{\boldsymbol{\alpha}_1, \boldsymbol{\alpha}_2, \ldots, \boldsymbol{\alpha}_n\} \begin{Bmatrix} x_1 \\ x_2 \\ \vdots \\ x_n \end{Bmatrix} \tag{1.1.11}$$

Substituting Eq. (1.1.8) into the above equation and applying the linear independence characteristics of $\{\boldsymbol{\alpha}_i\}$ yield

$$\boldsymbol{A} \begin{Bmatrix} x_1 \\ x_2 \\ \vdots \\ x_n \end{Bmatrix} = \mu \begin{Bmatrix} x_1 \\ x_2 \\ \vdots \\ x_n \end{Bmatrix} \tag{1.1.12}$$

or, equivalently, as

$$(\mu \boldsymbol{I}_n - \boldsymbol{A}) \begin{Bmatrix} x_1 \\ x_2 \\ \vdots \\ x_n \end{Bmatrix} = \mathbf{0} \tag{1.1.13}$$

where $\boldsymbol{I}_n$ is a n-order identity matrix, denoted briefly as $\boldsymbol{I}$. Equation (1.1.13) will have a nonzero (nontrivial) vector solution, if and only if, the determinant of the coefficient matrix is zero. Equivalently, μ is the root of

$$f(\mu) = |\mu \boldsymbol{I} - \boldsymbol{A}| = 0 \tag{1.1.14}$$

The eigenvalue problem of a linear transformation can be transformed to the eigenvalue problem of a matrix by the application of a basis. Equation (1.1.14) is the **characteristic polynomial** of matrix $\boldsymbol{A}$. Therefore, we will focus on the eigenvalue problem of a matrix in the following discussion on eigenvalue problems. The properties can be extended to any general linear transformation. An eigenvalue problem of a matrix has the following properties.

Theorem 1.1. Let $\mu_1, \mu_2, \ldots, \mu_t$ be t distinct eigenvalues of matrix $\mathbf{A}$, and $\boldsymbol{\alpha}_{i1}, \boldsymbol{\alpha}_{i2}, \ldots, \boldsymbol{\alpha}_{im_i}$ $(i = 1, \ldots, t)$ be the linearly independent eigenvectors corresponding to eigenvalue μ_i, then all eigenvectors $\boldsymbol{\alpha}_{11}, \boldsymbol{\alpha}_{12}, \ldots, \boldsymbol{\alpha}_{1m_1};\ \boldsymbol{\alpha}_{21}, \boldsymbol{\alpha}_{22}, \ldots, \boldsymbol{\alpha}_{2m_2}; \cdots; \boldsymbol{\alpha}_{t1}, \boldsymbol{\alpha}_{t2}, \ldots, \boldsymbol{\alpha}_{tm_t}$ of $\boldsymbol{A}$ are linearly independent.

Theorem 1.2. For every $n \times n$ matrix $\boldsymbol{A}$, there exists a nonsingular $n \times n$ matrix $\boldsymbol{X}$ (inclusive complex elements) such that matrix $\boldsymbol{A}$ can be transformed to the **Jordan canonical form**

$$\boldsymbol{X}^{-1}\boldsymbol{A}\boldsymbol{X} = \text{diag}(\boldsymbol{D}_1, \boldsymbol{D}_2, \ldots, \boldsymbol{D}_t) \tag{1.1.15}$$

where

$$\boldsymbol{D}_i = \begin{bmatrix} \mu_i & 1 & 0 & \cdots & 0 \\ 0 & \mu_i & 1 & \cdots & 0 \\ 0 & 0 & \mu_i & \cdots & 0 \\ \vdots & \vdots & \vdots & \ddots & \vdots \\ 0 & 0 & 0 & \cdots & \mu_i \end{bmatrix} \tag{1.1.16}$$

is the Jordan part, and $m_1 + \cdots + m_t = n$. Also, we have

$$\boldsymbol{X} = \left\{\boldsymbol{\psi}_1^{(0)}, \ldots, \boldsymbol{\psi}_i^{(0)}, \boldsymbol{\psi}_i^{(1)}, \ldots, \boldsymbol{\psi}_i^{(m_i-1)}, \ldots, \boldsymbol{\psi}_t^{(m_t-1)}\right\} \tag{1.1.17}$$

or, in other words, there are m_i vectors corresponding to the Jordan part $\boldsymbol{D}_i$ in matrix $\boldsymbol{X}$, i.e. the basic eigenvector $\boldsymbol{\psi}_i^{(0)}$ and the Jordan form eigenvectors $\boldsymbol{\psi}_i^{(k)}$ $(k = 1, 2, \ldots, m_i - 1)$ corresponding to eigenvalue μ_i where k denotes the kth-order Jordan form eigenvector.

Throughout this book, the order of Jordan form vectors is indicated as a superscript as above.

Equation (1.1.15) can be written as[1,2]

$$\boldsymbol{A}\boldsymbol{X} = \boldsymbol{X} \cdot \text{diag}(\boldsymbol{D}_1, \boldsymbol{D}_2, \ldots, \boldsymbol{D}_t) \tag{1.1.18}$$

Expanding Eq. (1.1.18) yields

$$\left.\begin{array}{l} \boldsymbol{A}\boldsymbol{\psi}_i^{(0)} = \mu_i\boldsymbol{\psi}_i^{(0)} \\ \boldsymbol{A}\boldsymbol{\psi}_i^{(1)} = \mu_i\boldsymbol{\psi}_i^{(1)} + \boldsymbol{\psi}_i^{(0)} \\ \cdots\cdots\cdots\cdots \\ \boldsymbol{A}\boldsymbol{\psi}_i^{(m_i-1)} = \mu_i\boldsymbol{\psi}_i^{(m_i-1)} + \boldsymbol{\psi}_i^{(m_i-2)} \end{array}\right\} \quad (i = 1, 2, \ldots, t) \tag{1.1.19}$$

Equation (1.1.19) shows the general method for solving the basic eigenvector and the Jordan form eigenvectors.

The discussion above is not restricted to a Cartesian coordinate system.

1.2. Euclidean Space

Vector addition and scalar multiplication are the two basic linear operations of vectors in a linear space. However, such linear operations cannot be used to describe the metric properties of vectors, such as length, orthogonality, etc. The metric concept can be introduced into linear space via the operation of inner product. This concept will be particularly discussed without proofs in this section.

Definition 1.4. Let V be a linear space defined in a real number filed R. For any two arbitrary vectors $\boldsymbol{\alpha}, \boldsymbol{\beta}$ in V, there exists a real number according to a specified rule, termed the **inner product** and denoted as $(\boldsymbol{\alpha}, \boldsymbol{\beta})$. The inner product operation satisfies the following four conditions:

$$(1)\quad (\boldsymbol{\alpha}, \boldsymbol{\alpha}) \geq 0, (\boldsymbol{\alpha}, \boldsymbol{\alpha}) = \mathbf{0} \text{ if and only if } \boldsymbol{\alpha} = \mathbf{0} \tag{1.2.1a}$$

$$(2)\quad (\boldsymbol{\alpha}, \boldsymbol{\beta}) = (\boldsymbol{\beta}, \boldsymbol{\alpha}) \tag{1.2.1b}$$

$$(3)\quad (\boldsymbol{\alpha} + \boldsymbol{\gamma}, \boldsymbol{\beta}) = (\boldsymbol{\alpha}, \boldsymbol{\beta}) + (\boldsymbol{\gamma}, \boldsymbol{\beta}), \boldsymbol{\gamma} \text{ is an arbitrary vector in } V \tag{1.2.1c}$$

$$(4)\quad (k\boldsymbol{\alpha}, \boldsymbol{\beta}) = k(\boldsymbol{\alpha}, \boldsymbol{\beta}), k \text{ is an arbitrary real number} \tag{1.2.1d}$$

A linear space satisfying the above conditions of inner product is called a **Euclidean space**.

Having defined inner product, it is possible to measure the length, orthogonality, unit vector, etc. of vectors related to metric concept in Euclidean space.

Let V be a Euclidean space. The **norm** of an arbitrary vector $\boldsymbol{\alpha}$ is defined as

$$\|\boldsymbol{\alpha}\| = \sqrt{(\boldsymbol{\alpha}, \boldsymbol{\alpha})} \tag{1.2.2}$$

$\boldsymbol{\alpha}$ is a **unit vector** if the norm of $\boldsymbol{\alpha}$ is $\|\boldsymbol{\alpha}\| = 1$.

Example 1.1. For arbitrary vectors $\boldsymbol{x} = \{x_1, x_2, \ldots, x_n\}^{\mathrm{T}}$, $\boldsymbol{y} = \{y_1, y_2, \ldots, y_n\}^{\mathrm{T}}$ in a n-dimensional real vector space R^n, the inner product is defined as

$$(\boldsymbol{x}, \boldsymbol{y}) = x_1 y_1 + x_2 y_2 + \cdots + x_n y_n = \boldsymbol{x}^{\mathrm{T}} \boldsymbol{y} (= \boldsymbol{x}^{\mathrm{T}} \boldsymbol{I} \boldsymbol{y}) \tag{1.2.3}$$

It is obvious that the operation satisfy the four conditions of inner product in Eqs. (1.2.1) and therefore they form an n-dimensional Euclidean space.

The magnitude of vector $\boldsymbol{x}$ in R^n is

$$\|\boldsymbol{x}\| = \sqrt{\sum_{i=1}^{n} x_i^2} \tag{1.2.4}$$

For a particular linear space, there are various definitions of inner product and therefore there exist different Euclidean spaces. The inner product in Example 1.1 is the normal inner product of a n-dimensional real vector space R^n. It is also the most common definition of inner product. The following discussion on R^n always refers to this normal inner product.

Definition 1.5. Vectors $\boldsymbol{\alpha}, \boldsymbol{\beta}$ are **orthogonal**, denoted as $\boldsymbol{\alpha} \perp \boldsymbol{\beta}$, if their inner product $(\boldsymbol{\alpha}, \boldsymbol{\beta}) = 0$.

If any two vectors in a nonzero vector set $\{\boldsymbol{\alpha}_i\}$ are orthogonal, then the vector set $\{\boldsymbol{\alpha}_i\}$ is called an **orthogonal vector set**. If the vectors are all unit vectors, then the vector set $\{\boldsymbol{\alpha}_i\}$ is called a **normal orthogonal vector set**. A basis formed by n (normal) orthogonal vectors in a n-dimensional Euclidean space is called a **(normal) orthogonal basis**.

From Definitions 1.4 and 1.5, it is obvious that

Theorem 1.3. A zero vector is orthogonal to any vector. Conversely, a vector which is orthogonal to any vector in the space must be a zero vector.

Theorem 1.4. An orthogonal vector set is a linearly independent vector set.

Theorem 1.5. Any arbitrary (normal) orthogonal vector set in a n-dimensional Euclidean space can be extended to a set of (normal) orthogonal bases.

Let V be a n-dimensional Euclidean space, $\{\boldsymbol{\alpha}_i\}$ be a set of normal orthogonal bases, then the coordinates $\{x_1, x_2, \ldots, x_n\}^{\mathrm{T}}$ of an arbitrary vector $\boldsymbol{\beta}$ referring to basis $\{\boldsymbol{\alpha}_i\}$ can be expressed (expansion theorem) as:

$$x_i = (\boldsymbol{\beta}, \boldsymbol{\alpha}_i) \quad (i = 1,\ 2, \ldots, n) \tag{1.2.5}$$

Let $\{y_1, y_2, \ldots, y_n\}^{\mathrm{T}}$ be the coordinates of another vector $\boldsymbol{\gamma}$ referring to basis $\{\boldsymbol{\alpha}_i\}$, then the inner product of $\boldsymbol{\beta}$ and $\boldsymbol{\gamma}$ is

$$(\boldsymbol{\beta}, \boldsymbol{\gamma}) = \sum_{i=1}^{n} x_i y_i = \boldsymbol{x}^{\mathrm{T}} \boldsymbol{y} (= \boldsymbol{x}^{\mathrm{T}} \boldsymbol{I} \boldsymbol{y}) = (\boldsymbol{\gamma}, \boldsymbol{\beta}) \tag{1.2.6}$$

where

$$\boldsymbol{x} = \{x_1, x_2, \ldots, x_n\}^{\mathrm{T}}, \quad \boldsymbol{y} = \{y_1, y_2, \ldots, y_n\}^{\mathrm{T}} \tag{1.2.7}$$

Through the use of a normal orthogonal basis, the inner product operation in a n-dimensional Euclidean space can be transformed to the normal inner product operation in a n-dimensional real vector space R^n.

To discuss on the transformation equation of bases for a normal orthogonal basis, we introduce.

Definition 1.6. If a $n \times n$ matrix $\boldsymbol{Q}$ satisfies

$$\boldsymbol{Q}^{\mathrm{T}}\boldsymbol{Q} = \boldsymbol{Q}\boldsymbol{Q}^{\mathrm{T}} = \boldsymbol{I} \tag{1.2.8}$$

then $\boldsymbol{Q}$ is an **orthogonal matrix**.

An orthogonal matrix has the following properties:

(1) The inverse matrix (i.e. transpose matrix) of an orthogonal matrix is an orthogonal matrix.
(2) The determinant of an orthogonal matrix is equal to either 1 or -1.
(3) The product of two orthogonal matrices is an orthogonal matrix.

From Definition 1.6, it is obviously that

Theorem 1.6. The transformation matrix for normal orthogonal bases is an orthogonal matrix.

The following discussion focuses on the most fundamental linear operator (transformation) in Euclidean space, i.e. the symmetrization operator.

Definition 1.7. Let V be a n-dimensional Euclidean space. If the linear transformation $\tilde{\boldsymbol{A}}$ on arbitrary vectors $\boldsymbol{\alpha}, \boldsymbol{\beta}$ in V satisfies

$$(\boldsymbol{\alpha}, \tilde{\boldsymbol{A}}\boldsymbol{\beta}) = (\boldsymbol{\beta}, \tilde{\boldsymbol{A}}\boldsymbol{\alpha}) \tag{1.2.9}$$

then $\tilde{\boldsymbol{A}}$ is a symmetric operator of the Euclidean space V.

Obviously, the matrix of symmetric operator $\tilde{\boldsymbol{A}}$ for any normal orthogonal basis $\{\boldsymbol{\alpha}_i\}$ is a $n \times n$ real symmetric matrix $\boldsymbol{A}$. A real symmetric matrix (**symmetric operator**) is self-adjoint. The eigenvalue problem of a self-adjoint operator is discussed in detail because of its needs in vibration theory and other problems in mathematical physics. There are some theorems as follows regarding the eigenvalue problem of a real symmetric matrix.

Theorem 1.7. The eigenvalues of a real symmetric matrix are all real numbers.

Theorem 1.8. Let $\boldsymbol{A}$ be a real symmetric matrix, then the eigenvectors corresponding to different eigenvalues of $\boldsymbol{A}$ in R^n are mutually orthogonal.

Theorem 1.9. For an arbitrary n order real symmetric matrix $\boldsymbol{A}$, there exists a n-order orthogonal matrix $\boldsymbol{Q}$ such that $\boldsymbol{Q}^{\mathrm{T}}\boldsymbol{A}\boldsymbol{Q} = \boldsymbol{Q}^{-1}\boldsymbol{A}\boldsymbol{Q}$ is diagonal.

Theorem 1.9 shows that for an arbitrary symmetric matrix (operator), there exists a normal orthogonal basis composed of eigenvectors. The corresponding eigenvalues are all real and there will be no Jordan form even if there are repeated eigenvalues. The eigenvectors are all mutually orthogonal. As a result, the eigen-solutions span to a complete space and every vector in this space can be constructed from a linear combination of the eigenvectors (expansion theorem).

1.3. Symplectic Space

All conservative real physical processes can be described by a suitable Hamiltonian system whose common mathematical fundamentals are the symplectic spaces. A symplectic space is different from a Euclidean space which studies the metric properties such as length, etc. It focuses on the study of area, or the study of work and this is a mathematical structure present throughout this book. Using a finite-dimensional symplectic space as an example in this section, the basic concepts and basic properties of a symplectic space are described and proved in detail[4,5]. The discussion lays a sound mathematical foundation for the study in the following chapters.

Definition 1.8. Let V be a n-dimensional linear space defined in a real number field R, and V' be the corresponding n-dimensional dual linear space. We define

$$W = V \times V' = \left\{ \begin{pmatrix} \boldsymbol{q} \\ \boldsymbol{p} \end{pmatrix} \middle| \boldsymbol{q} \in V, \boldsymbol{p} \in V' \right\} \tag{1.3.1}$$

where the linear space W is called a $2n$-dimensional **phase space** in a real number field R constructed by V and V'.

It is emphasized here that the linear spaces V and V' have absolutely different dimensions in actual problems, and there usually exists no direct relation between the spaces. However, the product of their corresponding components has a specific physical meaning. In this book, for instance, one is usually the displacement and the other the stress, and the product of their corresponding components has the dimension of work.

Definition 1.9. Let W be a $2n$-dimensional phase space in a real number field R. For any two vectors $\boldsymbol{\alpha}, \boldsymbol{\beta}$ in W, there exists a real number according to a specified rule. This real number is termed the **symplectic inner product**, denoted as $\langle \boldsymbol{\alpha}, \boldsymbol{\beta} \rangle$, and it satisfies the following four conditions:

$$\text{(1)}\quad \langle \boldsymbol{\alpha}, \boldsymbol{\beta} \rangle = -\langle \boldsymbol{\beta}, \boldsymbol{\alpha} \rangle \tag{1.3.2a}$$

$$\text{(2)}\quad \langle k\boldsymbol{\alpha}, \boldsymbol{\beta} \rangle = k\langle \boldsymbol{\alpha}, \boldsymbol{\beta} \rangle, k \text{ is an arbitrary real number} \tag{1.3.2b}$$

$$\text{(3)}\quad \langle \boldsymbol{\alpha} + \boldsymbol{\gamma}, \boldsymbol{\beta} \rangle = \langle \boldsymbol{\alpha}, \boldsymbol{\beta} \rangle + \langle \boldsymbol{\gamma}, \boldsymbol{\beta} \rangle, \boldsymbol{\gamma} \text{ is an arbitrary vector in } W \tag{1.3.2c}$$

$$\text{(4)}\quad \boldsymbol{\alpha} = \mathbf{0} \text{ if } \langle \boldsymbol{\alpha}, \boldsymbol{\beta} \rangle = 0 \text{ for every vector } \boldsymbol{\beta} \text{ in } W \tag{1.3.2d}$$

A phase space satisfying the above conditions of symplectic inner product is called a **symplectic space**.

From Eq. (1.3.2a), the symplectic self inner product of every vector must vanish, i.e. for every vector $\boldsymbol{\alpha}$

$$\langle \boldsymbol{\alpha}, \boldsymbol{\alpha} \rangle = 0 \tag{1.3.3}$$

Example 1.2. For any two vectors $\boldsymbol{x} = \{x_1, x_2\}^{\mathrm{T}}$, $\boldsymbol{y} = \{y_1, y_2\}^{\mathrm{T}}$ in a two-dimensional real vector space R^2, the symplectic inner product is defined as

$$\langle \boldsymbol{x}, \boldsymbol{y} \rangle = x_1 y_2 - x_2 y_1 \tag{1.3.4}$$

It is obvious that Eq. (1.3.4) satisfies the four conditions of symplectic inner product in Eqs. (1.3.2) and therefore it forms a 2-dimensional symplectic space. Here, the symplectic inner product (1.3.4) represents the area of a parallelogram constructed by $\boldsymbol{x}, \boldsymbol{y}$ as its adjacent sides.

Obviously, we can generalize Eq. (1.3.4) to a $2n$-dimensional real vector space R^{2n}. For any two vectors $\boldsymbol{x} = \{x_1, x_2, \ldots, x_{2n}\}^{\mathrm{T}}$, $\boldsymbol{y} = \{y_1, y_2, \ldots, y_{2n}\}^{\mathrm{T}}$, the symplectic inner product is defined as

$$\langle \boldsymbol{x}, \boldsymbol{y} \rangle \overset{\text{def}}{=} (\boldsymbol{x}, \boldsymbol{J}_{2n}\boldsymbol{y}) = \sum_{i=1}^{n} (x_i y_{n+i} - x_{n+i} y_i) = \boldsymbol{x}^{\mathrm{T}} \boldsymbol{J}_{2n} \boldsymbol{y} \tag{1.3.5}$$

where

$$J_{2n} = \begin{bmatrix} \mathbf{0} & \boldsymbol{I}_n \\ -\boldsymbol{I}_n & \mathbf{0} \end{bmatrix} \tag{1.3.6}$$

is called the **unit symplectic matrix**, denoted briefly as $\boldsymbol{J}$. It is obvious that Eq. (1.3.5) satisfies the four conditions of symplectic inner product in Eqs. (1.3.2), and therefore it forms a $2n$-dimensional symplectic space.

The determinant of a unit symplectic matrix is equal to 1. A unit symplectic matrix has the following properties:

$$\boldsymbol{J}^2 = -\boldsymbol{I}, \quad \boldsymbol{J}^{\mathrm{T}} = \boldsymbol{J}^{-1} = -\boldsymbol{J} \tag{1.3.7}$$

Similarly, there are various definitions of symplectic inner product for a phase space and therefore there exist different symplectic spaces. The symplectic inner product defined in Eq. (1.3.5) is called the normal symplectic inner product in a $2n$-dimensional real vector space R^{2n}. The following discussion on real vector space R^{2n} always refers to this normal symplectic inner product.

Definition 1.10. Vectors $\boldsymbol{\alpha}, \boldsymbol{\beta}$ are **symplectic orthogonal** if their symplectic inner product $\langle \boldsymbol{\alpha}, \boldsymbol{\beta} \rangle = 0$. Otherwise, they are **symplectic adjoint**.

Hence from Eq. (1.3.2d), there exists a symplectic adjoint nonzero vector for every nonzero vector. Virtually, $\boldsymbol{\alpha}$ and $\boldsymbol{J\alpha}$ must be symplectic adjoint if $\boldsymbol{\alpha} \neq \mathbf{0}$.

If the vectors in a vector set $\{\boldsymbol{\alpha}_1, \boldsymbol{\alpha}_2, \ldots, \boldsymbol{\alpha}_r, \boldsymbol{\beta}_1, \boldsymbol{\beta}_2, \ldots, \boldsymbol{\beta}_r\}$ $(r \leq n)$ satisfy

$$\left.\begin{aligned} &\langle \boldsymbol{\alpha}_i, \boldsymbol{\alpha}_j \rangle = \langle \boldsymbol{\beta}_i, \boldsymbol{\beta}_j \rangle = 0 \\ &\langle \boldsymbol{\alpha}_i, \boldsymbol{\beta}_j \rangle = \begin{cases} k_{ii} \neq 0 & (i = j) \\ 0 & (i \neq j) \end{cases} \end{aligned}\right\} \quad (i, j = 1,\ 2, \ldots, r) \tag{1.3.8}$$

then the vector set $\{\boldsymbol{\alpha}_1, \boldsymbol{\alpha}_2, \ldots, \boldsymbol{\alpha}_r, \boldsymbol{\beta}_1, \boldsymbol{\beta}_2, \ldots, \boldsymbol{\beta}_r\}$ is called an **adjoint symplectic orthonormal vector set**. If $k_{ii} \equiv 1 (i = 1, 2, \ldots, r)$ in Eq. (1.3.8), then the vector set $\{\boldsymbol{\alpha}_1, \boldsymbol{\alpha}_2, \ldots, \boldsymbol{\alpha}_r, \boldsymbol{\beta}_1, \boldsymbol{\beta}_2, \ldots, \boldsymbol{\beta}_r\}$ is called **normal adjoint symplectic orthonormal vector set.** From the definition, it is obvious that:

Theorem 1.10. An adjoint symplectic orthonormal vector set is a linearly independent vector set.

Proof. Using proof by contradiction, assume $\{\boldsymbol{\alpha}_1, \boldsymbol{\alpha}_2, \ldots, \boldsymbol{\alpha}_r, \boldsymbol{\beta}_1, \boldsymbol{\beta}_2, \ldots, \boldsymbol{\beta}_r\}$ is an adjoint symplectic orthonormal vector set and it is linearly dependent. Then there exists a vector, denoted as $\boldsymbol{\alpha}_1$, which can be expressed as a linear combination of the other vectors as

$$\boldsymbol{\alpha}_1 = s_2\boldsymbol{\alpha}_2 + \cdots + s_r\boldsymbol{\alpha}_r + t_1\boldsymbol{\beta}_1 + \cdots + t_r\boldsymbol{\beta}_r$$

Hence, we have

$$\begin{aligned}\langle\boldsymbol{\alpha}_1, \boldsymbol{\beta}_1\rangle &= s_2\langle\boldsymbol{\alpha}_2, \boldsymbol{\beta}_1\rangle + \cdots + s_r\langle\boldsymbol{\alpha}_r, \boldsymbol{\beta}_1\rangle \\ &\quad + t_1\langle\boldsymbol{\beta}_1, \boldsymbol{\beta}_1\rangle + \cdots + t_r\langle\boldsymbol{\beta}_r, \boldsymbol{\beta}_1\rangle = 0\end{aligned}$$

which contradicts $\langle\boldsymbol{\alpha}_1, \boldsymbol{\beta}_1\rangle \neq 0$. Therefore, the original proposition is tenable.

The basis formed by $2n$ (normal) adjoint symplectic orthonormal vectors in a $2n$-dimensional symplectic space is called a **(normal) adjoint symplectic orthonormal basis**.

Theorem 1.11. Every adjoint symplectic orthonormal vector set in a $2n$-dimensional symplectic space can be extended to an adjoint symplectic orthonormal basis.

Proof. Let $\{\boldsymbol{\alpha}_1, \boldsymbol{\alpha}_2, \ldots, \boldsymbol{\alpha}_r, \boldsymbol{\beta}_1, \boldsymbol{\beta}_2, \ldots, \boldsymbol{\beta}_r\}$ be an adjoint symplectic orthonormal vector set. A proof for $n - r$ by mathematical induction is established.

(1) For $n - r = 0$, $\{\boldsymbol{\alpha}_1, \boldsymbol{\alpha}_2, \ldots, \boldsymbol{\alpha}_r, \boldsymbol{\beta}_1, \boldsymbol{\beta}_2, \ldots, \boldsymbol{\beta}_r\}$ is an adjoint symplectic orthonormal basis. The theorem is tenable for $n - r = 0$.
(2) Assume that the theorem is tenable for $n - r = k$, then consider the case of $n - r = k + 1$.

As $r < n$, there exists a vector $\boldsymbol{\gamma}$ which cannot be expressed as a linear combination of $\{\boldsymbol{\alpha}_1, \boldsymbol{\alpha}_2, \ldots, \boldsymbol{\alpha}_r, \boldsymbol{\beta}_1, \boldsymbol{\beta}_2, \ldots, \boldsymbol{\beta}_r\}$. Denoting

$$\boldsymbol{\alpha}_{r+1} = \boldsymbol{\gamma} - \sum_{i=1}^{r} s_i\boldsymbol{\alpha}_i - \sum_{i=1}^{r} t_i\boldsymbol{\beta}_i$$

where

$$s_i = \frac{\langle\boldsymbol{\gamma}, \boldsymbol{\beta}_i\rangle}{\langle\boldsymbol{\alpha}_i, \boldsymbol{\beta}_i\rangle} \quad t_i = -\frac{\langle\boldsymbol{\gamma}, \boldsymbol{\alpha}_i\rangle}{\langle\boldsymbol{\alpha}_i, \boldsymbol{\beta}_i\rangle}$$

It is obvious that $\boldsymbol{\alpha}_{r+1}$ is symplectic orthogonal to $\{\boldsymbol{\alpha}_1, \boldsymbol{\alpha}_2, \ldots, \boldsymbol{\alpha}_r, \boldsymbol{\beta}_1, \boldsymbol{\beta}_2, \ldots, \boldsymbol{\beta}_r\}$. Besides, $\boldsymbol{\alpha}_{r+1}$ is a nonzero vector from the definition and, hence, there exists a symplectic adjoint nonzero vector $\tilde{\gamma}$, or $\langle \boldsymbol{\alpha}_{r+1}, \tilde{\gamma} \rangle \neq 0$. Obviously, $\tilde{\gamma}$ cannot be expressed as a linear combination of $\{\boldsymbol{\alpha}_1, \boldsymbol{\alpha}_2, \ldots, \boldsymbol{\alpha}_r, \boldsymbol{\beta}_1, \boldsymbol{\beta}_2, \ldots, \boldsymbol{\beta}_r\}$. Denoting

$$\boldsymbol{\beta}_{r+1} = \tilde{\gamma} - \sum_{i=1}^{r} \tilde{s}_i \boldsymbol{\alpha}_i - \sum_{i=1}^{r} \tilde{t}_i \boldsymbol{\beta}_i$$

where

$$\tilde{s}_i = \frac{\langle \tilde{\gamma}, \boldsymbol{\beta}_i \rangle}{\langle \boldsymbol{\alpha}_i, \boldsymbol{\beta}_i \rangle} \quad \tilde{t}_i = -\frac{\langle \tilde{\gamma}, \boldsymbol{\alpha}_i \rangle}{\langle \boldsymbol{\alpha}_i, \boldsymbol{\beta}_i \rangle}$$

It is obvious that $\boldsymbol{\beta}_{r+1}$ is symplectic orthogonal to $\{\boldsymbol{\alpha}_1, \boldsymbol{\alpha}_2, \ldots, \boldsymbol{\alpha}_r, \boldsymbol{\beta}_1, \boldsymbol{\beta}_2, \ldots, \boldsymbol{\beta}_r\}$, and also symplectic adjoint with $\boldsymbol{\alpha}_{r+1}$, or $\langle \boldsymbol{\alpha}_{r+1}, \boldsymbol{\beta}_{r+1} \rangle = \langle \boldsymbol{\alpha}_{r+1}, \tilde{\gamma} \rangle \neq 0$. Then $\{\boldsymbol{\alpha}_1, \ldots, \boldsymbol{\alpha}_r, \boldsymbol{\alpha}_{r+1}, \boldsymbol{\beta}_1, \ldots, \boldsymbol{\beta}_r, \boldsymbol{\beta}_{r+1}\}$ is an adjoint symplectic orthonormal vector set. According to the assumption of mathematical induction, the theorem is tenable for $n-(r+1) = k$. In other words, $\{\boldsymbol{\alpha}_1, \boldsymbol{\alpha}_2, \ldots, \boldsymbol{\alpha}_r, \boldsymbol{\alpha}_{r+1}, \boldsymbol{\beta}_1, \boldsymbol{\beta}_2, \ldots, \boldsymbol{\beta}_r, \boldsymbol{\beta}_{r+1}\}$ can be extended to an adjoint symplectic orthonormal basis and the theorem is tenable for $n - r = k + 1$. According to mathematical induction, therefore, the theorem is tenable for any $n - r$. Hence, the proposition is proven.

Deduction. Every normal adjoint symplectic orthonormal vector set in a $2n$-dimensional symplectic space can be extended to a normal adjoint symplectic orthonormal basis.

The theorem and deduction above indicate that there exists a normal adjoint symplectic orthonormal basis in a $2n$-dimensional symplectic space, but it is not unique. Based on a normal adjoint symplectic orthonormal basis and its properties, the expansion theorem of a symplectic space can be obtained directly.

Theorem 1.12. Let W be a $2n$-dimensional symplectic space and $\{\boldsymbol{\alpha}_i\}$ be a normal adjoint symplectic orthonormal basis, then the coordinates $\{x_1, \ldots, x_n, x_{n+1}, \ldots, x_{2n}\}^{\mathrm{T}}$ of any vector $\boldsymbol{\beta}$ in W referring to the basis $\{\boldsymbol{\alpha}_i\}$ can be expressed as:

$$x_i = \langle \boldsymbol{\beta}, \boldsymbol{\alpha}_{n+i} \rangle, x_{n+i} = -\langle \boldsymbol{\beta}, \boldsymbol{\alpha}_i \rangle \quad (i = 1, 2, \ldots, n) \tag{1.3.9}$$

Let $\{y_1, \ldots, y_n, y_{n+1}, \ldots, y_{2n}\}^{\mathrm{T}}$ be the coordinates of another vector $\boldsymbol{\gamma}$ referring to the basis $\{\boldsymbol{\alpha}_i\}$, then the symplectic inner product of $\boldsymbol{\beta}$ and $\boldsymbol{\gamma}$ is

$$\langle \boldsymbol{\beta}, \boldsymbol{\gamma} \rangle = \sum_{i=1}^{n} (x_i y_{n+i} - x_{n+i} y_i) = \boldsymbol{x}^{\mathrm{T}} \boldsymbol{J}_{2n} \boldsymbol{y} \tag{1.3.10}$$

where

$$\boldsymbol{x} = \{x_1, x_2, \ldots, x_{2n}\}^{\mathrm{T}}, \quad \boldsymbol{y} = \{y_1, y_2, \ldots, y_{2n}\}^{\mathrm{T}} \tag{1.3.11}$$

Through the use of a normal adjoint symplectic orthonormal basis, the symplectic inner product operation in a $2n$-dimensional symplectic space can be transformed to the matrix operation of ordinary vectors (matrices).

To discuss the transformation equation of bases for a normal adjoint symplectic orthonormal basis, we introduce

Definition 1.11. If a $2n \times 2n$ matrix $\boldsymbol{S}$ satisfies

$$\boldsymbol{S}^{\mathrm{T}}\boldsymbol{J}\boldsymbol{S} = \boldsymbol{J} \tag{1.3.12}$$

then $\boldsymbol{S}$ is a **symplectic matrix**, where $\boldsymbol{J}$ is the unit symplectic matrix.

A symplectic matrix has the following properties:

(1) The inverse matrix of a symplectic matrix is a symplectic matrix.
(2) The transpose matrix of a symplectic matrix is a symplectic matrix.
(3) The determinant of a symplectic matrix is equal to either 1 or -1.
(4) The product of two symplectic matrices is a symplectic matrix.

For a normal adjoint symplectic orthonormal basis, it is obviously that:

Theorem 1.13. The transformation matrix for normal adjoint symplectic orthonormal bases is a symplectic matrix.

The following discussion focuses on the most fundamental linear operator in a symplectic space, i.e. the Hamiltonian operator.

Definition 1.12. Let W be a $2n$-dimensional symplectic space. If a linear operator $\tilde{\boldsymbol{H}}$ acting on arbitrary vectors $\boldsymbol{\alpha}, \boldsymbol{\beta}$ satisfies

$$\langle \boldsymbol{\alpha}, \tilde{\boldsymbol{H}}\boldsymbol{\beta} \rangle = \langle \boldsymbol{\beta}, \tilde{\boldsymbol{H}}\boldsymbol{\alpha} \rangle \tag{1.3.13}$$

then linear transformation $\tilde{\boldsymbol{H}}$ is a **Hamiltonian operator** of the symplectic space W.

Definition 1.13. If a $2n \times 2n$ matrix $\boldsymbol{H}$ acting on arbitrary $2n$-dimensional vectors $\boldsymbol{x}, \boldsymbol{y}$ satisfies

$$\langle \boldsymbol{x}, \boldsymbol{H}\boldsymbol{y} \rangle = \langle \boldsymbol{y}, \boldsymbol{H}\boldsymbol{x} \rangle \tag{1.3.14}$$

then matrix $\boldsymbol{H}$ is a **Hamiltonian matrix**.

It is obvious that the definition of Hamiltonian matrix in Eq. (1.3.14) is equivalent to the following definition

$$(\boldsymbol{JH})^{\mathrm{T}} = \boldsymbol{JH}, \quad \text{or} \quad \boldsymbol{JHJ} = \boldsymbol{H}^{\mathrm{T}} \tag{1.3.14$'$}$$

Obviously, the matrix of Hamiltonian operator $\tilde{\boldsymbol{H}}$ referring to a normal adjoint symplectic orthonormal basis $\{\boldsymbol{\alpha}_i\}$ is a Hamiltonian matrix. The eigenvalue problem of a Hamiltonian matrix (Hamiltonian operator) is non-self-adjoint, hence it is possible to have complex eigenvalues or repeated eigenvalues. However, the said eigenvalue problem of a Hamiltonian matrix (Hamiltonian operator) has its specific characteristics. Hamiltonian matrices are used as examples in the following discussion. Certainly, the relevant conclusion can be directly generalized to the Hamiltonian operator of a finite-dimensional symplectic space.

Theorem 1.14. If μ is an eigenvalue of a Hamiltonian matrix with multiplicity m, then $-\mu$ is also an eigenvalue with multiplicity m. If zero is an eigenvalue of a Hamiltonian matrix $\boldsymbol{H}$, then the multiplicity number is even.

Proof. Let the characteristic polynomial of a Hamiltonian matrix $\boldsymbol{H}$ be

$$f(\mu) = |\mu\boldsymbol{I} - \boldsymbol{H}|$$

then according to the definitions of unit symplectic matrix in Eq. (1.3.6) and Hamiltonian matrix in Eq. (1.3.14$'$), we have

$$\begin{aligned} f(\mu) &= |\boldsymbol{J}(\mu\boldsymbol{I} - \boldsymbol{H})\boldsymbol{J}| = |\mu\boldsymbol{J}\boldsymbol{J} - \boldsymbol{J}\boldsymbol{H}\boldsymbol{J}| \\ &= |-\mu\boldsymbol{I} - \boldsymbol{H}^{\mathrm{T}}| = |-\mu\boldsymbol{I} - \boldsymbol{H}| = f(-\mu) \end{aligned}$$

As the expression above is true for every μ, the theorem is therefore proven.

Subsequently, the two eigenvalues $\pm\mu$ are called **mutually symplectic adjoint eigenvalues** of a Hamiltonian matrix. The nonzero eigenvalues of a Hamiltonian matrix are usually divided into two sets:

$$\left.\begin{array}{lll} (\alpha) & \mu_i, & \mathrm{Re}(\mu_i) < 0 \quad \text{or} \quad \mathrm{Re}(\mu_i) = 0 \wedge \mathrm{Im}(\mu_i) < 0 \\ (\beta) & \mu_{n+i} = -\mu_i & \end{array}\right\} \tag{1.3.15}$$

The eigenvalues in the (α)-set can be further arranged according to the absolute values of μ_i, in an ascending order, for instance. Note that Eq. (1.3.15) does not include the zero eigenvalue which is a special symplectic eigenvalue with itself the mutually symplectic adjoint eigenvalue.

Theorem 1.15. Let $\boldsymbol{H}$ be a Hamiltonian matrix, $\boldsymbol{\psi}_i^{(0)}, \boldsymbol{\psi}_i^{(1)}, \ldots, \boldsymbol{\psi}_i^{(m)}$ and $\boldsymbol{\psi}_j^{(0)}, \boldsymbol{\psi}_j^{(1)}, \ldots, \boldsymbol{\psi}_j^{(n)}$ be the basic eigenvectors and Jordan form eigenvectors corresponding to the eigenvalues μ_i, μ_j, respectively. For $\mu_i + \mu_j \neq 0$, there exists symplectic orthogonality between the eigenvectors as follows:

$$\langle \boldsymbol{\psi}_i^{(s)}, \boldsymbol{\psi}_j^{(t)} \rangle = \boldsymbol{\psi}_i^{(s)\mathrm{T}} \boldsymbol{J} \boldsymbol{\psi}_j^{(t)} = 0 \quad (s = 0, 1, \ldots, m;\ t = 0,\ 1, \ldots, n) \tag{1.3.16}$$

Proof. A proof for $r = s + t$ by mathematical induction is established.

(1) For $r = 0$, i.e. $s = 0$ and $t = 0$, we have

$$\boldsymbol{H}\boldsymbol{\psi}_i^{(0)} = \mu_i \boldsymbol{\psi}_i^{(0)}, \quad \boldsymbol{H}\boldsymbol{\psi}_j^{(0)} = \mu_j \boldsymbol{\psi}_j^{(0)}$$

because $\boldsymbol{\psi}_i^{(0)}, \boldsymbol{\psi}_j^{(0)}$ are the basic eigenvectors corresponding to the eigenvalues μ_i, μ_j, respectively. Hence,

$$\langle \boldsymbol{\psi}_i^{(0)}, \boldsymbol{H}\boldsymbol{\psi}_j^{(0)} \rangle = \langle \boldsymbol{\psi}_i^{(0)}, \mu_j \boldsymbol{\psi}_j^{(0)} \rangle = \mu_j \langle \boldsymbol{\psi}_i^{(0)}, \boldsymbol{\psi}_j^{(0)} \rangle$$

Similarly,

$$\langle \boldsymbol{\psi}_j^{(0)}, \boldsymbol{H}\boldsymbol{\psi}_i^{(0)} \rangle = \mu_i \langle \boldsymbol{\psi}_j^{(0)}, \boldsymbol{\psi}_i^{(0)} \rangle = -\mu_i \langle \boldsymbol{\psi}_i^{(0)}, \boldsymbol{\psi}_j^{(0)} \rangle$$

As $\boldsymbol{H}$ is a Hamiltonian matrix, the left-hand-sides of the two expressions above are equal. After substituting and rearranging the expressions, we have

$$(\mu_i + \mu_j) \langle \boldsymbol{\psi}_i^{(0)}, \boldsymbol{\psi}_j^{(0)} \rangle = 0$$

Since $\mu_i + \mu_j \neq 0$, Eq. (1.3.16) is tenable for $r = 0$.

(2) Assume that Eq. (1.3.16) is tenable for $r = k$, consider the case of $r = s + t = k + 1$.

Firstly, from Eq. (1.1.19), the eigenvectors $\boldsymbol{\psi}_i^{(s)}$ and $\boldsymbol{\psi}_j^{(t)}$ satisfy these equations

$$\boldsymbol{H}\boldsymbol{\psi}_i^{(s)} = \mu_i \boldsymbol{\psi}_i^{(s)} + \boldsymbol{\psi}_i^{(s-1)} \quad \text{and} \quad \boldsymbol{H}\boldsymbol{\psi}_j^{(t)} = \mu_j \boldsymbol{\psi}_j^{(t)} + \boldsymbol{\psi}_j^{(t-1)}$$

respectively. Hence

$$\langle \boldsymbol{\psi}_i^{(s)}, \boldsymbol{H}\boldsymbol{\psi}_j^{(t)} \rangle = \mu_j \langle \boldsymbol{\psi}_i^{(s)}, \boldsymbol{\psi}_j^{(t)} \rangle + \langle \boldsymbol{\psi}_i^{(s)}, \boldsymbol{\psi}_j^{(t-1)} \rangle$$

In accordance with the assumption of mathematical induction, Eq. (1.3.16) is tenable for $r = k$. As a result, we have

$$\langle \boldsymbol{\psi}_i^{(s)}, \boldsymbol{\psi}_j^{(t-1)} \rangle = 0$$

hence,

$$\langle \boldsymbol{\psi}_i^{(s)}, \boldsymbol{H}\boldsymbol{\psi}_j^{(t)} \rangle = \mu_j \langle \boldsymbol{\psi}_i^{(s)}, \boldsymbol{\psi}_j^{(t)} \rangle$$

Similarly,

$$\langle \boldsymbol{\psi}_j^{(t)}, \boldsymbol{H}\boldsymbol{\psi}_i^{(s)} \rangle = \mu_i \langle \boldsymbol{\psi}_j^{(t)}, \boldsymbol{\psi}_i^{(s)} \rangle = -\mu_i \langle \boldsymbol{\psi}_i^{(s)}, \boldsymbol{\psi}_j^{(t)} \rangle$$

As $\boldsymbol{H}$ is a Hamiltonian matrix, the left-hand-sides of the two expressions above are equal. After substituting and rearranging the expression, we have

$$(\mu_i + \mu_j) \langle \boldsymbol{\psi}_i^{(s)}, \boldsymbol{\psi}_j^{(t)} \rangle = 0$$

Since $\mu_i + \mu_j \neq 0$, Eq. (1.3.16) is tenable for $r = k + 1$ too. In accordance with mathematical induction, Eq. (1.3.16) is tenable for any s and t, and the proposition is proven.

The theorem above indicates that the basic eigenvectors and Jordan form eigenvectors corresponding to the non-symplectic adjoint eigenvalues are symplectic orthogonal. Subsequently, we discuss the relations among the eigenvectors corresponding to the mutually symplectic adjoint eigenvalues. For brevity of proofs in the remaining parts in this section, only one Jordan chain is assumed to exist for each eigenvalue.

Theorem 1.16. Let $\pm\mu \neq 0$ be a pair of mutually symplectic adjoint eigenvalues of a Hamiltonian matrix $\boldsymbol{H}$ with multiplicity m, then there exists an adjoint symplectic orthonormal vector set

$$\{\boldsymbol{\psi}^{(0)}, \boldsymbol{\psi}^{(1)}, \ldots, \boldsymbol{\psi}^{(m-1)}, \boldsymbol{\phi}^{(m-1)}, \boldsymbol{\phi}^{(m-2)}, \ldots, \boldsymbol{\phi}^{(0)}\}$$

such that

$$\langle \boldsymbol{\psi}^{(i)}, \boldsymbol{\phi}^{(j)} \rangle \begin{cases} = (-1)^i a \neq 0; & \text{when} \quad i + j = m - 1 \\ = 0; & \text{when} \quad i + j \neq m - 1 \end{cases} \tag{1.3.17}$$

where $\{\boldsymbol{\psi}^{(0)}, \boldsymbol{\psi}^{(1)}, \ldots, \boldsymbol{\psi}^{(m-1)}\}$ and $\{\boldsymbol{\phi}^{(0)}, \boldsymbol{\phi}^{(1)}, \ldots, \boldsymbol{\phi}^{(m-1)}\}$ are, respectively, the basic eigenvectors and Jordan form eigenvectors corresponding to μ and $-\mu$.

Proof. A proof for i by mathematical induction is established.

(1) For $i = 0$, let $\boldsymbol{\psi}^{(0)}$ be the basic eigenvector corresponding to the eigenvalue μ, and $\{\boldsymbol{\phi}^{(0)}, \ldots, \boldsymbol{\phi}^{(m-1)}\}$ be a set of arbitrary eigenvectors corresponding to the eigenvalue $-\mu$.

Firstly, for any $j \leq m - 2$, we have

$$\langle \boldsymbol{\psi}^{(0)}, \boldsymbol{H}\boldsymbol{\phi}^{(j+1)} \rangle = -\mu \langle \boldsymbol{\psi}^{(0)}, \boldsymbol{\phi}^{(j+1)} \rangle + \langle \boldsymbol{\psi}^{(0)}, \boldsymbol{\phi}^{(j)} \rangle$$

and

$$\langle \boldsymbol{\phi}^{(j+1)}, \boldsymbol{H}\boldsymbol{\psi}^{(0)} \rangle = \mu \langle \boldsymbol{\phi}^{(j+1)}, \boldsymbol{\psi}^{(0)} \rangle = -\mu \langle \boldsymbol{\psi}^{(0)}, \boldsymbol{\phi}^{(j+1)} \rangle$$

As $\boldsymbol{H}$ is a Hamiltonian matrix, the left-hand-sides of the two expressions above are equal. Hence, we have

$$\langle \boldsymbol{\psi}^{(0)}, \boldsymbol{\phi}^{(j)} \rangle = 0 \quad (j \leq m-2)$$

Next, according to Theorem 1.15, $\boldsymbol{\psi}^{(0)}$ is symplectic orthogonal to all other basic eigenvectors and Jordan form eigenvectors except that corresponding to the eigenvalue $-\mu$. Hence, $\boldsymbol{\psi}^{(0)}$ must be symplectic adjoint with $\boldsymbol{\phi}^{(m-1)}$. Otherwise $\boldsymbol{\psi}^{(0)}$ will be symplectic orthogonal to all eigenvectors, then $\boldsymbol{\psi}^{(0)} \equiv \mathbf{0}$. The conclusion is contradictory, i.e.

$$\langle \boldsymbol{\psi}^{(0)}, \boldsymbol{\phi}^{(m-1)} \rangle = a \neq 0$$

Hence, Eq. (1.3.17) is true for $i = 0$.

(2) Assuming there exists a set of basic eigenvectors and Jordan form eigenvectors $\{\boldsymbol{\psi}^{(0)}, \ldots, \boldsymbol{\psi}^{(m-1)}\}$ and $\{\boldsymbol{\phi}^{(0)}, \ldots, \boldsymbol{\phi}^{(m-1)}\}$ corresponding to μ, $-\mu$ such that Eq. (1.3.17) is true for $i \leq k$. Now consider the case of $i \leq k+1$.

Firstly, denote that

$$t = -\frac{1}{a} \langle \boldsymbol{\psi}^{(k+1)}, \boldsymbol{\phi}^{(m-1)} \rangle$$

and

$$\tilde{\boldsymbol{\psi}}^{(k+1+p)} = \boldsymbol{\psi}^{(k+1+p)} + t\boldsymbol{\psi}^{(p)} \quad (p = 0, 1, \ldots, (m-k-2))$$

Obviously, $\{\boldsymbol{\psi}^{(0)}, \ldots, \boldsymbol{\psi}^{(k)}, \tilde{\boldsymbol{\psi}}^{(k+1)}, \ldots, \tilde{\boldsymbol{\psi}}^{(m-1)}\}$ remains as a set of basic eigenvectors and Jordan form eigenvectors corresponding to the eigenvalue μ, and we have

$$\langle \tilde{\boldsymbol{\psi}}^{(k+1)}, \boldsymbol{\phi}^{(m-1)} \rangle = \langle \boldsymbol{\psi}^{(k+1)}, \boldsymbol{\phi}^{(m-1)} \rangle + t \langle \boldsymbol{\psi}^{(0)}, \boldsymbol{\phi}^{(m-1)} \rangle = 0$$

Next, for every $j < m-1$, we have

$$\langle \tilde{\boldsymbol{\psi}}^{(k+1)}, \boldsymbol{H}\boldsymbol{\phi}^{(j+1)} \rangle = -\mu \langle \tilde{\boldsymbol{\psi}}^{(k+1)}, \boldsymbol{\phi}^{(j+1)} \rangle + \langle \tilde{\boldsymbol{\psi}}^{(k+1)}, \boldsymbol{\phi}^{(j)} \rangle$$

and

$$\begin{aligned} \langle \boldsymbol{\phi}^{(j+1)}, \boldsymbol{H}\tilde{\boldsymbol{\psi}}^{(k+1)} \rangle &= \mu \langle \boldsymbol{\phi}^{(j+1)}, \tilde{\boldsymbol{\psi}}^{(k+1)} \rangle + \langle \boldsymbol{\phi}^{(j+1)}, \boldsymbol{\psi}^{(k)} \rangle \\ &= -\mu \langle \tilde{\boldsymbol{\psi}}^{(k+1)}, \boldsymbol{\phi}^{(j+1)} \rangle - \langle \boldsymbol{\psi}^{(k)}, \boldsymbol{\phi}^{(j+1)} \rangle \end{aligned}$$

As $\boldsymbol{H}$ is a Hamiltonian matrix, the left-hand-sides of the two expressions above are equal. Hence, we have

$$\langle \tilde{\boldsymbol{\psi}}^{(k+1)}, \boldsymbol{\phi}^{(j)} \rangle = -\langle \tilde{\boldsymbol{\psi}}^{(k)}, \boldsymbol{\phi}^{(j+1)} \rangle$$

In accordance with the assumption of mathematical induction, Eq. (1.3.17) is tenable for $r = k$. Hence, we have

$$\langle \tilde{\boldsymbol{\psi}}^{(k+1)}, \boldsymbol{\phi}^{(j)} \rangle \begin{cases} = (-1)^{k+1} a \neq 0; & \text{for } k+1+j = m-1 \\ = 0; & \text{for } k+1+j \neq m-1 \end{cases}$$

i.e. there exists a set of basic eigenvectors and Jordan form eigenvectors $\{\boldsymbol{\psi}^{(0)}, \ldots, \boldsymbol{\psi}^{(k)}, \tilde{\boldsymbol{\psi}}^{(k+1)}, \ldots, \tilde{\boldsymbol{\psi}}^{(m-1)}\}$ and $\{\boldsymbol{\phi}^{(0)}, \boldsymbol{\phi}^{(1)}, \ldots, \boldsymbol{\phi}^{(m-1)}\}$ corresponding to μ, $-\mu$ such that Eq. (1.3.17) is true for $i \leq k+1$. Hence, the theorem is proven.

Theorem 1.16 merely indicates that there exists an adjoint symplectic orthonormal vector set $\{\boldsymbol{\psi}^{(0)}, \boldsymbol{\psi}^{(1)}, \ldots, \boldsymbol{\psi}^{(m-1)}, \boldsymbol{\phi}^{(m-1)}, \ldots, \boldsymbol{\phi}^{(1)}, \boldsymbol{\phi}^{(0)}\}$ formed by the eigenvectors corresponding to eigenvalues $\pm\mu \neq 0$. If we take

$$\tilde{\boldsymbol{\phi}}^{(j)} = \frac{(-1)^{m-1-j}}{a} \boldsymbol{\phi}^{(j)} \quad (j = 0, 1, \ldots, m-1) \tag{1.3.18}$$

then $\{\boldsymbol{\psi}^{(0)}, \boldsymbol{\psi}^{(1)}, \ldots, \boldsymbol{\psi}^{(m-1)}, \tilde{\boldsymbol{\phi}}^{(m-1)}, \tilde{\boldsymbol{\phi}}^{(m-2)}, \ldots, \tilde{\boldsymbol{\phi}}^{(0)}\}$ forms a normal adjoint symplectic orthonormal vector set. It should be noted that the Jordan part corresponding to eigenvalue μ remains as the form of Eq. (1.1.16), while the Jordan part corresponding to eigenvalue $-\mu$ should be rewritten as

$$\boldsymbol{D}_{-\mu} = \begin{bmatrix} -\mu & 0 & 0 & \cdots & 0 \\ -1 & -\mu & 0 & \cdots & 0 \\ 0 & -1 & -\mu & \cdots & 0 \\ \vdots & \vdots & \vdots & \ddots & \vdots \\ 0 & 0 & 0 & \cdots & -\mu \end{bmatrix} \tag{1.3.19}$$

Then, we have

$$\left.\begin{aligned} \boldsymbol{H}\tilde{\boldsymbol{\phi}}^{(m-1)} &= -\mu\tilde{\boldsymbol{\phi}}^{(m-1)} - \tilde{\boldsymbol{\phi}}^{(m-2)} \\ \boldsymbol{H}\tilde{\boldsymbol{\phi}}^{(m-2)} &= -\mu\tilde{\boldsymbol{\phi}}^{(m-2)} - \tilde{\boldsymbol{\phi}}^{(m-3)} \\ &\cdots\cdots\cdots \\ \boldsymbol{H}\tilde{\boldsymbol{\phi}}^{(1)} &= -\mu\tilde{\boldsymbol{\phi}}^{(1)} - \tilde{\boldsymbol{\phi}}^{(0)} \\ \boldsymbol{H}\tilde{\boldsymbol{\phi}}^{(0)} &= -\mu\tilde{\boldsymbol{\phi}}^{(0)} \end{aligned}\right\} \tag{1.3.19'}$$

In such a way, a Hamiltonian matrix can be ensured when the Hamiltonian matrix $\boldsymbol{H}$ is transformed into a block diagonal matrix. Hence, we establish symplectic orthonormalization which is the basis of the expansion theorem.

Subsequently, we assume that the Jordan form eigenvectors corresponding to the (α)-set eigenvalues in Eq. (1.3.15) is again determined by Eq. (1.1.16), while the Jordan form eigenvectors corresponding to the (β)-set eigenvalues is determined by Eq. (1.3.19).

The discussion above concerns the adjoint symplectic orthonormality between the eigenvectors of the adjoint nonzero eigenvalues. From Theorem 1.14, there must be an even multiple root if a Hamiltonian matrix $\boldsymbol{H}$ has a zero eigenvalue. There usually exists the Jordan form for zero eigenvalue, and their solutions for physical problems have special interpretation in physics. The various physical problems will be introduced in due course in the following chapters.

Due to the special property of zero eigenvalue $\mu = -\mu = 0$, the corresponding basic eigenvectors and Jordan form eigenvectors form an adjoint symplectic orthonormal vector set. In order to discuss their nature of symplectic orthogonality, the following lemma is introduced.

Lemma. Let a Hamiltonian matrix $\boldsymbol{H}$ has a zero eigenvalue, and $\{\boldsymbol{\psi}^{(0)}, \boldsymbol{\psi}^{(1)}, \ldots, \boldsymbol{\psi}^{(2m-1)}\}$ be a set of arbitrary basic eigenvectors and Jordan form eigenvectors corresponding to the zero eigenvalue. Then for any $1 \le p \le 2m-1, 0 \le q \le 2m-2$, we have

$$\langle \boldsymbol{\psi}^{(p)}, \boldsymbol{\psi}^{(q)} \rangle = -\langle \boldsymbol{\psi}^{(p-1)}, \boldsymbol{\psi}^{(q+1)} \rangle \tag{1.3.20}$$

and for even $p+q$, we have

$$\langle \boldsymbol{\psi}^{(p)}, \boldsymbol{\psi}^{(q)} \rangle = 0 \tag{1.3.21}$$

Proof. Firstly, the eigenvectors $\boldsymbol{\psi}^{(p)}, \boldsymbol{\psi}^{(q+1)}$ corresponding to the zero eigenvalue satisfies

$$\langle \boldsymbol{\psi}^{(p)}, \boldsymbol{H}\boldsymbol{\psi}^{(q+1)} \rangle = \langle \boldsymbol{\psi}^{(p)}, \boldsymbol{\psi}^{(q)} \rangle$$

and

$$\langle \boldsymbol{\psi}^{(q+1)}, \boldsymbol{H}\boldsymbol{\psi}^{(p)} \rangle = \langle \boldsymbol{\psi}^{(q+1)}, \boldsymbol{\psi}^{(p-1)} \rangle = -\langle \boldsymbol{\psi}^{(p-1)}, \boldsymbol{\psi}^{(q+1)} \rangle$$

in accordance with Eq. (1.1.19). As $\boldsymbol{H}$ is a Hamiltonian matrix, the left-hand-sides of the two expressions above are equal. Hence, we have Eq. (1.3.20).

Next, we can assume without the loss of generality that $p = q + 2k$ for even $p + q$ where k is a nonnegative integer. Repeatedly applying Eq. (1.3.20) and using Eq. (1.3.3) yield

$$\begin{aligned}\langle \boldsymbol{\psi}^{(q+2k)}, \boldsymbol{\psi}^{(q)} \rangle &= -\langle \boldsymbol{\psi}^{(q+2k-1)}, \boldsymbol{\psi}^{(q+1)} \rangle \\ &= \cdots\cdots = (-1)^k \langle \boldsymbol{\psi}^{(q+k)}, \boldsymbol{\psi}^{(q+k)} \rangle = 0\end{aligned}$$

Hence the proposition is proven.

Theorem 1.17. If there exists a zero eigenvalue with multiplicity $2m$ for a Hamiltonian matrix $\boldsymbol{H}$, then there also exists a set of basic eigenvectors and Jordan form eigenvectors $\{\boldsymbol{\psi}^{(0)}, \boldsymbol{\psi}^{(1)}, \ldots, \boldsymbol{\psi}^{(2m-1)}\}$ corresponding to the zero eigenvalue such that these eigenvectors have adjoint symplectic orthonormality as follows

$$\langle \boldsymbol{\psi}^{(i)}, \boldsymbol{\psi}^{(j)} \rangle \begin{cases} = (-1)^i a \neq 0 & \text{for } i + j = 2m - 1 \\ = 0 & \text{for } i + j \neq 2m - 1 \end{cases} \tag{1.3.22}$$

Proof. A proof for i by mathematical induction is established.

(1) For $i = 0$, let $\{\boldsymbol{\psi}^{(0)}, \boldsymbol{\psi}^{(1)}, \ldots, \boldsymbol{\psi}^{(2m-1)}\}$ be a set of basic eigenvectors and Jordan form eigenvectors corresponding to a zero eigenvalue.

Firstly, using Eq. (1.1.19) for any $j \leq 2m - 2$ yields

$$\langle \boldsymbol{\psi}^{(0)}, \boldsymbol{H}\boldsymbol{\psi}^{(j+1)} \rangle = \langle \boldsymbol{\psi}^{(0)}, \boldsymbol{\psi}^{(j)} \rangle$$

On the other hand, as $\boldsymbol{H}$ is a Hamiltonian matrix, we have

$$\langle \boldsymbol{\psi}^{(0)}, \boldsymbol{H}\boldsymbol{\psi}^{(j+1)} \rangle = \langle \boldsymbol{\psi}^{(j+1)}, \boldsymbol{H}\boldsymbol{\psi}^{(0)} \rangle = 0$$

hence

$$\langle \boldsymbol{\psi}^{(0)}, \boldsymbol{\psi}^{(j)} \rangle = 0 \quad (j \leq 2m - 2)$$

Next, $\boldsymbol{\psi}^{(0)}$ is symplectic orthogonal to all eigenvectors of nonzero eigenvalues as well as the Jordan form eigenvectors in accordance to Theorem 1.15, hence $\boldsymbol{\psi}^{(0)}$ must be symplectic adjoint with $\boldsymbol{\psi}^{(2m-1)}$. Otherwise it will be symplectic orthogonal to all eigenvectors and this conclusion is contradictory. Hence we have

$$\langle \boldsymbol{\psi}^{(0)}, \boldsymbol{\psi}^{(2m-1)} \rangle = a \neq 0$$

Hence Eq. (1.3.22) is true for $i = 0$.

(2) Assume that there exists a set of basic eigenvectors and Jordan form eigenvectors $\{\boldsymbol{\psi}^{(0)}, \boldsymbol{\psi}^{(1)}, \ldots, \boldsymbol{\psi}^{(2m-1)}\}$ corresponding to a zero eigenvalue

such that Eq. (1.3.22) is true for $i \le k (k \ge 0)$. Now consider the case of $i \le k+1$.

(a) For odd k, we denote

$$t = -\frac{1}{2a}\langle \psi^{(k+1)}, \psi^{(2m-1)} \rangle$$

and take

$$\tilde{\psi}^{(k+1+p)} = \psi^{(k+1+p)} + t\psi^{(p)} \quad (p = 0, 1, \ldots, (2m-k-2))$$

Obviously, $\{\psi^{(0)}, \ldots, \psi^{(k)}, \tilde{\psi}^{(k+1)}, \ldots, \tilde{\psi}^{(2m-1)}\}$ is also a set of basic eigenvectors and Jordan form eigenvectors corresponding to the zero eigenvalue. Furthermore, it still holds for $i \le k (k \ge 0)$ with respect to Eq. (1.3.22). But here we have

$$\begin{aligned}\langle \tilde{\psi}^{(k+1)}, \tilde{\psi}^{(2m-1)} \rangle &= \langle \psi^{(k+1)}, \psi^{(2m-1)} \rangle + t\langle \psi^{(0)}, \psi^{(2m-1)} \rangle \\ &\quad + t\langle \psi^{(k+1)}, \psi^{(2m-k-2)} \rangle + t^2\langle \psi^{(0)}, \psi^{(2m-k-2)} \rangle \\ &= -2ta + ta + (-1)t\langle \psi^{(k)}, \psi^{(2m-k-1)} \rangle = 0\end{aligned}$$

(b) For even k, from the lemma in Eq. (1.3.21), we have

$$\langle \psi^{(k+1)}, \psi^{(2m-1)} \rangle = 0$$

Combining (a) and (b) yields a set of basic eigenvectors and Jordan form eigenvectors $\{\psi^{(0)}, \ldots, \psi^{(2m-1)}\}$ corresponding to the zero eigenvalue such that Eq. (1.3.22) is true for $i \le k$ $(k \ge 0)$. Furthermore, they satisfy $\langle \psi^{(k+1)}, \psi^{(2m-1)} \rangle = 0$. From Eq. (1.3.20), for any $j \le 2m-2$, we have

$$\langle \psi^{(k+1)}, \psi^{(j)} \rangle = -\langle \psi^{(k)}, \psi^{(j+1)} \rangle$$

According to the assumption of mathematical induction, Eq. (1.3.22) is tenable for $i \le k$ $(k \ge 0)$. Hence we have

$$\langle \psi^{(k+1)}, \psi^{(j)} \rangle \begin{cases} = (-1)^{k+1}a \ne 0; & \text{when } k+1+j = 2m-1 \\ = 0; & \text{when } k+1+j \ne 2m-1 \end{cases}$$

Therefore, Eq. (1.3.22) is tenable for $i \le k+1$ with respect to $\{\psi^{(0)}, \psi^{(1)}, \ldots, \psi^{(2m-1)}\}$.

In other words, there exists a set of eigenvectors $\{\psi^{(0)}, \psi^{(1)}, \ldots, \psi^{(2m-1)}\}$ corresponding to the zero eigenvalue such that Eq. (1.3.22) is tenable for $i \le k+1$. Therefore according to mathematical induction, the theorem is proven.

Theorems 1.14 to 1.17 indicate that there exists an adjoint symplectic orthonormal basis composed of the basic eigenvectors and Jordan form eigenvectors of the Hamiltonian matrix $\boldsymbol{H}$ in a $2n$-dimensional symplectic space. Through normalization, a normal adjoint symplectic orthonormal basis can be formed. The matrix formed by the column vectors is indeed a symplectic matrix. The properties of Hamiltonian matrix above are verified by a specific example as follows.

Example 1.3. Construct a normal adjoint symplectic orthonormal basis composed of the basic eigenvectors and Jordan form eigenvectors of the following Hamiltonian matrix

$$\boldsymbol{H} = \begin{bmatrix} 1 & -2 & 0 & 2 & 1 & 3 \\ 0 & -1 & 0 & 1 & 1 & 1 \\ 0 & -1 & -1 & 3 & 1 & 0 \\ 0 & 0 & 0 & -1 & 0 & 0 \\ 0 & -1 & 0 & 2 & 1 & 1 \\ 0 & 0 & 0 & 0 & 0 & 1 \end{bmatrix}$$

Solution: First determine the eigenvalue of the Hamiltonian matrix $\boldsymbol{H}$. From

$$|\mu \boldsymbol{I} - \boldsymbol{H}| = \mu^2(\mu - 1)^2(\mu + 1)^2$$

hence $\mu = 0, \pm 1$ are the eigenvalues of Hamiltonian matrix $\boldsymbol{H}$. All of them are eigenvalues with multiplicity 2.

Then we solve the eigenvectors. For the eigenvectors corresponding to eigenvalue $\mu = 0$, from

$$\boldsymbol{H}\boldsymbol{\psi} = 0$$

we obtain

$$\boldsymbol{\psi}_0^{(0)} = \{1, 1, 0, 0, 1, 0\}^{\mathrm{T}}$$

As there is only a single Jordan form chain corresponding to the zero eigenvalue, a Jordan form solution exists. From

$$\boldsymbol{H}\boldsymbol{\psi} = \boldsymbol{\psi}_0^{(0)}$$

we obtain

$$\boldsymbol{\psi}_0^{(1)} = \{0, 0, 1, 0, 1, 0\}^{\mathrm{T}}$$

Obviously, $\boldsymbol{\psi}_0^{(0)}$ is symplectic adjoint with $\boldsymbol{\psi}_0^{(1)}$

$$\langle \boldsymbol{\psi}_0^{(0)}, \boldsymbol{\psi}_0^{(1)} \rangle = 1$$

For the eigenvectors corresponding to eigenvalue $\mu = 1$, from

$$\boldsymbol{H}\boldsymbol{\psi} = \boldsymbol{\psi}$$

we obtain

$$\boldsymbol{\psi}_1^{(0)} = \{\, 1, 0, 0, 0, 0, 0 \,\}^{\mathrm{T}}$$

Similarly, there exists the Jordan form eigenvectors corresponding to eigenvalue $\mu = 1$. From

$$\boldsymbol{H}\boldsymbol{\psi} = \boldsymbol{\psi} + \boldsymbol{\psi}_1^{(0)}$$

we obtain

$$\boldsymbol{\psi}_1^{(1)} = \left\{ 0,\ \frac{1}{2},\ 0,\ 0,\ \frac{1}{2},\ \frac{1}{2} \right\}^{\mathrm{T}}$$

For the eigenvectors corresponding to eigenvalue $\mu = -1$, from

$$\boldsymbol{H}\boldsymbol{\psi} = -\boldsymbol{\psi}$$

we obtain

$$\boldsymbol{\psi}_{-1}^{(0)} = \{\, 0\ 0, 1, 0, 0, 0 \,\}^{\mathrm{T}}$$

Similarly, there exists the Jordan form eigenvectors corresponding to eigenvalue $\mu = -1$. From

$$\boldsymbol{H}\boldsymbol{\psi} = -\boldsymbol{\psi} + \boldsymbol{\psi}_{-1}^{(0)}$$

we obtain

$$\boldsymbol{\psi}_{-1}^{(1)} = \left\{ -\frac{1}{4},\ 0,\ 0,\ \frac{1}{2},\ -\frac{1}{2},\ 0 \right\}^{\mathrm{T}}$$

As there are only two vectors for the Jordan form chains corresponding to eigenvalue $\mu = \pm 1$, hence $\boldsymbol{\psi}_1^{(0)}$ and $\boldsymbol{\psi}_{-1}^{(1)}$, $\boldsymbol{\psi}_{-1}^{(0)}$ and $\boldsymbol{\psi}_1^{(1)}$ must be symplectic adjoint

$$a = \langle \boldsymbol{\psi}_{-1}^{(0)}, \boldsymbol{\psi}_1^{(1)} \rangle = -\langle \boldsymbol{\psi}_{-1}^{(1)}, \boldsymbol{\psi}_1^{(0)} \rangle = \frac{1}{2}$$

Here $\boldsymbol{\psi}_1^{(1)}$ and $\boldsymbol{\psi}_{-1}^{(1)}$ are not symplectic orthogonal. Referring to the proof of Theorem 1.16, denote

$$t = -2\langle \boldsymbol{\psi}_1^{(1)}, \boldsymbol{\psi}_{-1}^{(1)} \rangle = \frac{1}{2}$$

and take

$$\tilde{\boldsymbol{\psi}}_1^{(1)} = \boldsymbol{\psi}_1^{(1)} + t\boldsymbol{\psi}_1^{(0)} = \left\{ \frac{1}{2},\ \frac{1}{2},\ 0,\ 0,\ \frac{1}{2},\ \frac{1}{2} \right\}^{\mathrm{T}}$$

then the eigenvectors $\boldsymbol{\psi}_{-1}^{(1)}$ and $\tilde{\boldsymbol{\psi}}_1^{(1)}$ are symplectic orthogonal. According to Theorems 1.15 and 1.16, the other eigenvectors satisfy symplectic orthogonality. In this way an adjoint symplectic orthonormal basis formed by the eigenvectors of Hamiltonian matrix $\boldsymbol{H}$ is

$$\boldsymbol{\psi}_{-1}^{(0)},\quad \boldsymbol{\psi}_{-1}^{(1)},\quad \boldsymbol{\psi}_0^{(0)};\quad \tilde{\boldsymbol{\psi}}_1^{(1)},\quad \boldsymbol{\psi}_1^{(0)},\quad \boldsymbol{\psi}_0^{(1)}$$

In addition, from Eq. (1.3.18) and via symplectic normalization, a normal adjoint symplectic orthonormal basis

$$\boldsymbol{\psi}_{-1}^{(0)},\quad \boldsymbol{\psi}_{-1}^{(1)},\quad \boldsymbol{\psi}_0^{(0)};\quad \hat{\boldsymbol{\psi}}_1^{(1)},\quad \hat{\boldsymbol{\psi}}_1^{(0)},\quad \boldsymbol{\psi}_0^{(1)}$$

can be composed where

$$\begin{aligned} \hat{\boldsymbol{\psi}}_1^{(0)} &= -2\boldsymbol{\psi}_1^{(0)} = \{\,-2,\, 0,\, 0,\, 0,\, 0,\, 0\,\}^{\mathrm{T}} \\ \hat{\boldsymbol{\psi}}_1^{(1)} &= 2\tilde{\boldsymbol{\psi}}_1^{(1)} = \{\,1,\, 1,\, 0,\, 0,\, 1,\, 1\,\}^{\mathrm{T}} \end{aligned}$$

It should be clearly stated here that the arrangement of $\{\boldsymbol{\alpha}_1, \ldots, \boldsymbol{\alpha}_r, \boldsymbol{\beta}_1, \ldots, \boldsymbol{\beta}_r\}$ for all adjoint symplectic orthonormal vector sets expressed as Eq. (1.3.8) is adopted in this book. It is absolutely valid to have another arrangement as $\{\boldsymbol{\alpha}_1, \boldsymbol{\beta}_1, \boldsymbol{\alpha}_2, \boldsymbol{\beta}_2, \ldots, \boldsymbol{\alpha}_r, \boldsymbol{\beta}_r\}$, but in this case the definition of unit symplectic matrix Eq. (1.3.6) should be rewritten as

$$\boldsymbol{J}'_{2n} = \begin{bmatrix} \boldsymbol{J}'_2 & \boldsymbol{0} & \cdots & \boldsymbol{0} \\ \boldsymbol{0} & \boldsymbol{J}'_2 & \cdots & \boldsymbol{0} \\ \vdots & \vdots & \ddots & \vdots \\ \boldsymbol{0} & \boldsymbol{0} & \cdots & \boldsymbol{J}'_2 \end{bmatrix} \quad \text{where } \boldsymbol{J}'_2 = \boldsymbol{J}_2 = \begin{bmatrix} 0 & 1 \\ -1 & 0 \end{bmatrix} \tag{1.3.23}$$

Furthermore, other relevant definitions such as symplectic matrix, Hamiltonian matrix, etc. should be modified accordingly. This arrangement is comparatively more convenient for numerical analysis. The details are omitted here.

Table 1.1. The correlation between Euclidean space and symplectic space.

Euclidean space	**Symplectic space**
inner product $(\boldsymbol{\alpha}, \boldsymbol{\beta})$ — {length}	symplectic inner product $\langle \boldsymbol{\alpha}, \boldsymbol{\beta} \rangle$ — {area}
unit matrix $\boldsymbol{I}$	unit symplectic matrix $\boldsymbol{J}$
orthogonality $(\boldsymbol{x}, \boldsymbol{y}) = \boldsymbol{x}^{\mathrm{T}}\boldsymbol{y} (= \boldsymbol{x}^{\mathrm{T}}\boldsymbol{I}\boldsymbol{y}) = 0$	symplectic orthogonality $\langle \boldsymbol{x}, \boldsymbol{y} \rangle = \boldsymbol{x}^{\mathrm{T}}\boldsymbol{J}\boldsymbol{y} = 0$
(normal) orthogonal basis	(normal) adjoint symplectic orthonormal basis
orthogonal matrix $\boldsymbol{Q}^{\mathrm{T}}\boldsymbol{Q} = (\boldsymbol{Q}^{\mathrm{T}}\boldsymbol{I}\boldsymbol{Q} =)\boldsymbol{I}$	symplectic matrix $\boldsymbol{S}^{\mathrm{T}}\boldsymbol{J}\boldsymbol{S} = \boldsymbol{J}$
symmetry transformation $(\boldsymbol{\alpha}, \tilde{\boldsymbol{A}}\boldsymbol{\beta}) = (\boldsymbol{\beta}, \tilde{\boldsymbol{A}}\boldsymbol{\alpha})$	Hamiltonian transformation $\langle \boldsymbol{\alpha}, \tilde{\boldsymbol{H}}\boldsymbol{\beta} \rangle = \langle \boldsymbol{\beta}, \tilde{\boldsymbol{H}}\boldsymbol{\alpha} \rangle$
symmetric matrix $\boldsymbol{A}^{\mathrm{T}} = \boldsymbol{A} (= \boldsymbol{I}\boldsymbol{A}\boldsymbol{I})$	Hamiltonian matrix $\boldsymbol{H}^{\mathrm{T}} = \boldsymbol{JHJ}$
The eigenvalues of real symmetric matrix are real	If μ is an eigenvalue of a Hamiltonian matrix, $-\mu$ is also its eigenvalue
The eigenvectors corresponding to distinct eigenvalues of a real symmetric matrix are orthogonal	The eigenvectors corresponding to non-symplectic-adjoint eigenvalues of a Hamiltonian matrix are symplectic orthogonal
The eigenvectors of a real symmetric matrix can form a normal orthogonal basis	The eigenvectors of a Hamiltonian matrix can form a normal adjoint symplectic orthonormal basis

In this section, the basic concepts of a finite-dimensional symplectic space are elaborated and some fundamental properties are briefly introduced. It is of certainty that the many concepts and properties can be directly generalized to an infinite-dimensional symplectic space. Towards the end of this section, the correlation between a Euclidean space and a symplectic space is presented in order to better describe the relevant concepts and properties of a symplectic space for the benefit of readers.

1.4. Legengre's Transformation

Legendre's transformation in the scope of mathematics is introduced in this section. It is the key to realize a transformation from the Lagrange system to the Hamiltonian system.

Consider Legendre's transformation in two variables[6]. Let $f = f(x, y)$, then

$$\mathrm{d}f = u\mathrm{d}x + v\mathrm{d}y \tag{1.4.1}$$

where

$$u = \frac{\partial f}{\partial x}, \quad v = \frac{\partial f}{\partial y} \tag{1.4.2}$$

Here we choose x, y as the independent variables. In reality, we may choose any two of x, y, u, v as the independent variables to suit the problem under consideration. If we choose u, y as the independent variables, we obtain from Eq. (1.4.2):

$$x = x(u, y), \quad v = v(u, y) \tag{1.4.3}$$

while the function f can be expressed in terms of u, y as:

$$\bar{f}(u, y) = f[x(u, y), y] \tag{1.4.4}$$

then

$$\left.\begin{aligned} \frac{\partial \bar{f}}{\partial y} &= \frac{\partial f}{\partial x}\frac{\partial x}{\partial y} + \frac{\partial f}{\partial y} = u\frac{\partial x}{\partial y} + v \\ \frac{\partial \bar{f}}{\partial u} &= \frac{\partial f}{\partial x}\frac{\partial x}{\partial u} = u\frac{\partial x}{\partial u} = \frac{\partial}{\partial u}(ux) - x \end{aligned}\right\} \tag{1.4.5}$$

Equation (1.4.5) can be expressed as

$$\left.\begin{aligned} v &= -\frac{\partial}{\partial y}\left(ux - \bar{f}\right) = -\frac{\partial g}{\partial y} \\ x &= \frac{\partial}{\partial u}\left(ux - \bar{f}\right) = \frac{\partial g}{\partial u} \end{aligned}\right\} \tag{1.4.6}$$

where $g(u, y) = ux - \bar{f} = x\partial f/\partial x - f$. It shows that when the independent variables change from x, y to u, y and using function $\bar{f}$, then x, v cannot be directly expressed in terms of the partial derivative of $\bar{f}$ with respect to u and y as in Eq. (1.4.2). Instead, we should use function g which is equal to the variable to be eliminated x multiplied by the partial derivative of the former function with respect to this variable $u = \partial f/\partial x$ and minus the original function f. Hence, x, v can then be expressed in terms of the partial derivative of g with respect to u and y. This is the basic principle of Legendre's transformation.

The discussion above is merely for the Legendre's transformation on variable x. Of course, we may also perform Legendre's transformation on two variables x, y simultaneously, i.e. we may choose u, v as the independent variables. In a similar way, using (1.4.2) yields:

$$x = x(u, v), \quad y = y(u, v) \tag{1.4.7}$$

and function f can be alternatively expressed in terms of u, v, as

$$\tilde{f}(u,v) = f[x(u,v), y(u,v)] \tag{1.4.8}$$

Introducing transformation function

$$\tilde{g}(u,v) = ux + vy - \tilde{f}(u,v) \tag{1.4.9}$$

it is obvious that following relation exists

$$\left.\begin{aligned} \frac{\partial \tilde{g}}{\partial u} &= x + u\frac{\partial x}{\partial u} + v\frac{\partial y}{\partial u} - \frac{\partial f}{\partial x}\frac{\partial x}{\partial u} - \frac{\partial f}{\partial y}\frac{\partial y}{\partial u} = x \\ \frac{\partial \tilde{g}}{\partial v} &= u\frac{\partial x}{\partial v} + y + v\frac{\partial y}{\partial v} - \frac{\partial f}{\partial x}\frac{\partial x}{\partial v} - \frac{\partial f}{\partial y}\frac{\partial y}{\partial v} = y \end{aligned}\right\} \tag{1.4.10}$$

i.e. x, y can be expressed in terms of the partial derivative of $\tilde{g}$ with respect to u and v.

In this section, we introduce the Legendre's transformation for two variables. The approach can be directly generalized from two variables to multiple variables. The details are omitted here.

1.5. The Hamiltonian Principle and the Hamiltonian Canonical Equations

"The nature always chooses the simplest and most possible way." This is a famous principle of Fermat. The minimum action principle in classical mechanics is originated from the Hamiltonian principle. It is often described in terms of the n-dimensional general displacement $q_i (i = 1, 2, \ldots, n)$ with finite degrees of freedom or in terms of vector $\boldsymbol{q}$. Using $\dot{q}_i$ to indicate differentiation with respect to time, the **Lagrange function** of a dynamic system (kinetic energy minus potential energy) is

$$\mathscr{L}(\boldsymbol{q}, \dot{\boldsymbol{q}}) \quad \text{or}\ \mathscr{L}(q_1, q_2, \ldots, q_n; \dot{q}_1, \dot{q}_2, \ldots, \dot{q}_n) \tag{1.5.1}$$

The Hamiltonian principle states that the actual path of a conservative system from the initial point $(\boldsymbol{q}_0, t_0)$ to the terminal point $(\boldsymbol{q}_\mathrm{e}, t_\mathrm{e})$ is such that the action A is a stationary value,

$$A = \int_{t_0}^{t_\mathrm{e}} \mathscr{L}(\boldsymbol{q}, \dot{\boldsymbol{q}})\mathrm{d}t; \quad \delta A = 0 \tag{1.5.2}$$

Performing variation of Eq. (1.5.2) and integrating by parts yield

$$\delta A = \int_{t_0}^{t_e} \left[\frac{\partial \mathscr{L}}{\partial \boldsymbol{q}} - \frac{\mathrm{d}}{\mathrm{d}t} \left(\frac{\partial \mathscr{L}}{\partial \dot{\boldsymbol{q}}} \right) \right]^{\mathrm{T}} \cdot \delta \boldsymbol{q} \mathrm{d}t = 0 \tag{1.5.3}$$

As $\delta \boldsymbol{q}$ is arbitrary, the **Lagrange equation**:

$$\frac{\mathrm{d}}{\mathrm{d}t} \left(\frac{\partial \mathscr{L}}{\partial \dot{\boldsymbol{q}}} \right) = \frac{\partial \mathscr{L}}{\partial \boldsymbol{q}} \tag{1.5.4}$$

is derived. Hence, the Hamiltonian principle (1.5.2) corresponds to the Lagrange equation (1.5.4), which is a system of second-order ordinary differential equations. The expression above only includes one class of variables like displacement, and therefore it is a variational principle with a single class of variables.

The Hamiltonian canonical system has already been developed in classical analytical mechanics. It transforms a class of independent variables $\dot{\boldsymbol{q}}$ (generalized velocity) of Lagrange function $\mathscr{L}$ into $\boldsymbol{p}$ (generalized momentum, i.e. dual variable) by Legendre's transformation

$$\boldsymbol{p} = \frac{\partial \mathscr{L}}{\partial \dot{\boldsymbol{q}}} \tag{1.5.5}$$

From Eq. (1.5.5), we may solve $\dot{\boldsymbol{q}}$ such that $\dot{\boldsymbol{q}}$ is a function of $\boldsymbol{p}, \boldsymbol{q}$, or

$$\dot{\boldsymbol{q}} = \dot{\boldsymbol{q}}(\boldsymbol{p}, \boldsymbol{q}) \tag{1.5.6}$$

According to the principle of Legendre's transformation, we introduce a transformation function, i.e. the **Hamiltonian function** (kinetic energy plus potential energy)

$$\mathscr{H}(\boldsymbol{q}, \boldsymbol{p}) = \boldsymbol{p}^{\mathrm{T}} \dot{\boldsymbol{q}} - \mathscr{L}(\boldsymbol{q}, \dot{\boldsymbol{q}}(\boldsymbol{p}, \boldsymbol{q})) \tag{1.5.7}$$

Hence, Eq. (1.4.6) yields

$$\frac{\partial \mathscr{L}}{\partial \boldsymbol{q}} = -\frac{\partial \mathscr{H}}{\partial \boldsymbol{q}}; \quad \dot{\boldsymbol{q}} = \frac{\partial \mathscr{H}}{\partial \boldsymbol{p}} \tag{1.5.8}$$

On the other hand, from Eq. (1.5.4)

$$\frac{\partial \mathscr{L}}{\partial \boldsymbol{q}} = \frac{\mathrm{d}}{\mathrm{d}t} \left(\frac{\partial \mathscr{L}}{\partial \dot{\boldsymbol{q}}} \right) = \dot{\boldsymbol{p}} \tag{1.5.9}$$

we obtain

$$\dot{\boldsymbol{q}} = \frac{\partial \mathscr{H}}{\partial \boldsymbol{p}}; \quad \dot{\boldsymbol{q}} = -\frac{\partial \mathscr{H}}{\partial \boldsymbol{q}} \tag{1.5.10}$$

Equation (1.5.10) are the **Hamiltonian canonical equations**, in which there are two classes of variables: the generalized displacement $\boldsymbol{q}$ and the generalized momentum $\boldsymbol{p}$. The variational principle corresponding to the Hamiltonian equations (1.5.10) is

$$\delta \int_{t_0}^{t_e} [\boldsymbol{p}^{\mathrm{T}} \dot{\boldsymbol{q}} - \mathscr{H}(\boldsymbol{q}, \boldsymbol{p})] \, \mathrm{d}t = 0 \tag{1.5.11}$$

where $\boldsymbol{q}$ and $\boldsymbol{p}$ should be regarded as two unrelated variables with independent variation. Equation (1.5.10) can be obtained directly when the variational formula (1.5.11) is expanded.

The process of transforming from the variational principle with a single class of variables in Eq. (1.5.2) to the variational principle with two classes of variables in Eq. (1.5.11) bears a classical feature realized through Legendre's transformation.

1.6. The Reciprocal Theorems

An elastic system is a system without energy dissipation. Hence the strain energy stored during the process of deformation must be equal to the work done by the external force during the process. If applied in a quasi-static process, the work done by the external force only depends on the state of displacement state at the particular moment. If a linear infinitesimal deformation problem is considered, the principle of superposition is applicable. A series of reciprocal theorems can therefore be derived by combining the principles of superposition and energy conservation.

1.6.1. *The Reciprocal Theorem for Work*[7]

Consider a linear system under the action of two sets of forces F_A and F_B at two different positions as illustrated in Fig. 1.1. Because it is an elastic system, the work done by the external forces during the process of deformation is independent of the sequence of loading. It only depends on the final state of the external forces. Now consider the outcome of two different loading routes.

For the first loading route, the action of F_A occurs first in a quasi-static progressive manner and it is followed by the action of F_B. F_B is not in action when F_A is applied. Hence the only work done is by F_A with magnitude 0.5 $F_A u_A$, where u_A is the displacement at A caused by F_A.

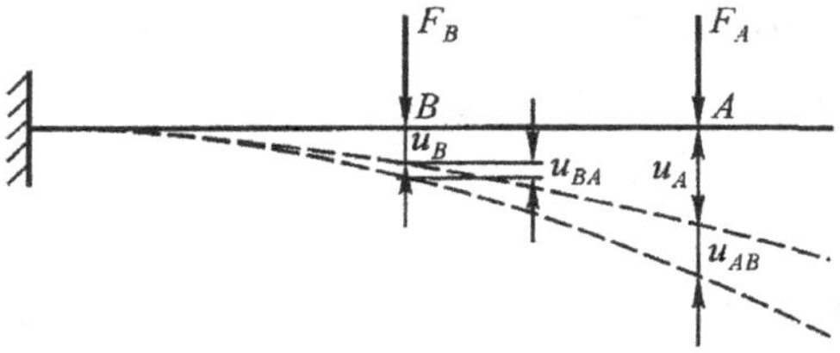

Fig. 1.1. Two different load processes.

After applying F_A, the action of force F_B on the system begins. The work done by F_B is 0.5 $F_B u_B$ where u_B is the displacement at B caused by F_B. During the process of loading F_B, a displacement u_{AB} at A (i.e. the displacement at A caused by F_B) occurs. As F_A applied on, the system remains, the additional work done by F_A is $F_A u_{AB}$. Hence the total work done by the external forces is

$$W_1 = \frac{1}{2}F_A u_A + \frac{1}{2}F_B u_B + F_A u_{AB} \tag{1.6.1}$$

For the second loading route, F_B acts first on the system and follows by F_A. Following the same analytical procedure, we obtain the total work done by the external forces as

$$W_2 = \frac{1}{2}F_B u_B + \frac{1}{2}F_A u_A + F_B u_{BA} \tag{1.6.2}$$

where u_{BA} is the displacement at B caused by F_A.

Because the final state due to different loading sequences does not change, the ultimate deformation of structure for the two varying loading sequences are identical. Therefore, we have $W_1 = W_2$ and

$$F_A u_{AB} = F_B u_{BA} \tag{1.6.3}$$

Hence we establish **the reciprocal theorem for work: in a system undergoing arbitrary linear elastic deformation, the work done by a first set of external forces due to the displacement caused by a second set of external forces is equal to the work done by the second set of external forces due to the displacement caused by the first set of external forces.** The reciprocal theorem for work in elastic mechanics and structural mechanics is very useful. From this theorem it is deduced the reciprocal theorem for displacement and the reciprocal theorem for reaction.

1.6.2. *The Reciprocal Theorem for Displacement*

Supposing $F_A = F_B = 1$ in Eq. (1.6.3), we obtain

$$u_{AB} = u_{BA} \tag{1.6.4}$$

Hence, we establish **the reciprocal theorem for displacement: in a system undergoing linear elastic deformation, the displacement at B caused by unit force acting at A is equal to the displacement at A caused by unit force acting at B.**

1.6.3. *The Reciprocal Theorem for Reaction*

Consider the settlement problem of supports as shown in Fig. 1.2. Let s_A and s_B are the settlements at supports A and B, respectively. For a linear elastic system, the state of system is independent of the order of settlement of supports. It only depends on the final settlement shape. Now we consider settlement in two different cases.

In the first case, s_A takes place first and follows by s_B. During the settlement of s_A, there are reactions F_A and F_{BA} at A and B, respectively. During the process, there is no settlement at support B and therefore the only work done is by the reaction F_A at A, which is 0.5 $F_A s_A$. Then a settlement s_B at support B takes place and it generates the reactions F_B and F_{AB} at B and A, respectively. There is no work done at support A because A does not move. As reaction F_{BA} at B exists during the process, the work done is $F_{BA} s_B$. In addition, the work done by the gradually

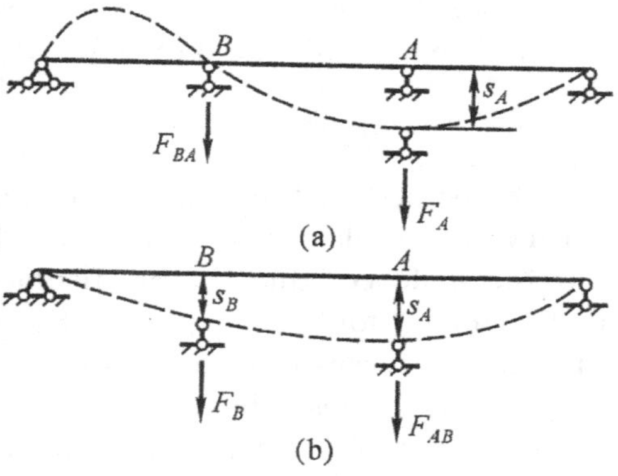

Fig. 1.2. Support settlement.

increasing reaction F_B is 0.5 $F_B s_B$. Hence the total work is

$$W_1 = \frac{1}{2}F_A s_A + \frac{1}{2}F_B s_B + F_{BA} s_B \tag{1.6.5}$$

In the second case, s_B takes place first and follows by s_A. Following the same procedure, we obtain the total work as

$$W_2 = \frac{1}{2}F_B s_B + \frac{1}{2}F_A s_A + F_{AB} s_A \tag{1.6.6}$$

Because the final settlements states are identical, hence $W_1 = W_2$ and

$$F_{BA} s_B = F_{AB} s_A \tag{1.6.7}$$

Further, assuming $s_A = s_B = 1$ yields

$$F_{AB} = F_{BA} \tag{1.6.8}$$

Hence, we establish **the reciprocal theorem for reaction: in a system undergoing linear elastic deformation, the reaction at B caused by unit displacement at A is equal to the reaction at A caused by unit displacement at B.**

1.6.4. *The Reciprocal Theorem for Displacement and Negative Reaction*

Suppose there is a unit external force[a] acting at point A of an elastic system, while there is unit support displacement at support B as illustrated in Fig. 1.3. As the strain energy of a linear elastic system depends only on the final states of loads and support displacements and is independent of orders they are applied, we may consider the following two cases with different orders of application.

[a]The magnitude of unit external force and unit displacement is 1. These quantities are omitted in the following equations.

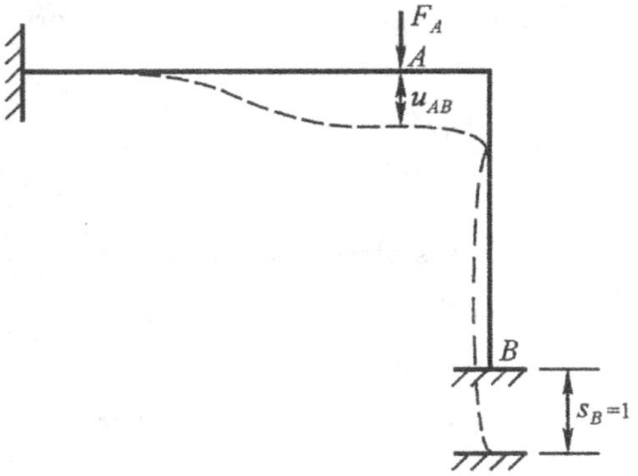

Fig. 1.3. The reciprocal for displacement and negative reaction.

In the first case, we apply a unit displacement $s_B = 1$ at support B and the displacement at A is u_{AB}. There is no work at A because the unit force at A has not been applied. The only work $0.5F_Bs_B$ is done by the reaction F_B at B. Then we apply a unit external force $F_A = 1$ at A, which results in a reaction F_{BA} at B. Again the reaction at B does no work because there is no support displacement during this process. The only work $u_A \times 1/2$ is due to the unit external force where u_A is the displacement at A resulting from the unit force at A. The work done by external forces corresponding to this order of application is

$$W_1 = \frac{1}{2}u_A + \frac{1}{2}F_B \tag{1.6.9}$$

In the second case, we apply a unit external force at A first and the work done by this external force is $u_A \times 1/2$. Then we apply a unit displacement at support B. As then there exists a reaction F_{BA} at B, the work done by the external forces during the latter process is $u_{AB} \times 1 + F_{BA} \times 1 + F_B \times 1/2$. Hence the work done by the external forces corresponding to this order of application sums up to

$$W_2 = \frac{1}{2}u_A + \frac{1}{2}F_B + u_{AB} + F_{BA} \tag{1.6.10}$$

These two orders of application correspond the identical final state of external influences and structural deformation. Therefore $W_1 = W_2$ and

$$F_{BA} = -u_{AB} \tag{1.6.11}$$

As a result, we establish **the reciprocal theorem for displacement and negative reaction: in an arbitrary system undergoing linear**

elastic deformation, the reaction at support B caused by a unit force acting at A is equal to the negative value of the displacement at A caused by a unit displacement at support B.

When the hybrid method (i.e. partially the unknown quantities of some displacements and partially the unknown quantities of redundant internal force) is adopted to solve a system of structural mechanics, the symmetry or antisymmetry of the canonical equation is reflected in the reciprocal theorems for displacement, for reaction and for displacement and negative reaction. The Hamiltonian system corresponds to the adoption of the hybrid method, and the properties of Hamiltonian matrix reflects the reciprocal theorems for displacement, for reaction, and for displacement and negative reaction.

The chapter discusses some basic concepts, such as symplectic space, Hamiltonian system, etc. from various aspects of mathematics and analytical mechanics. We will introduce in detail the solutions of symplectic Hamiltonian systems through some elasticity problems in the following chapters.

References

1. Rongchang Shi, *Matrix Analysis*, Beijing: Beijing Institute of Technology Press, 1996 (in Chinese).
2. Yizhong Lan and Chunlai Zhao, *Theory of Linear Algebra*, Beijing: Peking University Press, 1998 (in Chinese).
3. Department of Mathematical Mechanics of Peking University, *Advanced Algebra*, Beijing: People's Education Press, 1978 (in Chinese).
4. J. Koszul and Yiming Zou, *Theory of Symplectic Geometry*, Beijing: Science Press, 1986 (in Chinese).
5. MengZhao Qin, Symplectic geometry and computing Hamiltonian mechanics, *Mechanics and Practice*, 1990, 12(6): 1–20 (in Chinese).
6. Yanbai Zhou, *Lectures on Theoretical Mechanics*, Beijing: People's Education Press, 1979 (in Chinese).
7. Lingxi Qian, *Statically Indeterminate Structural Analysis*, Shanghai: Shanghai Scientific and Technical Publishers, 1951 (in Chinese).

Chapter 2

Fundamental Equations of Elasticity and Variational Principle

Beginning with some fundamental concepts such as stress, strain, stress-strain relationship, etc., this chapter briefly introduces the fundamental equations of elasticity and the variational principle. The conventions of the mechanical quantities are then discussed generally.

2.1. Stress Analysis

In a rectangular Cartesian coordinate system $Oxyz$, the state of stress at an arbitrary point P in an elastic body can be described by six independent components of stress[1], which can be written as a vector

$$\boldsymbol{\sigma} = \left\{ \sigma_x, \quad \sigma_y, \quad \sigma_z, \quad \tau_{xy}, \quad \tau_{xz}, \quad \tau_{yz} \right\}^{\mathrm{T}} \tag{2.1.1}$$

The positive directions of the stress components are shown in Fig. 2.1. The three pairs of shear stresses are $\tau_{xy} = \tau_{yx}$, $\tau_{xz} = \tau_{zx}$, $\tau_{zy} = \tau_{yz}$.

If these components of stress are known, the stress acting at point P on any inclined plane can be determined as

$$\left.\begin{aligned} F_{nx} &= \sigma_x l + \tau_{xy} m + \tau_{xz} n \\ F_{ny} &= \tau_{xy} l + \sigma_y m + \tau_{yz} n \\ F_{nz} &= \tau_{xz} l + \tau_{yz} m + \sigma_z n \end{aligned}\right\} \tag{2.1.2}$$

where l, m, n are the direction cosines of the exterior normal vector $\boldsymbol{n}$ with respect to the inclined plane, and F_{nx}, F_{ny}, F_{nz} are, respectively, the projections of the force per unit area acting on the inclined plane on x, y, z.

Equation (2.1.2) can be expressed in the form of matrix as

$$\boldsymbol{F}_n = \boldsymbol{E}(\boldsymbol{n})\boldsymbol{\sigma} \tag{2.1.2$'$}$$

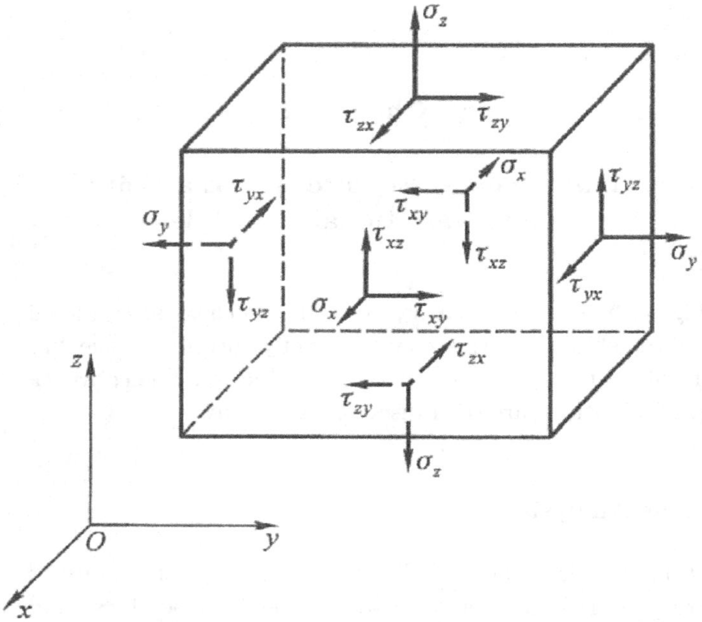

Fig. 2.1. Senses of positive stress components.

where

$$\boldsymbol{E}(\boldsymbol{n}) = \begin{bmatrix} l & 0 & 0 & m & n & 0 \\ 0 & m & 0 & l & 0 & n \\ 0 & 0 & n & 0 & l & m \end{bmatrix} \tag{2.1.3}$$

Using Eq. (2.1.2), the expressions of stress components under rotation transformation of coordinate systems can be obtained. Consider two sets of rectangular Cartesian frames of reference $Ox'y'z'$ and $Oxyz$. Let the direction cosines of the axes in $Ox'y'z'$ relative to the axes in $Oxyz$ be

axes	x	y	z
x'	l_1	m_1	n_1
y'	l_2	m_2	n_2
z'	l_3	m_3	n_3

Hence, by virtue of Eq. (2.1.2), the stresses acting on the plane with the new x'-axis as the outer normal direction are

$$\left.\begin{aligned} F_{1x} &= \sigma_x l_1 + \tau_{xy} m_1 + \tau_{xz} n_1 \\ F_{1y} &= \tau_{xy} l_1 + \sigma_y m_1 + \tau_{yz} n_1 \\ F_{1z} &= \tau_{xz} l_1 + \tau_{yz} m_1 + \sigma_z n_1 \end{aligned}\right\} \tag{2.1.4}$$

Then by projecting F_{1x}, F_{1y}, F_{1z} on axes x', y' and z', respectively, the normal stresses and shear stresses acting on this plane are

$$\begin{aligned} \sigma_{x'} &= F_{1x} l_1 + F_{1y} m_1 + F_{1z} n_1 \\ &= \sigma_x l_1^2 + \sigma_y m_1^2 + \sigma_z n_1^2 + 2\tau_{xy} l_1 m_1 \\ &\quad + 2\tau_{xz} l_1 n_1 + 2\tau_{yz} m_1 n_1 \end{aligned} \tag{2.1.5a}$$

$$\begin{aligned} \tau_{x'y'} &= F_{1x} l_2 + F_{1y} m_2 + F_{1z} n_2 \\ &= \sigma_x l_1 l_2 + \sigma_y m_1 m_2 + \sigma_z n_1 n_2 + \tau_{xy}(l_1 m_2 + l_2 m_1) \\ &\quad + \tau_{xz}(l_1 n_2 + l_2 n_1) + \tau_{yz}(m_1 n_2 + m_2 n_1) \end{aligned} \tag{2.1.5b}$$

$$\begin{aligned} \tau_{x'z'} &= F_{1x} l_3 + F_{1y} m_3 + F_{1z} n_3 \\ &= \sigma_x l_1 l_3 + \sigma_y m_1 m_3 + \sigma_z n_1 n_3 + \tau_{xy}(l_1 m_3 + l_3 m_1) \\ &\quad + \tau_{xz}(l_1 n_3 + l_3 n_1) + \tau_{yz}(m_1 n_3 + m_3 n_1) \end{aligned} \tag{2.1.5c}$$

Similarly, the normal stresses and shear stresses acting on the planes with the new y'- and z'-axis as the outer normal directions are

$$\begin{aligned} \sigma_{y'} &= \sigma_x l_2^2 + \sigma_y m_2^2 + \sigma_z n_2^2 + 2\tau_{xy} l_2 m_2 \\ &\quad + 2\tau_{xz} l_2 n_2 + 2\tau_{yz} m_2 n_2 \end{aligned} \tag{2.1.5d}$$

$$\begin{aligned} \sigma_{z'} &= \sigma_x l_3^2 + \sigma_y m_3^2 + \sigma_z n_3^2 + 2\tau_{xy} l_3 m_3 \\ &\quad + 2\tau_{xz} l_3 n_3 + 2\tau_{yz} m_3 n_3 \end{aligned} \tag{2.1.5e}$$

$$\begin{aligned} \tau_{y'z'} &= \sigma_x l_2 l_3 + \sigma_y m_2 m_3 + \sigma_z n_2 n_3 + \tau_{xy}(l_2 m_3 + l_3 m_2) \\ &\quad + \tau_{xz}(l_2 n_3 + l_3 n_2) + \tau_{yz}(m_2 n_3 + m_3 n_2) \end{aligned} \tag{2.1.5f}$$

The expressions in Eq. (2.1.5) are the formula of stress components under rotation transformation of coordinate systems.

As a coordinate system rotates, there exists a specific coordinate system on an inclined plane with its outer normal direction as one of the coordinate axes such that the shear stresses acting on that plane vanish and only the normal stresses $\sigma_{x'}$, $\sigma_{y'}$ and $\sigma_{z'}$ remain. This particular direction is called the **principal direction of stress** and the corresponding normal stresses are called the **principal stresses**. The normal stress σ and the direction cosines of the principal direction l, m, n fulfill the following equations

$$\left.\begin{aligned} &(\sigma_x - \sigma)l + \tau_{xy}m + \tau_{xz}n = 0 \\ &\tau_{xy}l + (\sigma_y - \sigma)m + \tau_{yz}n = 0 \\ &\tau_{xz}l + \tau_{yz}m + (\sigma_z - \sigma)n = 0 \end{aligned}\right\} \tag{2.1.6}$$

Because the direction cosines do not vanish simultaneously, the determinant of the coefficient matrix vanishes; i.e.

$$\begin{vmatrix} \sigma_x - \sigma & \tau_{xy} & \tau_{xz} \\ \tau_{xy} & \sigma_y - \sigma & \tau_{yz} \\ \tau_{xz} & \tau_{yz} & \sigma_z - \sigma \end{vmatrix} = 0 \tag{2.1.7}$$

On expanding the determinant, we have the eigenvalue equation of σ, which can be written in a compact form as

$$\sigma^3 - I_1\sigma^2 + I_2\sigma - I_3 = 0 \tag{2.1.8}$$

where

$$\begin{aligned} I_1 &= \sigma_x + \sigma_y + \sigma_z \\ I_2 &= \sigma_x\sigma_y + \sigma_x\sigma_z + \sigma_y\sigma_z - \tau_{xy}^2 - \tau_{xz}^2 - \tau_{yz}^2 \\ I_3 &= \begin{vmatrix} \sigma_x & \tau_{xy} & \tau_{xz} \\ \tau_{xy} & \sigma_y & \tau_{yz} \\ \tau_{xz} & \tau_{yz} & \sigma_z \end{vmatrix} \end{aligned} \tag{2.1.9}$$

I_1, I_2, I_3 are independent of the choice of coordinate systems and they are called the first, the second, and the third stress invariants, respectively. The invariant I_2 is often used in the Mises criteria in plastic yield.

As the stress components form a real symmetric matrix, there exist three real roots of Eq. (2.1.7) denoted as $\sigma_1, \sigma_2, \sigma_3$, which are the principal stresses. Furthermore, there must exist three mutually orthogonal principal directions.

The state of stress at an arbitrary point in an elastic body has been discussed above.

Consider an elastic body acted upon by some external forces where F_x, F_y, F_z denote the body force components per unit volume. Then the stress components fulfill the following equations of equilibrium

$$\frac{\partial \sigma_x}{\partial x} + \frac{\partial \tau_{xy}}{\partial y} + \frac{\partial \tau_{xz}}{\partial z} + F_x = 0 \tag{2.1.10a}$$

$$\frac{\partial \tau_{xy}}{\partial x} + \frac{\partial \sigma_y}{\partial y} + \frac{\partial \tau_{yz}}{\partial z} + F_y = 0 \tag{2.1.10b}$$

$$\frac{\partial \tau_{xz}}{\partial x} + \frac{\partial \tau_{yz}}{\partial y} + \frac{\partial \sigma_z}{\partial z} + F_z = 0 \tag{2.1.10c}$$

Equations (2.1.10) can be expressed by an operator matrix $\boldsymbol{E}(\nabla)^2$ as

$$\boldsymbol{E}(\nabla)\boldsymbol{\sigma} + \boldsymbol{F} = \boldsymbol{0} \tag{2.1.10$'$}$$

where

$$\boldsymbol{E}(\nabla) = \begin{bmatrix} \partial/\partial x & 0 & 0 & \partial/\partial y & \partial/\partial z & 0 \\ 0 & \partial/\partial y & 0 & \partial/\partial x & 0 & \partial/\partial z \\ 0 & 0 & \partial/\partial z & 0 & \partial/\partial x & \partial/\partial y \end{bmatrix} \tag{2.1.11}$$

$$\boldsymbol{F} = \begin{bmatrix} F_x, & F_y, & F_z \end{bmatrix}^{\mathrm{T}} \tag{2.1.12}$$

2.2. Strain Analysis

The state of strain at every point in an elastic body can be described by six independent components of strain in a rectangular Cartesian coordinate system $Oxyz$. They can be written as a vector

$$\boldsymbol{\varepsilon} = \{\varepsilon_x, \quad \varepsilon_y, \quad \varepsilon_z, \quad \gamma_{xy}, \quad \gamma_{xz}, \quad \gamma_{yz}\}^{\mathrm{T}} \tag{2.2.1}$$

The sign convention of strain corresponds to that of stress described in the previous section. Here, extension is regarded as positive while contraction as negative. A positive shear strain implies a decrease in the angle between the vectors along the positive axes; a negative shear strain otherwise. The state of strain can be described by the displacement along the coordinate axes $\boldsymbol{u} = \{u, v, w\}^{\mathrm{T}}$. The strain-displacement relations for infinitesimal deformation, are

$$\left.\begin{aligned} \varepsilon_x &= \frac{\partial u}{\partial x}; & \varepsilon_y &= \frac{\partial v}{\partial y}; & \varepsilon_z &= \frac{\partial w}{\partial z}; \\ \gamma_{xy} &= \frac{\partial u}{\partial y} + \frac{\partial v}{\partial x}; & \gamma_{xz} &= \frac{\partial u}{\partial z} + \frac{\partial w}{\partial x}; & \gamma_{yz} &= \frac{\partial v}{\partial z} + \frac{\partial w}{\partial y} \end{aligned}\right\} \tag{2.2.2}$$

or in the form of a vector as

$$\boldsymbol{\varepsilon} = \hat{\boldsymbol{E}}(\nabla)\boldsymbol{u} \tag{2.2.2$'$}$$

where

$$\hat{\boldsymbol{E}}(\nabla) = \begin{bmatrix} \partial/\partial x & 0 & 0 \\ 0 & \partial/\partial y & 0 \\ 0 & 0 & \partial/\partial z \\ \partial/\partial y & \partial/\partial x & 0 \\ \partial/\partial z & 0 & \partial/\partial x \\ 0 & \partial/\partial z & \partial/\partial y \end{bmatrix} \tag{2.2.3}$$

Obviously, the six components of strain are not independent. They satisfy the equations of compatibility

$$\begin{aligned} &\frac{\partial^2 \varepsilon_x}{\partial y^2} + \frac{\partial^2 \varepsilon_y}{\partial x^2} = \frac{\partial^2 \gamma_{xy}}{\partial x \partial y}, \quad \frac{\partial}{\partial x}\left(\frac{\partial \gamma_{xz}}{\partial y} + \frac{\partial \gamma_{xy}}{\partial z} - \frac{\partial \gamma_{yz}}{\partial x}\right) = 2\frac{\partial^2 \varepsilon_x}{\partial y \partial z} \\ &\frac{\partial^2 \varepsilon_x}{\partial z^2} + \frac{\partial^2 \varepsilon_z}{\partial x^2} = \frac{\partial^2 \gamma_{xz}}{\partial x \partial z}, \quad \frac{\partial}{\partial y}\left(\frac{\partial \gamma_{xy}}{\partial z} + \frac{\partial \gamma_{yz}}{\partial x} - \frac{\partial \gamma_{xz}}{\partial y}\right) = 2\frac{\partial^2 \varepsilon_y}{\partial x \partial z} \\ &\frac{\partial^2 \varepsilon_y}{\partial z^2} + \frac{\partial^2 \varepsilon_z}{\partial y^2} = \frac{\partial^2 \gamma_{yz}}{\partial y \partial z}, \quad \frac{\partial}{\partial z}\left(\frac{\partial \gamma_{yz}}{\partial x} + \frac{\partial \gamma_{xz}}{\partial y} - \frac{\partial \gamma_{xy}}{\partial z}\right) = 2\frac{\partial^2 \varepsilon_z}{\partial x \partial y} \end{aligned} \tag{2.2.4}$$

These equations are also called **Saint–Venant equations**. If the region occupied by the object is multiple-connected, Eq. (2.2.4) ensure the existence of the solution of Eq. (2.2.2), but the solution, namely, the displacement function may turn out to be multiple-valued.

Similar to stress, the strain components transform under rotation of coordinate systems[3].

The displacement components in the new coordinate system follow the transformation

$$\left.\begin{aligned} u' &= ul_1 + vm_1 + wn_1 \\ v' &= ul_2 + vm_2 + wn_2 \\ w' &= ul_3 + vm_3 + wn_3 \end{aligned}\right\} \tag{2.2.5}$$

The corresponding strain components are

$$\left.\begin{aligned}
&\varepsilon_{x'}=\frac{\partial u'}{\partial x'}; &&\varepsilon_{y'}=\frac{\partial v'}{\partial y'}; &&\varepsilon_{z'}=\frac{\partial w'}{\partial z'};\\
&\gamma_{x'y'}=\frac{\partial u'}{\partial y'}+\frac{\partial v'}{\partial x'}; &&\gamma_{x'z'}=\frac{\partial u'}{\partial z'}+\frac{\partial w'}{\partial x'}; &&\gamma_{y'z'}=\frac{\partial v'}{\partial z'}+\frac{\partial w'}{\partial y'}
\end{aligned}\right\} \tag{2.2.6}$$

Substituting Eq. (2.2.5) into Eq. (2.2.6) and using the formula of partial derivative as follows

$$\frac{\partial}{\partial s}()=\cos(s,x)\frac{\partial}{\partial x}()+\cos(s,y)\frac{\partial}{\partial y}()+\cos(s,z)\frac{\partial}{\partial z}() \tag{2.2.7}$$

We obtain the expressions of strain components under rotation transformation of coordinates systems

$$\varepsilon_{x'}=\varepsilon_x l_1^2+\varepsilon_y m_1^2+\varepsilon_z n_1^2+\gamma_{xy}l_1m_1+\gamma_{xz}l_1n_1+\gamma_{yz}m_1n_1 \tag{2.2.8a}$$

$$\varepsilon_{y'}=\varepsilon_x l_2^2+\varepsilon_y m_2^2+\varepsilon_z n_2^2+\gamma_{xy}l_2m_2+\gamma_{xz}l_2n_2+\gamma_{yz}m_2n_2 \tag{2.2.8b}$$

$$\varepsilon_{z'}=\varepsilon_x l_3^2+\varepsilon_y m_3^2+\varepsilon_z n_3^2+\gamma_{xy}l_3m_3+\gamma_{xz}l_3n_3+\gamma_{yz}m_3n_3 \tag{2.2.8c}$$

$$\begin{aligned}\gamma_{x'y'}&=2\varepsilon_x l_1l_2+2\varepsilon_y m_1m_2+2\varepsilon_z n_1n_2+\gamma_{xy}(l_1m_2+l_2m_1)\\&\quad+\gamma_{xz}(l_1n_2+l_2n_1)+\gamma_{yz}(m_1n_2+m_2n_1)\end{aligned} \tag{2.2.8d}$$

$$\begin{aligned}\gamma_{x'z'}&=2\varepsilon_x l_1l_3+2\varepsilon_y m_1m_3+2\varepsilon_z n_1n_3+\gamma_{xy}(l_1m_3+l_3m_1)\\&\quad+\gamma_{xz}(l_1n_3+l_3n_1)+\gamma_{yz}(m_1n_3+m_3n_1)\end{aligned} \tag{2.2.8e}$$

$$\begin{aligned}\gamma_{y'z'}&=2\varepsilon_x l_2l_3+2\varepsilon_y m_2m_3+2\varepsilon_z n_2n_3+\gamma_{xy}(l_2m_3+l_3m_2)\\&\quad+\gamma_{xz}(l_2n_3+l_3n_2)+\gamma_{yz}(m_2n_3+m_3n_2)\end{aligned} \tag{2.2.8f}$$

It is observed that the expressions of strain components under rotation transformation of coordinates system are the same as the corresponding stress components if we denote strain components as $\varepsilon_{xy}=0.5\gamma_{xy}$, $\varepsilon_{xz}=0.5\gamma_{xz}$, $\varepsilon_{yz}=0.5\gamma_{yz}$. Hence, there exist three mutually orthogonal principal directions and the corresponding principal strains ε_1, $\varepsilon_2,\varepsilon_3$. The principal strain ε and its direction cosines l,m,n fulfill the following equations

$$\left.\begin{aligned}
(\varepsilon_x-\varepsilon)l+\varepsilon_{xy}m+\varepsilon_{xz}n=0\\
\varepsilon_{xy}l+(\varepsilon_y-\varepsilon)m+\varepsilon_{yz}n=0\\
\varepsilon_{xz}l+\varepsilon_{yz}m+(\varepsilon_z-\varepsilon)n=0
\end{aligned}\right\} \tag{2.2.9}$$

Because the direction cosines do not vanish simultaneously, the determinant of the coefficient matrix vanishes. On expanding the determinant, we have the cubic equation of ε

$$\varepsilon^3 - J_1\varepsilon^2 + J_2\varepsilon - J_3 = 0 \tag{2.2.10}$$

where

$$\begin{aligned} J_1 &= \varepsilon_x + \varepsilon_y + \varepsilon_z \\ J_2 &= \varepsilon_x\varepsilon_y + \varepsilon_x\varepsilon_z + \varepsilon_y\varepsilon_z - \varepsilon_{xy}^2 - \varepsilon_{xz}^2 - \varepsilon_{yz}^2 \\ J_3 &= \begin{vmatrix} \varepsilon_x & \varepsilon_{xy} & \varepsilon_{xz} \\ \varepsilon_{xy} & \varepsilon_y & \varepsilon_{yz} \\ \varepsilon_{xz} & \varepsilon_{yz} & \varepsilon_z \end{vmatrix} \end{aligned} \tag{2.2.11}$$

J_1, J_2, J_3 are independent of the choice of coordinate systems and they are called the first, the second, and the third strain invariants, respectively. The first strain invariant J_1 represents the expansion of a unit volume due to strain and it is called the volume strain. J_2 is often used in the plastic yield criteria of isotropic materials in terms of strain.

2.3. Stress-Strain Relations

The equations of equilibrium (2.1.10) and the strain-displacement relations (2.2.2) are independent of the material properties. Hence, they are applicable to any cases having linear elastic infinitesimal deformation. The **generalized Hooke's law**, on the other hand, couples the characteristics of elasticity and material properties via the stress-strain relations. The generalized linear stress-strain relations[4] are given by

$$\left.\begin{aligned} \sigma_x &= c_{11}\varepsilon_x + c_{12}\varepsilon_y + c_{13}\varepsilon_z + c_{14}\gamma_{xy} + c_{15}\gamma_{xz} + c_{16}\gamma_{yz} \\ \sigma_y &= c_{21}\varepsilon_x + c_{22}\varepsilon_y + c_{23}\varepsilon_z + c_{24}\gamma_{xy} + c_{25}\gamma_{xz} + c_{26}\gamma_{yz} \\ \sigma_z &= c_{31}\varepsilon_x + c_{32}\varepsilon_y + c_{33}\varepsilon_z + c_{34}\gamma_{xy} + c_{35}\gamma_{xz} + c_{36}\gamma_{yz} \\ \tau_{xy} &= c_{41}\varepsilon_x + c_{42}\varepsilon_y + c_{43}\varepsilon_z + c_{44}\gamma_{xy} + c_{45}\gamma_{xz} + c_{46}\gamma_{yz} \\ \tau_{xz} &= c_{51}\varepsilon_x + c_{52}\varepsilon_y + c_{53}\varepsilon_z + c_{54}\gamma_{xy} + c_{55}\gamma_{xz} + c_{56}\gamma_{yz} \\ \tau_{yz} &= c_{61}\varepsilon_x + c_{62}\varepsilon_y + c_{63}\varepsilon_z + c_{64}\gamma_{xy} + c_{65}\gamma_{xz} + c_{66}\gamma_{yz} \end{aligned}\right\} \tag{2.3.1}$$

where $c_{ij} = c_{ji} (i, j = 1, 2, \ldots, 6)$. The equations above can be expressed concisely in vector form as

$$\boldsymbol{\sigma} = \boldsymbol{C}\boldsymbol{\varepsilon} \tag{2.3.1$'$}$$

where $\boldsymbol{C}$ is the **stiffness coefficient matrix** of the material.

Using the inverse expressions of Eq. (2.3.1), the generalized Hooke's law can also be given in the following form

$$\left.\begin{aligned}
\varepsilon_x &= s_{11}\sigma_x + s_{12}\sigma_y + s_{13}\sigma_z + s_{14}\tau_{xy} + s_{15}\tau_{xz} + s_{16}\tau_{yz} \\
\varepsilon_y &= s_{21}\sigma_x + s_{22}\sigma_y + s_{23}\sigma_z + s_{24}\tau_{xy} + s_{25}\tau_{xz} + s_{26}\tau_{yz} \\
\varepsilon_z &= s_{31}\sigma_x + s_{32}\sigma_y + s_{33}\sigma_z + s_{34}\tau_{xy} + s_{35}\tau_{xz} + s_{36}\tau_{yz} \\
\gamma_{xy} &= s_{41}\sigma_x + s_{42}\sigma_y + s_{43}\sigma_z + s_{44}\tau_{xy} + s_{45}\tau_{xz} + s_{46}\tau_{yz} \\
\gamma_{xz} &= s_{51}\sigma_x + s_{52}\sigma_y + s_{53}\sigma_z + s_{54}\tau_{xy} + s_{55}\tau_{xz} + s_{56}\tau_{yz} \\
\gamma_{yz} &= s_{61}\sigma_x + s_{62}\sigma_y + s_{63}\sigma_z + s_{64}\tau_{xy} + s_{65}\tau_{xz} + s_{66}\tau_{yz}
\end{aligned}\right\} \tag{2.3.2}$$

where $s_{ij} = s_{ji}$ $(i, j = 1, 2, \ldots, 6)$ are the flexibility coefficients representing the elastic nature of materials, and they form the **flexibility coefficient matrix** $\boldsymbol{S}$. The above expression can be rewritten in a compact form as

$$\boldsymbol{\varepsilon} = \boldsymbol{S}\boldsymbol{\sigma} = \boldsymbol{C}^{-1}\boldsymbol{\sigma} \tag{2.3.2$'$}$$

During the deformation of an elastic body, the work done by an external force is converted into strain energy stored in the elastic body. For linear stress-strain relation, the strain energy per unit volume or the **strain energy density** can be expressed as

$$v_\varepsilon(\boldsymbol{\varepsilon}) = \frac{1}{2}\boldsymbol{\varepsilon}^{\mathrm{T}}\boldsymbol{C}\boldsymbol{\varepsilon} \tag{2.3.3}$$

Hence the stress-strain relation (2.3.1) can also be expressed in terms of the strain energy density as

$$\boldsymbol{\sigma} = \frac{\partial v_\varepsilon(\boldsymbol{\varepsilon})}{\partial \boldsymbol{\varepsilon}} \tag{2.3.4}$$

Apply Legendre's transformation to all the independent variables $\boldsymbol{\varepsilon}$ of the strain energy density v_ε, i.e. by introducing a function (**strain complementary energy density**)

$$v_{\mathrm{c}}(\boldsymbol{\sigma}) = \boldsymbol{\sigma}^{\mathrm{T}}\boldsymbol{\varepsilon} - v_\varepsilon(\boldsymbol{\varepsilon}) = \frac{1}{2}\boldsymbol{\sigma}^{\mathrm{T}}\boldsymbol{S}\boldsymbol{\sigma} = \frac{1}{2}\boldsymbol{\sigma}^{\mathrm{T}}\boldsymbol{C}^{-1}\boldsymbol{\sigma} \tag{2.3.5}$$

Then we can express strain $\boldsymbol{\varepsilon}$ in terms of stress $\boldsymbol{\sigma}$ as

$$\boldsymbol{\varepsilon} = \frac{\partial v_{\mathrm{c}}(\boldsymbol{\sigma})}{\partial \boldsymbol{\sigma}} \tag{2.3.6}$$

Obviously, as long as stress or strain are not all zero, we have

$$v_\varepsilon(\boldsymbol{\varepsilon}) > 0 \quad \text{and} \quad v_{\mathrm{c}}(\boldsymbol{\sigma}) > 0 \tag{2.3.7}$$

From Eq. (2.3.1), there are 21 independent elastic constants for the most general form of an anisotropic elastic body. If the elastic body possesses symmetric interior structure, so does its elastic property. The generalized Hooke's law can then be simplified.

If an elastic body exhibits one plane of elastic symmetry at any interior point, then the elastic properties for any two mirror points on opposite sides of the plane are identical. The axis (or direction) perpendicular to the plane of elastic symmetry is called the **elastic principal axis** (or direction) of the material. Supposing the z-axis is an elastic principal axis of the material, then the generalized Hooke's law (2.3.1) simplifies to

$$\left.\begin{aligned} \sigma_x &= c_{11}\varepsilon_x + c_{12}\varepsilon_y + c_{13}\varepsilon_z + c_{14}\gamma_{xy} \\ \sigma_y &= c_{12}\varepsilon_x + c_{22}\varepsilon_y + c_{23}\varepsilon_z + c_{24}\gamma_{xy} \\ \sigma_z &= c_{13}\varepsilon_x + c_{23}\varepsilon_y + c_{33}\varepsilon_z + c_{34}\gamma_{xy} \\ \tau_{xy} &= c_{14}\varepsilon_x + c_{24}\varepsilon_y + c_{34}\varepsilon_z + c_{44}\gamma_{xy} \\ \tau_{xz} &= c_{55}\gamma_{xz} + c_{56}\gamma_{yz} \\ \tau_{yz} &= c_{56}\gamma_{xz} + c_{66}\gamma_{yz} \end{aligned}\right\} \tag{2.3.8}$$

Now the number of independent elastic constants reduces to 13.

If the elastic body has three mutually perpendicular planes of elastic symmetry at any interior point, it is called an **orthotropic** solid. It has three orthogonal elastic principal axes and the number of independent elastic constants becomes 9. Suppose the coordinate planes coincide with the planes of elastic symmetry, then the generalized Hooke's law (2.3.8) further simplifies to

$$\left.\begin{aligned} \sigma_x &= c_{11}\varepsilon_x + c_{12}\varepsilon_y + c_{13}\varepsilon_z \\ \sigma_y &= c_{12}\varepsilon_x + c_{22}\varepsilon_y + c_{23}\varepsilon_z \\ \sigma_z &= c_{13}\varepsilon_x + c_{23}\varepsilon_y + c_{33}\varepsilon_z \\ \tau_{xy} &= c_{44}\gamma_{xy} \\ \tau_{xz} &= c_{55}\gamma_{xz} \\ \tau_{yz} &= c_{66}\gamma_{yz} \end{aligned}\right\} \tag{2.3.9}$$

In other words, all the elastic coefficients representing the coupling effects of tension-shearing and shearing–shearing along the elastic principal directions vanish. This is an important elastic property of an orthotropic material. But when the coordinate axes do not coincide with the principal axes of material, the matrix of elastic coefficients becomes a fully populated matrix with various coupling effects. However the number of independent elastic constants is still 9 and such property is called **generalized orthotropy**.

If the elastic body has an isotropic plane at any interior point, namely the elastic properties are identical in all directions with respect to the plane, it is called a **transversely isotropic** solid. The axis perpendicular to that plane is the elastic rotation axis of symmetry of material, and there are only five independent elastic constants. If the z-axis coincides with the elastic rotation axis of symmetry, then Eq. (2.3.9) further simplifies to

$$\left.\begin{aligned} \sigma_x &= c_{11}\varepsilon_x + c_{12}\varepsilon_y + c_{13}\varepsilon_z \\ \sigma_y &= c_{12}\varepsilon_x + c_{11}\varepsilon_y + c_{13}\varepsilon_z \\ \sigma_z &= c_{13}\varepsilon_x + c_{13}\varepsilon_y + c_{33}\varepsilon_z \\ \tau_{xy} &= \frac{1}{2}(c_{11} - c_{12})\gamma_{xy} \\ \tau_{xz} &= c_{55}\gamma_{xz} \\ \tau_{yz} &= c_{55}\gamma_{yz} \end{aligned}\right\} \tag{2.3.10}$$

Finally, for an isotropic elastic material, there are only two independent elastic constants. The generalized Hooke's law can then be stated as

$$\left.\begin{aligned} \sigma_x &= \lambda\theta + 2G\varepsilon_x; \quad \tau_{xy} = G\gamma_{xy} \\ \sigma_y &= \lambda\theta + 2G\varepsilon_y; \quad \tau_{xz} = G\gamma_{xz} \\ \sigma_z &= \lambda\theta + 2G\varepsilon_z \quad\;\; \tau_{yz} = G\gamma_{yz} \end{aligned}\right\} \tag{2.3.11}$$

where λ and G are the **Lame' constants**. The **volume strain**, θ, is

$$\theta = \varepsilon_x + \varepsilon_y + \varepsilon_z = \frac{\partial u}{\partial x} + \frac{\partial v}{\partial y} + \frac{\partial w}{\partial z} \tag{2.3.12}$$

For isotropic materials, the stress-strain relation is always expressed in terms of the **modulus of elasticity** (or **Young's modulus**) E and **Poisson's ratio** ν as

$$\left.\begin{aligned} \varepsilon_x &= \frac{1}{E}[\sigma_x - \nu(\sigma_y + \sigma_z)]; \quad \gamma_{xy} = \frac{2(1+\nu)}{E}\tau_{xy} \\ \varepsilon_y &= \frac{1}{E}[\sigma_y - \nu(\sigma_x + \sigma_z)]; \quad \gamma_{xz} = \frac{2(1+\nu)}{E}\tau_{xz} \\ \varepsilon_z &= \frac{1}{E}[\sigma_z - \nu(\sigma_x + \sigma_y)]; \quad \gamma_{yz} = \frac{2(1+\nu)}{E}\tau_{yz} \end{aligned}\right\} \tag{2.3.13}$$

in which the relations between the modulus of elasticity E, Poisson's ν and Lame' constants are given by

$$E = \frac{G(3\lambda + 2G)}{\lambda + G}; \quad \nu = \frac{\lambda}{2(\lambda + G)} \tag{2.3.14}$$

or

$$\lambda = \frac{E\nu}{(1+\nu)(1-2\nu)}; \quad G = \frac{E}{2(1+\nu)} \tag{2.3.15}$$

where Poisson's ν must satisfy

$$0 \leq \nu \leq 0.5 \tag{2.3.16}$$

For $\nu = 0.5$, the material is incompressible.

For isotropic materials, the strain energy density can be expressed as

$$\begin{aligned} v_\varepsilon(\boldsymbol{\varepsilon}) &= \frac{1}{2}(\lambda + 2G)(\varepsilon_x^2 + \varepsilon_y^2 + \varepsilon_z^2) \\ &\quad + \lambda(\varepsilon_x\varepsilon_y + \varepsilon_x\varepsilon_z + \varepsilon_y\varepsilon_z) + \frac{1}{2}G(\gamma_{xy}^2 + \gamma_{xz}^2 + \gamma_{yz}^2) \end{aligned} \tag{2.3.17}$$

and the complementary strain energy density is given by

$$\begin{aligned} v_c(\boldsymbol{\sigma}) &= \frac{1}{2E}[\sigma_x^2 + \sigma_y^2 + \sigma_z^2 - 2\nu(\sigma_x\sigma_y + \sigma_x\sigma_z + \sigma_y\sigma_z) \\ &\quad + 2(1+\nu)(\tau_{xy}^2 + \tau_{xz}^2 + \tau_{yz}^2)] \end{aligned} \tag{2.3.18}$$

Although an isotropic continuum is restricted by Eq. (2.3.16), the restriction may not be mathematically necessary. For instance, the fundamental equations for plate bending to be discussed later are equivalent to the case with a negative Poisson's ν.

2.4. The Fundamental Equations of Elasticity

From the analyses of stress, strain and the stress-strain relations in the previous three sections, there are 15 fundamental equations of equilibrium in elasticity in the region V:

(1) Equations of equilibrium (2.1.10).
(2) Strain-displacement relations (2.2.2.).
(3) Stress-strain relations (2.3.1).

In addition to satisfying the above fundamental equations within the region, the solution to elasticity problems must also fulfill the corresponding boundary conditions. The boundary Γ of region V is usually separated into two parts: Γ_σ for specified surface boundary tractions and Γ_u for specified displacements.

The boundary conditions on Γ_σ are

$$\left.\begin{aligned} F_{nx} &= \sigma_x l + \tau_{xy} m + \tau_{xz} n = \overline{F}_{nx} \\ F_{ny} &= \tau_{xy} l + \sigma_y m + \tau_{yz} n = \overline{F}_{ny} \\ F_{nz} &= \tau_{xz} l + \tau_{yz} m + \sigma_z n = \overline{F}_{nz} \end{aligned}\right\} \quad \text{on } \Gamma_\sigma \tag{2.4.1}$$

where $\overline{F}_{nx}, \overline{F}_{ny}, \overline{F}_{nz}$ are the projections of the known force per unit area acting on the surface boundary on the coordinate axes x, y and z, respectively, and l, m, n are the direction cosines of the exterior normal vector $\boldsymbol{n}$ of the surface boundary. The boundary conditions (2.4.1) can be expressed in matrix as

$$\boldsymbol{F}_n = \boldsymbol{E}(\boldsymbol{n})\boldsymbol{\sigma} = \overline{\boldsymbol{F}}_n \quad \text{on } \Gamma_\sigma \tag{2.4.1$'$}$$

The boundary conditions on Γ_u are

$$u = \overline{u}, \quad v = \overline{v}, \quad w = \overline{w} \quad \text{on } \Gamma_u \tag{2.4.2}$$

where $\overline{u}, \overline{v}, \overline{w}$ are the specified displacements. Similarly, the boundary conditions (2.4.2) can also be expressed in vector as

$$\boldsymbol{u} = \overline{\boldsymbol{u}} \quad \text{on } \Gamma_u \tag{2.4.2$'$}$$

Having had the governing partial differential equations and the boundary conditions, the question remains for obtaining the solution. The traditional approach for solving partial differential equations is to eliminate as many unknown functions as possible in order to reduce the number of basic unknown functions in the equations. This approach results in relatively higher-order partial differential equations. Hence, two ways of traditional solution methodology for elasticity problems are formed, i.e. the displacement solution method and the stress solution method. Although a hybrid solution method exists in theory, it has been rarely applied except for some cases in shell analysis.

For displacement solution method, the basic unknown functions are referred to the displacements of each point in the elastic body. By using strain-displacement relations (2.2.2) and generalized Hooke's law (2.3.11) and eliminating the components of strain and stress, we obtain three equations of equilibrium involving only three unknown functions u, v, w, which belong to the same kind of variables. For isotropic materials, the equations of equilibrium in terms of displacement functions u, v, w are

$$\left.\begin{aligned}(\lambda+G)\frac{\partial\theta}{\partial x}+G\nabla^2u+F_x=0\\(\lambda+G)\frac{\partial\theta}{\partial y}+G\nabla^2v+F_y=0\\(\lambda+G)\frac{\partial\theta}{\partial z}+G\nabla^2w+F_z=0\end{aligned}\right\}\tag{2.4.3}$$

where ∇^2 is the Laplace operator.

$$\nabla^2=\frac{\partial^2}{\partial x^2}+\frac{\partial^2}{\partial y^2}+\frac{\partial^2}{\partial z^2}\tag{2.4.4}$$

The volume strain θ should be substituted by Eq. (2.3.12). These equations are called **Lame equations**.

We are now ready to impose the boundary conditions. On boundary Γ_u for specified displacements, the boundary conditions are (2.4.2) while on boundary Γ_σ for specified boundary tractions, the boundary conditions in terms of displacements are

$$\left.\begin{aligned}\lambda\theta l+G\left(\frac{\partial u}{\partial x}l+\frac{\partial u}{\partial y}m+\frac{\partial u}{\partial z}n\right)+G\left(\frac{\partial u}{\partial x}l+\frac{\partial v}{\partial x}m+\frac{\partial w}{\partial x}n\right)=\overline{F}_{nx}\\\lambda\theta m+G\left(\frac{\partial v}{\partial x}l+\frac{\partial v}{\partial y}m+\frac{\partial v}{\partial z}n\right)+G\left(\frac{\partial u}{\partial y}l+\frac{\partial v}{\partial y}m+\frac{\partial w}{\partial y}n\right)=\overline{F}_{ny}\\\lambda\theta n+G\left(\frac{\partial w}{\partial x}l+\frac{\partial w}{\partial y}m+\frac{\partial w}{\partial z}n\right)+G\left(\frac{\partial u}{\partial z}l+\frac{\partial v}{\partial z}m+\frac{\partial w}{\partial z}n\right)=\overline{F}_{nz}\end{aligned}\right\}\tag{2.4.5}$$

Equation (2.4.3) and boundary conditions (2.4.2), (2.4.5) are the fundamental differential equations and boundary conditions for displacement solution method.

Contrary to the displacement solution method, the stress solution method considers the stress components as the basic unknown functions by eliminating displacement and strain components. Thus the six Beltrami–Michell partial differential equations[2,3] are formed and they are the strain compatibility equations in terms of stress. For three-dimensional elasticity problems, it is rather difficult to solve these equations simultaneously with the equations of equilibrium. The expressions of equations are thus omitted here. For the stress solution method, the stress functions adopted usually satisfy the equations of equilibrium spontaneously. Hence the solution is merely determined by the equations of compatibility. The stress solution method, especially the solution method by stress functions, is always applied to problems in plane elasticity and axisymmetric revolution body. Again, only one kind of variables is involved in this solution methodology.

2.5. The Principle of Virtual Work[5,6]

According to general definition in mechanics, the displacement $\boldsymbol{u}_d$ and the corresponding strain $\boldsymbol{\varepsilon}_d$ satisfying the strain-displacement relations (2.2.2) and displacement boundary conditions (2.4.2) are called, respectively, the admissible deformation displacement and the admissible deformation strain, or **admissible displacement** and **admissible strain** in short. Obviously, the admissible displacement and its corresponding admissible strain are not unique. There are infinite sets of admissible displacement and admissible strain and the ones which satisfy the equations of equilibrium are the true displacement and the true strain, respectively. On the other hand, the stress $\boldsymbol{\sigma}_s$ satisfying the equations of equilibrium (2.1.10) and traction boundary condition (2.4.1) is called statically admissible stress, or **admissible stress** in short. Similarly, there are infinite sets of admissible stress and the one which satisfies the strain compatibility equations is the true stress. When solving elasticity problems, it is possible to search for the true stress field in the entire statically admissible stress field (stress solution), or to search for the true displacement field in the entire admissible displacement field (displacement solution).

Firstly, we obtain the following identity via integration by parts

$$\int_V \boldsymbol{\sigma}^{\mathrm{T}} \hat{\boldsymbol{E}}(\nabla)\boldsymbol{u}\mathrm{d}V = -\int_V [\boldsymbol{E}(\nabla)\boldsymbol{\sigma}]^{\mathrm{T}}\boldsymbol{u}\mathrm{d}V + \int_\Gamma [\boldsymbol{E}(\boldsymbol{n})\boldsymbol{\sigma}]^{\mathrm{T}}\boldsymbol{u}\mathrm{d}\Gamma \qquad (2.5.1)$$

Hence, the statically admissible stress field $\boldsymbol{\sigma}_s$ and the admissible deformation displacement field $\boldsymbol{u}_d$ fulfill

$$\int_V \boldsymbol{F}^{\mathrm{T}}\boldsymbol{u}_d\mathrm{d}V + \int_\Gamma \boldsymbol{F}_n^{\mathrm{T}}\boldsymbol{u}_d\mathrm{d}\Gamma = \int_V \boldsymbol{\sigma}_s^{\mathrm{T}}\boldsymbol{\varepsilon}_d\mathrm{d}V \qquad (2.5.2)$$

which expresses **the principle of virtual work: for an elastic body, the work done by an external force with respect to an arbitrary set of admissible deformation displacement is equal to the work done by an arbitrary set of statically admissible stress with respect to the admissible strain corresponding to the stated admissible deformation displacement.**

We denote

$$\boldsymbol{u}_d = \boldsymbol{u} + \delta\boldsymbol{u} \tag{2.5.3}$$

where $\boldsymbol{u}$ denotes the actual displacement field and $\delta\boldsymbol{u}$ denotes the virtual displacement. The latter is a quantity with a small deviation to the actual displacement field consistent with the displacement constrains. Hence the virtual displacement vanishes over the boundary Γ_u for specified displacements,

$$\delta\boldsymbol{u} = \boldsymbol{0} \tag{2.5.4}$$

As $\boldsymbol{u}_d$ and $\boldsymbol{u}$ are both admissible displacement fields and if we take the actual stress field as the statically admissible stress field, by means of Eq. (2.5.2), we derive

$$\int_V \boldsymbol{F}^{\mathrm{T}}\delta\boldsymbol{u}\mathrm{d}V + \int_{\Gamma_\sigma} \overline{\boldsymbol{F}}_n^{\mathrm{T}}\delta\boldsymbol{u}\mathrm{d}\Gamma = \int_V \boldsymbol{\sigma}^{\mathrm{T}}\delta\boldsymbol{\varepsilon}\mathrm{d}V \tag{2.5.5}$$

which states **the principle of virtual displacement: for an elastic body in equilibrium, the external virtual work done by an external force with respect to a virtual displacement is equal to the internal virtual work of the elastic body.** Conversely, if the external virtual work with respect to an arbitrary virtual displacement is equal to the internal virtual work, the elastic body must be in equilibrium. The principle of virtual displacement is sometimes called the principle of virtual work.

2.6. The Principle of Minimum Total Potential Energy

Irrespective of displacement or stress solution methods, it has been enormously difficult to seek analytical solutions for elasticity problems. Hence, various approximate theories, such as the theories of beam, thin-walled bar, plate shell, etc. were established. Even with practical theories, analytical solutions are limited to a certain extent due to the complexity of partial differential equations. Particularly, it is almost impossible to obtain analytical solutions if the boundary conditions are complicated. Hence, the seeking of solutions for elasticity problems has long been a "bottleneck" in the development of theory of elasticity. Similarly, research in the solution methodology has also been a main subject of interest in elasticity.

Because exact solutions in elasticity are difficult, approximate solutions become significant. The method of variation is one of the most effective methods among the approximate methods. It is the theoretical foundation of some numerical methods, for instance, the finite element method. This method has gained enormous success. The functional of the variational principle has close relations with energy, hence, it is also called the variational principle of energy.

Since the external forces are invariant during the process of virtual displacements, based on Eq. (2.3.4), Eq. (2.5.5) can be written as

$$\delta E_{\mathrm{p}}(\boldsymbol{u}) = \delta(V_\varepsilon(\boldsymbol{u}) + E_{\mathrm{w}}(\boldsymbol{u})) = 0 \tag{2.6.1}$$

where

$$V_\varepsilon(\boldsymbol{u}) = \int_V v_\varepsilon(\boldsymbol{\varepsilon})\mathrm{d}V \tag{2.6.2}$$

$$E_{\mathrm{w}}(\boldsymbol{u}) = -\int_V \boldsymbol{F}^{\mathrm{T}}\boldsymbol{u}\mathrm{d}V - \int_{\Gamma_\sigma} \overline{\boldsymbol{F}}_n^{\mathrm{T}}\boldsymbol{u}\mathrm{d}\Gamma \tag{2.6.3}$$

in which V_ε and E_{w} are the strain energy and the potential energy of external forces of the elastic body, respectively, and E_{p}, a functional of displacement, is the total potential energy.

Equation (2.6.1) only indicates the vanishing of the first variation of total potential energy. It is also a condition for equilibrium. It is observed that the potential energy of external forces is a first-order functional of displacement $\boldsymbol{u}$, and the strain energy is a second-order functional of displacement. Hence, we have

$$\left.\begin{aligned} V_\varepsilon(\boldsymbol{u}+\delta\boldsymbol{u}) &= V_\varepsilon(\boldsymbol{u}) + \delta V_\varepsilon(\boldsymbol{u}) + V_\varepsilon(\delta\boldsymbol{u}) \\ E_{\mathrm{w}}(\boldsymbol{u}+\delta\boldsymbol{u}) &= E_{\mathrm{w}}(\boldsymbol{u}) + \delta E_{\mathrm{w}}(\boldsymbol{u}) \end{aligned}\right\} \tag{2.6.4}$$

Using Eqs. (2.6.1) and (2.3.7), we obtain

$$E_{\mathrm{p}}(\boldsymbol{u}+\delta\boldsymbol{u}) = E_{\mathrm{p}}(\boldsymbol{u}) + V_\varepsilon(\delta\boldsymbol{u}) \geq E_{\mathrm{p}}(\boldsymbol{u}) \tag{2.6.5}$$

It states that the actual solution corresponds to minimum potential energy. Hence, we have **the principle of minimum total potential energy: for all admissible displacement fields, the actual displacement field of an elastic system corresponds to minimum total potential energy**

of the elastic body. The minimum of total potential energy indicates a state of stable equilibrium.

2.7. The Principle of Minimum Total Complementary Energy

For an elastic body in equilibrium, the statically admissible stress field is

$$\boldsymbol{\sigma}_s = \boldsymbol{\sigma} + \delta\boldsymbol{\sigma} \tag{2.7.1}$$

where $\boldsymbol{\sigma}$ is the actual stress field and $\delta\boldsymbol{\sigma}$ is the virtual stress which is a quantity with a small deviation to the actual stress field and yet making $\boldsymbol{\sigma}_s$ a statically admissible stress field. Hence on the boundary Γ_σ for specified forces, we have

$$\delta\boldsymbol{F}_n = \boldsymbol{E}(\boldsymbol{n})\delta\boldsymbol{\sigma} = \boldsymbol{0} \tag{2.7.2}$$

As $\boldsymbol{\sigma}_s$ and $\boldsymbol{\sigma}$ are both statically admissible stress fields and if the actual displacement field is taken as the admissible deformation displacement field, by means of Eq. (2.5.2), we derive

$$\int_V \boldsymbol{\varepsilon}^{\mathrm{T}}\delta\boldsymbol{\sigma}\mathrm{d}V - \int_{\Gamma_u} \overline{\boldsymbol{u}}^{\mathrm{T}}\mathbf{E}(\boldsymbol{n})\delta\boldsymbol{\sigma}\mathrm{d}\Gamma = 0 \tag{2.7.3}$$

Since displacements are invariable during the generation of virtual stress, and based on Eq. (2.3.6), Eq. (2.7.3) can be written as

$$\delta E_{\mathrm{pc}}(\boldsymbol{\sigma}) = \delta[V_{\mathrm{c}}(\boldsymbol{\sigma}) + E_{\mathrm{c}}(\boldsymbol{\sigma})] = 0 \tag{2.7.4}$$

where

$$V_{\mathrm{c}}(\boldsymbol{\sigma}) = \int_V v_{\mathrm{c}}(\boldsymbol{\sigma})\mathrm{d}V \tag{2.7.5}$$

$$E_{\mathrm{c}}(\boldsymbol{\sigma}) = -\int_{\Gamma_u} \overline{\boldsymbol{u}}^{\mathrm{T}}\boldsymbol{E}(\boldsymbol{n})\boldsymbol{\sigma}\mathrm{d}\Gamma \tag{2.7.6}$$

V_{c} and E_{c} are the strain complementary energy and the complementary energy for support displacements, respectively, and E_{pc}, a functional of stress, is the total complementary energy of elastic body.

Similarly, Eq. (2.7.4) only indicates the vanishing of the first variation of total complementary energy. It also represents the deformation compatibility conditions. It is observed that the complementary energy for support displacements is a first-order functional of reaction $\boldsymbol{E}(\boldsymbol{n})\boldsymbol{\sigma}$, and the strain complementary energy is second-order functional of stress $\boldsymbol{\sigma}$. Hence, we have

$$\left.\begin{aligned} V_c(\boldsymbol{\sigma}+\delta\boldsymbol{\sigma}) &= V_c(\boldsymbol{\sigma})+\delta V_c(\boldsymbol{\sigma})+V_c(\delta\boldsymbol{\sigma}) \\ E_c(\boldsymbol{\sigma}+\delta\boldsymbol{\sigma}) &= E_c(\boldsymbol{\sigma})+\delta E_c(\boldsymbol{\sigma}) \end{aligned}\right\} \tag{2.7.7}$$

According to Eqs. (2.7.4) and (2.3.7), we obtain

$$E_{pc}(\boldsymbol{\sigma}+\delta\boldsymbol{\sigma}) = E_{pc}(\boldsymbol{\sigma})+V_c(\delta\boldsymbol{\sigma}) \geq E_{pc}(\boldsymbol{\sigma}) \tag{2.7.8}$$

It states that the actual solution corresponds to minimum complementary energy. Hence, we have **the principle of minimum total complementary energy**[7]**: for all admissible stress fields, the actual stress field of an elastic system corresponds to minimum total complementary energy of the elastic body.** The minimum of total complementary energy indicates a state of stable equilibrium.

For exact solutions, the relation between the total potential energy and the total complementary energy of an elastic system is

$$E_p + E_{pc} = 0 \tag{2.7.9}$$

Associating with Eqs. (2.6.5) and (2.7.8), we obtain a series of inequalities

$$-E_p(\boldsymbol{u}+\delta\boldsymbol{u}) \leq -E_p(\boldsymbol{u}) = E_{pc}(\boldsymbol{\sigma}) \leq E_{pc}(\boldsymbol{\sigma}+\delta\boldsymbol{\sigma}) \tag{2.7.10}$$

2.8. The Hellinger–Reissner Variational Principle with Two Kinds of Variables

Both the principle of minimum potential energy and the principle of minimum complementary energy are subjected to specified constraints. From the viewpoint of multiple kinds of variables, they are the principles of conditional extremum[8].

For the principle of minimum complementary energy, for instance, the independent variable function $\boldsymbol{\sigma}$ satisfies the equation of equilibrium (2.1.10) and the boundary conditions for specified surface tractions (2.4.1). In order to get rid of these constraints, we introduce the Lagrange multipliers $\boldsymbol{\xi}, \boldsymbol{\eta}$. Substituting Eqs. (2.1.10) and (2.4.1) into the variation formula

(2.7.4), a new functional

$$\Pi_2 = \int_V \{v_c(\boldsymbol{\sigma}) + \boldsymbol{\xi}^{\mathrm{T}}[\boldsymbol{E}(\nabla)\boldsymbol{\sigma} + \boldsymbol{F}]\}\mathrm{d}V - \int_{\Gamma_u} \overline{\boldsymbol{u}}^{\mathrm{T}}\boldsymbol{E}(\boldsymbol{n})\boldsymbol{\sigma}\mathrm{d}\Gamma - \int_{\Gamma_\sigma} \boldsymbol{\eta}^{\mathrm{T}}(\boldsymbol{E}(\boldsymbol{n})\boldsymbol{\sigma} - \overline{\boldsymbol{F}}_n)\mathrm{d}\Gamma \tag{2.8.1}$$

is obtained. For functional Π_2, $\boldsymbol{\sigma}$ is considered as an independent variable function without any constraints. The variation of Π_2 with respect to $\boldsymbol{\sigma}$ is

$$\int_V \{\delta\boldsymbol{\sigma}^{\mathrm{T}}[\boldsymbol{\varepsilon} - \hat{\boldsymbol{E}}(\nabla)\boldsymbol{\xi}]\}\mathrm{d}V + \int_{\Gamma_u} (\boldsymbol{\xi} - \overline{\boldsymbol{u}})^{\mathrm{T}}\boldsymbol{E}(\boldsymbol{n})\delta\boldsymbol{\sigma}\mathrm{d}\Gamma + \int_{\Gamma_\sigma} (\boldsymbol{\xi} - \boldsymbol{\eta})^{\mathrm{T}}\boldsymbol{E}(\boldsymbol{n})\delta\boldsymbol{\sigma}\mathrm{d}\Gamma \tag{2.8.2}$$

As $\delta\boldsymbol{\sigma}$ is arbitrary, we have

$$\boldsymbol{\varepsilon} = \hat{\boldsymbol{E}}(\nabla)\boldsymbol{\xi} \tag{2.8.3}$$

$$\boldsymbol{\xi} = \overline{\boldsymbol{u}} \quad \text{on } \Gamma_u \tag{2.8.4}$$

$$\boldsymbol{\xi} = \boldsymbol{\eta} \quad \text{on } \Gamma_\sigma \tag{2.8.5}$$

It states that the Lagrange multipliers $\boldsymbol{\xi}, \boldsymbol{\eta}$ are the displacement $\boldsymbol{u}$. Substituting $\boldsymbol{u}$ for $\boldsymbol{\xi}, \boldsymbol{\eta}$ in the functional Π_2, **the Hellinger–Reissner variational principle with two kinds of variables** is

$$\delta\Pi_2 = 0 \tag{2.8.6}$$

where

$$\Pi_2 = \int_V \{v_c(\boldsymbol{\sigma}) + \boldsymbol{u}^{\mathrm{T}}[\boldsymbol{E}(\nabla)\boldsymbol{\sigma} + \boldsymbol{F}]\}\mathrm{d}V - \int_{\Gamma_u} \overline{\boldsymbol{u}}^{\mathrm{T}}\boldsymbol{E}(\boldsymbol{n})\boldsymbol{\sigma}\mathrm{d}\Gamma - \int_{\Gamma_\sigma} \boldsymbol{u}^{\mathrm{T}}(\boldsymbol{E}(\boldsymbol{n})\boldsymbol{\sigma} - \overline{\boldsymbol{F}}_n)\mathrm{d}\Gamma \tag{2.8.7}$$

By means of the identity (2.5.1), another form of variational principle with two kinds of variables is obtained.

$$\delta\Pi_2' = 0 \tag{2.8.8}$$

where

$$\Pi_2' = \int_V \{\boldsymbol{\sigma}^{\mathrm{T}}\hat{\boldsymbol{E}}(\nabla)\boldsymbol{u} - v_c(\boldsymbol{\sigma}) - \boldsymbol{F}^{\mathrm{T}}\boldsymbol{u}\}\mathrm{d}V - \int_{\Gamma_u} (\boldsymbol{u} - \overline{\boldsymbol{u}})^{\mathrm{T}}\boldsymbol{E}(\boldsymbol{n})\boldsymbol{\sigma}\mathrm{d}\Gamma - \int_{\Gamma_\sigma} \overline{\boldsymbol{F}}_n^{\mathrm{T}}\boldsymbol{u}\mathrm{d}\Gamma \tag{2.8.9}$$

2.9. The Hu–Washizu Variational Principle with Three Kinds of Variables

For the Hellinger–Reissner variational principle with two kinds of variables, the strain $\boldsymbol{\varepsilon}$ in terms of the displacement $\boldsymbol{u}$ is derived from Eq. (2.2.2). If this constraint is relaxed, we obtain the Hu–Washizu variational principle with three kinds of variables. It was published by Haichang Hu in China in 1954 and by Washizu in USA in 1955, respectively. Although the Hu–Washizu variational principle can also be derived by introducing Lagrange multipliers, only derivation by Legendre's transformation is employed here.

First of all, we apply Legendre's transformation on all the independent variables $\boldsymbol{\sigma}$ of strain complementary energy density $v_{\rm c}(\boldsymbol{\sigma})$ in Hellinger–Reissner variational principle. By introducing a function (strain energy density)

$$v_\varepsilon(\boldsymbol{\varepsilon}) = \boldsymbol{\sigma}^{\rm T}\boldsymbol{\varepsilon} - v_{\rm c}(\boldsymbol{\sigma}) = \frac{1}{2}\boldsymbol{\varepsilon}^{\rm T}\boldsymbol{C}\boldsymbol{\varepsilon} \tag{2.9.1}$$

stress $\boldsymbol{\sigma}$ can be expressed in terms of strain $\boldsymbol{\varepsilon}$ as

$$\boldsymbol{\sigma} = \frac{\partial v_\varepsilon(\boldsymbol{\varepsilon})}{\partial \boldsymbol{\varepsilon}} \tag{2.9.2}$$

Obviously, Eqs. (2.9.1), (2.9.2) and (2.3.5), Eq. (2.3.6) are in dual. Equation (2.9.1) can be rewritten as

$$v_{\rm c}(\boldsymbol{\sigma}) = \max_{\boldsymbol{\sigma}}[\boldsymbol{\sigma}^{\rm T}\boldsymbol{\varepsilon} - v_\varepsilon(\boldsymbol{\varepsilon})] \tag{2.9.3}$$

Substitution of the above expression into the formula of the Hellinger–Reissner variational principle in Eqs. (2.8.6) and (2.8.8), respectively, results in two forms of **the Hu–Washizu variational principle with three kinds of variables**

$$\delta\Pi_3 = 0 \tag{2.9.4}$$

$$\begin{aligned}\Pi_3 = &\int_V \{\boldsymbol{\sigma}^{\rm T}\boldsymbol{\varepsilon} - v_\varepsilon(\boldsymbol{\varepsilon}) + \boldsymbol{u}^{\rm T}[\boldsymbol{E}(\nabla)\boldsymbol{\sigma} + \boldsymbol{F}]\}{\rm d}V \\ &- \int_{\Gamma_u} \overline{\boldsymbol{u}}^{\rm T}\boldsymbol{E}(\boldsymbol{n})\boldsymbol{\sigma}{\rm d}\Gamma - \int_{\Gamma_\sigma} \boldsymbol{u}^{\rm T}(\boldsymbol{E}(\boldsymbol{n})\boldsymbol{\sigma} - \overline{\boldsymbol{F}}_n){\rm d}\Gamma\end{aligned} \tag{2.9.5}$$

and

$$\delta\Pi_3' = 0 \tag{2.9.6}$$

$$\begin{aligned}\Pi_3' = &\int_V \{\boldsymbol{\sigma}^{\mathrm{T}}\hat{E}(\nabla)\boldsymbol{u} - \boldsymbol{\sigma}^{\mathrm{T}}\boldsymbol{\varepsilon} + v_\varepsilon(\boldsymbol{\varepsilon}) - \boldsymbol{F}^{\mathrm{T}}\boldsymbol{u}\}\mathrm{d}V \\ &- \int_{\Gamma_u} (\boldsymbol{u} - \overline{\boldsymbol{u}})^{\mathrm{T}}\boldsymbol{E}(\boldsymbol{n})\boldsymbol{\sigma}\mathrm{d}\Gamma - \int_{\Gamma_\sigma} \overline{\boldsymbol{F}}_n^{\mathrm{T}}\boldsymbol{u}\mathrm{d}\Gamma\end{aligned} \tag{2.9.7}$$

It is not necessary to apply Legendre's transformation on all stress components of $\boldsymbol{\sigma}$. Depending on the circumstances, it is completely valid to apply Legendre's transformation on a portion of the stress components while keeping the other stress components unchanged. Hence, the **mixed energy** and its corresponding variational principle can be derived. Applying the **mixed energy variational principle** to the analysis of chains of substructures yields a theory similar to the theory in control optimization. If it is applied with respect to the longitudinal direction of an elastic column, a Hamiltonian system is derived. Using the methods of mathematical physics, such as separation of variables and expansion of eigenvector, a breakthrough on the restriction of Sturm–Liouville's problem (self-adjoint operator) can be established. This approach is applicable for solving the Saint–Venant problem.

In the Hu–Washizu variational principle with three kinds of variables: displacement, stress and strain are the independent variables. It encompasses the equations of equilibrium (2.1.10), the strain-displacement relations (2.2.2.) and the stress–strain relations (2.3.1). Although such development created satisfaction among researchers, however, many stress functions widely used in solutions of classical stress analysis are not included. From the viewpoint of similarity of plate bending and plane elasticity problems[9], it is not correct to consider the stress function as a less known variable. We can also introduce the **multi-variable variational principle of three-dimensional elasticity**[10] for residual deformation. It does not only include displacement, stress and strain, but also stress function and residual strain, a total of five kinds of fundamental variables. It covers five kinds of fundamental equations: the equations of equilibrium, strain-displacement relations, stress–strain relations, compatibility equations of deformation and stress–stress function. The variational principle can be used to derive the various aforementioned classical variational principles, and therefore, it can be regarded as the most general variational principle hitherto. In Chapter 8, the multi-variable variational principle for plate bending and plane elasticity problems will be discussed in detail, while the multi-variable variational principle for three-dimensional elasticity problems will be omitted.

2.10. The Principle of Superposition and the Uniqueness Theorem

An important, specific character of the elasticity systems above is the linearity of the fundamental equations and boundary conditions. As they are linear system, the superposition principle is applicable. Let $\boldsymbol{u}, \boldsymbol{\varepsilon}, \boldsymbol{\sigma}$ be the solutions of an elastic system subject to body forces $\boldsymbol{F}$ and surface tractions $\overline{\boldsymbol{F}}_n$, and $\boldsymbol{u}', \boldsymbol{\varepsilon}', \boldsymbol{\sigma}'$ be the solutions corresponding to $\boldsymbol{F}'$and $\overline{\boldsymbol{F}}'_n$. Then $\boldsymbol{u}+\boldsymbol{u}', \boldsymbol{\varepsilon}+\boldsymbol{\varepsilon}', \boldsymbol{\sigma}+\boldsymbol{\sigma}'$ are the solutions corresponding to body forces $\boldsymbol{F}+\boldsymbol{F}'$ and surface tractions $\overline{\boldsymbol{F}}_n + \overline{\boldsymbol{F}}'_n$.

The principle of superposition is very useful. It can divide a complicated problem into the solutions of several simpler problems. In this book we only discuss the solutions of linear elasticity systems. Large displacement and large deformation are not considered and, therefore, the principle of superposition is absolutely applicable.

The uniqueness of solution is discussed below. Suppose there exist two sets of solutions

$$\boldsymbol{\sigma}^{(1)}, \quad \boldsymbol{\varepsilon}^{(1)}, \quad \boldsymbol{u}^{(1)} \quad \text{and} \quad \boldsymbol{\sigma}^{(2)}, \quad \boldsymbol{\varepsilon}^{(2)}, \quad \boldsymbol{u}^{(2)} \tag{2.10.1}$$

for an elastic system with prescribed body forces $\boldsymbol{F}$ and surface tractions $\overline{\boldsymbol{F}}_n$. Both solutions satisfy the equations of equilibrium (2.1.10), strain-displacement relations (2.2.2.) and generalized Hooke's law (2.3.1), as well as the identical boundary conditions for specified surface tractions (2.4.1) and specified displacements (2.4.2). According to the principle of superposition, the difference of the two solutions

$$\tilde{\boldsymbol{\sigma}} = \boldsymbol{\sigma}^{(1)} - \boldsymbol{\sigma}^{(2)}, \quad \tilde{\boldsymbol{\varepsilon}} = \boldsymbol{\varepsilon}^{(1)} - \boldsymbol{\varepsilon}^{(2)}, \quad \tilde{\boldsymbol{u}} = \boldsymbol{u}^{(1)} - \boldsymbol{u}^{(2)} \tag{2.10.2}$$

are also the solutions for the elastic system. However, these solutions correspond to an elastic body without any body forces and surface tractions. Obviously, the external virtual work done by external forces for any virtual displacements vanishes. According to the principle of virtual displacement (2.5.5), the internal virtual work vanishes when an elastic body is in a state of equilibrium, or

$$\int_V \tilde{\boldsymbol{\sigma}}^{\mathrm{T}} \delta \tilde{\boldsymbol{\varepsilon}} \mathrm{d}V = 0 \tag{2.10.3}$$

Further, as the virtual strain $\delta\tilde{\varepsilon}$ is arbitrary, we have $\tilde{\boldsymbol{\sigma}} = \mathbf{0}$, and subsequently $\tilde{\varepsilon} = \mathbf{0}$. If the elastic body is constrained completely by support without any rigid displacement, we have the displacement $\tilde{\boldsymbol{u}} = \mathbf{0}$. Hence, the two sets of solutions $\boldsymbol{\sigma}^{(1)}, \boldsymbol{\varepsilon}^{(1)}, \boldsymbol{u}^{(1)}$ and $\boldsymbol{\sigma}^{(2)}$, $\boldsymbol{\varepsilon}^{(2)}, \boldsymbol{u}^{(2)}$ are identical. It states **the uniqueness theorem** of solution for elasticity problems: **the stress state and strain state satisfying all equations of elasticity and boundary conditions are unique. Furthermore, if rigid displacement of the elastic body is constrained completely by support, the displacement is also unique.**

2.11. Saint–Venant Principle

As far as exact solution of elasticity problems are concerned, the solutions are different for different boundary conditions. However, in many practical circumstances, it is rather difficult to accurately determine the actual distribution of surface tractions. Even if the exact distribution of surface tractions is known, there is formidable mathematical complexity if the boundary conditions are to be satisfied strictly.

In 1855, French scientist Saint–Venant presented the famous **Saint–Venant principle** to overcome the problem. It states that **if a system of forces in equilibrium acts on any portion of the surface of an elastic body, the resultant stress and deformation thus caused are localized in the vicinity of the applied forces, and they decrease rapidly with the increase of distance from the region of application**. For instance, consider the bending of bar with uniform cross section. The exact distribution of forces acting at the end of the bar can be neglected and replaced by a statically equivalent system of forces to be solved. The corresponding approximate solution is called the solution of Saint–Venant's problems. According to the principle, the state of stress and deformation at regions far enough from the end of bar are little affected. Thus, the Saint–Venant solution is applicable.

References

1. S.P. Timoshenko and J.N. Goodier, *Theory of Elasticity*, 3rd edn., New York: McGraw-Hill, 1970.
2. Haichang Hu, *Variational Principle of Elasticity and Its Application*, Beijing: Science Press, 1981 (in Chinese).

3. Longfu Wang, *Theory of Elasticity*, Beijing: Science Press, 1984 (in Chinese).
4. Zudao Luo and Sijian Li, *Mechanics of Anisotropic Materials*, Shanghai: Shanghai Jiaotong University Press, 1994 (in Chinese).
5. Qinghua Du, Zhuhua Xiong and Xuewen Tao, *Fundamentals of Applied Solid Mechanics*, Vol. 2, Beijing: Higher Education Press, 1996 (in Chinese).
6. Jialong Wu, *Elasticity*, Shanghai: Tongji University Press, 1993 (in Chinese).
7. L.H. Tsien, *Principle of Complementary Energy*, Scientia Sinica, 1950, 1: 449–456 (in Chinese).
8. Xueren Zhao, Xusheng Zhao and Zhaoxiong Cheng, *Solid Mechanics*, Beijing: Chinese Scientific and Technical Publishers, 1993 (in Chinese).
9. Wanxie Zhong and Weian Yao, New solution system for plate bending and its application, *ACTA Mechanica Sinica*, 1999, 31(2): 173–184 (in Chinese).
10. Wanxie Zhong and Weian Yao, Multi-variable variational principle for three dimensional elasticity, *ACTA Mechanica Solida Sinica*, 2000, 21(SI): 126–129 (in Chinese).

Chapter 3

The Timoshenko Beam Theory and Its Extension

Using the Timoshenko beam, an elastic system in a single continuous coordinate system, as an example, the derivation of a Hamiltonian system from the fundamental equations of elasticity is introduced in this chapter. Subsequently, some effective methods of mathematical physics such as the methods of separation of variables and expansion of eigenvector are used to obtain the solutions. At the same time, the physical interpretation of symplectic orthogonality is given. The problem of wave propagation in a Timoshenko beam is also discussed.

3.1. The Timoshenko Beam Theory

It is well known the classical theory of Euler–Bernoulli beam assumes that (1) the cross-sectional plane perpendicular to the axis of the beam remains plane after deformation (assumption of a rigid cross-sectional plane); (2) the deformed cross-sectional plane is still perpendicular to the axis after deformation. The classical theory of beam neglects transverse shearing deformation where the transverse shear stress is determined by the equations of equilibrium. It is applicable to a thin beam. For a beam with short effective length or composite beams, plates and shells, it is inapplicable to neglect the transverse shear deformation. In 1921, Timoshenko presented a revised beam theory considering shear deformation[1] which retains the first assumption and satisfies the stress-strain relation of shear.

Let the x-axis be along the beam axis before deformation and the xz-plane be the deflection plane as shown in Fig. 3.1. The bending problem of a Timoshenko beam is considered. The displacements $\hat{u}(x,z)$, $\hat{w}(x,z)$ at any point (x,z) in the beam along the x- and z-axis, respectively, can be expressed in terms of two generalized displacements, i.e. the deflection of beam axis $\tilde{w}(x)$ and the rotational angle of the cross section $\tilde{\theta}(x)$

$$\hat{u}(x,z) = -z\tilde{\theta}(x), \quad \hat{w}(x,z) = \tilde{w}(x) \tag{3.1.1}$$

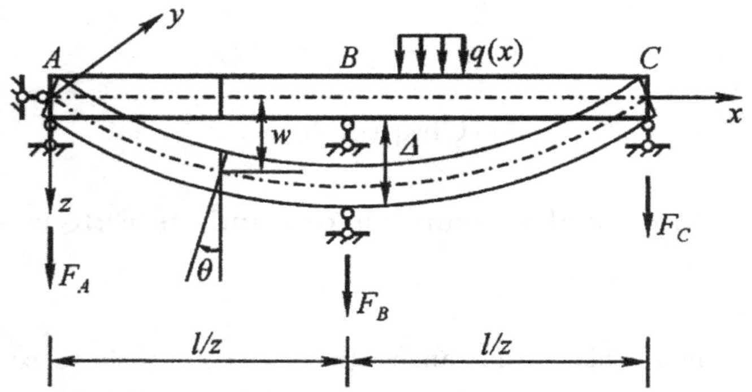

Fig. 3.1. Bending of Timoshenko beam.

Then according to the strain-displacement relation (2.2.2), the strains components are

$$\varepsilon_x = -z\frac{\mathrm{d}\tilde{\theta}}{\mathrm{d}x}, \quad \gamma_{xz} = \frac{\mathrm{d}\tilde{w}}{\mathrm{d}x} - \tilde{\theta} \tag{3.1.2}$$

Substituting Eq. (3.1.2) into the stress-strain relation (2.3.13) and neglecting the effect of higher-order small stress quantities σ_y, σ_z yield

$$-z\frac{\mathrm{d}\tilde{\theta}}{\mathrm{d}x} = \frac{1}{E}\sigma_x, \quad \frac{\mathrm{d}\tilde{w}}{\mathrm{d}x} - \tilde{\theta} = \frac{1}{G}\tau_{xz} \tag{3.1.3}$$

Multiplying both sides of the first equation in Eq. (3.1.3) by z and integrating over the cross-sectional area; and further integrating both sides of the second equation directly over the cross-sectional area, we obtain

$$-I\frac{\mathrm{d}\tilde{\theta}}{\mathrm{d}x} = \frac{1}{E}\int_A z\sigma_x \mathrm{d}A = -\frac{1}{E}\tilde{M} \tag{3.1.4a}$$

$$A\left(\frac{\mathrm{d}\tilde{w}}{\mathrm{d}x} - \tilde{\theta}\right) = \frac{1}{G}\int_A \tau_{xz}\mathrm{d}A = \frac{1}{G}\tilde{F}_{\mathrm{s}} \tag{3.1.4b}$$

where I is the moment of inertia about the y-axis of the cross section of beam, A is the cross-sectional area, and $\tilde{M}, \tilde{F}_{\mathrm{s}}$ are the bending moment and shear force on the cross section, respectively. When deducing the above equations, we assume constant shear stresses on the cross section which, however, are not true in actual situations. Hence a **shear correction factor** k of the cross section is always introduced in Eq. (3.1.4b) to rectify the inappropriate assumption. Equation (3.1.4) can then be expressed as

$$\tilde{M} = EI \cdot \tilde{\kappa}, \quad \tilde{F}_{\mathrm{s}} = kGA \cdot \tilde{\gamma} \tag{3.1.5}$$

where

$$\tilde{\kappa}(x) = \frac{\mathrm{d}\tilde{\theta}}{\mathrm{d}x}, \quad \tilde{\gamma} = \frac{\mathrm{d}\tilde{w}}{\mathrm{d}x} - \tilde{\theta} \tag{3.1.6}$$

are the generalized strains corresponding to the bending moment $\tilde{M}$ and shear force $\tilde{F}_s$. Physically, they are referred as the relative rotational angle between two adjacent cross sections and the shear angle of cross section, respectively. To determine the shear correction factor k, it is usually necessary to assume beforehand the type of shear stress distribution on various shapes of cross sections. Different assumption results in different numerical values. For example, $k = 1 \sim 1.2$ for a rectangular cross section while others may be referred to Ref. 2. An infinite shear correction factor, $k \to \infty$, implies negligible effects of transverse shear deformation and the model degenerates to the classical theory of Euler–Bernoulli beam.

The equation of motion of Timoshenko beam is derived as follows[3]. Consider an infinitesimal segment of the beam as shown in Fig. 3.2. The loads acting on the beam include the transverse distributed force $\tilde{q}(x)$ and the distributed external moment $\tilde{m}(x)$. Besides, there are the transverse inertia force due to the deflection of beam $\tilde{w}(x)$

$$\tilde{q}_{\mathrm{I}}(x) = -\rho A \frac{\partial^2 \tilde{w}}{\partial t^2} \tag{3.1.7}$$

and the inertia moment due to the rotational angle of cross section $\tilde{\theta}(x)$

$$\tilde{m}_{\mathrm{I}}(z) = -\rho I \frac{\partial^2 \tilde{\theta}}{\partial t^2} \tag{3.1.8}$$

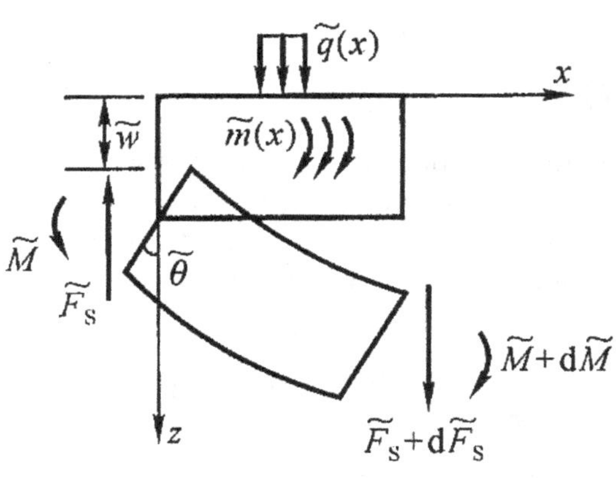

Fig. 3.2. Stress analysis on a segment of Timoshenko beam.

The dynamic equations are then

$$\left.\begin{aligned}\left(\tilde{F}_s+\frac{\partial \tilde{F}_S}{\partial x}\mathrm{d}x\right)-\tilde{F}_s+(\tilde{q}+\tilde{q}_I)\mathrm{d}x=0\\ \left(\tilde{M}+\frac{\partial \tilde{M}}{\partial x}\mathrm{d}x\right)-\tilde{M}+\tilde{F}_s\mathrm{d}x+(\tilde{m}+\tilde{m}_I)\mathrm{d}x=0\end{aligned}\right\} \tag{3.1.9}$$

or

$$\frac{\partial \tilde{F}_s}{\partial x}+\tilde{q}=\rho A\frac{\partial^2\tilde{w}}{\partial t^2},\quad \frac{\partial \tilde{M}}{\partial x}+\tilde{F}_s+\tilde{m}=\rho I\frac{\partial^2\tilde{\theta}}{\partial t^2} \tag{3.1.10}$$

Substituting Eqs. (3.1.5) and (3.1.6) into Eq. (3.1.10) and eliminating the internal forces and strain, the equations of motion in terms of displacements are

$$\left.\begin{aligned}\frac{\partial}{\partial x}\left[kGA\left(\frac{\partial \tilde{w}}{\partial x}-\tilde{\theta}\right)\right]+\tilde{q}=\rho A\frac{\partial^2\tilde{w}}{\partial t^2}\\ \frac{\partial}{\partial x}\left(EI\frac{\partial\tilde{\theta}}{\partial x}\right)+kGA\left(\frac{\partial \tilde{w}}{\partial x}-\tilde{\theta}\right)+\tilde{m}=\rho I\frac{\partial^2\tilde{\theta}}{\partial t^2}\end{aligned}\right\} \tag{3.1.11}$$

The solution of these differential equations requires the relevant boundary conditions and initial conditions. The initial conditions are

$$\tilde{w}=w_0(x)\quad\text{and}\quad\tilde{\theta}=\theta_0(x)\qquad\text{for } t=0 \tag{3.1.12}$$

while there are normally three types of boundary conditions:

(a) Simply supported boundary $\tilde{w}=0, \tilde{M}=0$ (3.1.13a)

(b) Clamped boundary $\tilde{w}=0, \tilde{\theta}=0$ (3.1.13b)

(c) Free boundary $\tilde{M}=0, \tilde{F}_s=0$ (3.1.13c)

To analyze vibration problems, we usually use Fourier expansion for time t and transform the time domain into the frequency domain. These quantities then have a multiplier $\mathrm{e}^{\mathrm{i}\omega t}$ resulting in $\tilde{w}=w\mathrm{e}^{\mathrm{i}\omega t}, \tilde{\theta}=\theta\mathrm{e}^{\mathrm{i}\omega t}$ etc.

In the frequency domain, the equations of motion (3.1.11) are simplified to

$$\left.\begin{aligned}\frac{\mathrm{d}}{\mathrm{d}x}\left[kGA\left(\frac{\mathrm{d}w}{\mathrm{d}x}-\theta\right)\right]+\rho A\omega^2 w+q=0\\ \frac{\mathrm{d}}{\mathrm{d}x}\left(EI\frac{\mathrm{d}\theta}{\mathrm{d}x}\right)+kGA\left(\frac{\mathrm{d}w}{\mathrm{d}x}-\theta\right)+\rho I\omega^2\theta+m=0\end{aligned}\right\} \tag{3.1.14}$$

The following discussion commences with the equations of motion (3.1.14). First, the x-coordinate is modeled as the time coordinate of the Lagrangian system and the Hamiltonian system. A dot is used to indicate differentiation with respect to the x-coordinate, i.e. $(\dot{}) = \mathrm{d}/\mathrm{d}x$, and further denote

$$\boldsymbol{q} = \{w, \theta\}^{\mathrm{T}}, \quad \dot{\boldsymbol{q}} = \{\dot{w}, \dot{\theta}\}^{\mathrm{T}} \tag{3.1.15}$$

Then the principle of potential energy corresponding to the equation of motion (3.1.14) is

$$\delta \int_0^L \mathscr{L}(\boldsymbol{q}, \dot{\boldsymbol{q}})\mathrm{d}x = 0 \tag{3.1.16}$$

and **Lagrange density function** is

$$\mathscr{L}(\boldsymbol{q}, \dot{\boldsymbol{q}}) = \frac{1}{2}\dot{\boldsymbol{q}}^{\mathrm{T}}\boldsymbol{K}_{22}\dot{\boldsymbol{q}} + \dot{\boldsymbol{q}}^{\mathrm{T}}\boldsymbol{K}_{21}\boldsymbol{q} + \frac{1}{2}\boldsymbol{q}^{\mathrm{T}}\boldsymbol{K}_{11}\boldsymbol{q} - \boldsymbol{g}^{\mathrm{T}}\boldsymbol{q} \tag{3.1.17}$$

where

$$\boldsymbol{K}_{22} = \begin{bmatrix} kGA & 0 \\ 0 & EI \end{bmatrix}, \quad \boldsymbol{K}_{21} = \begin{bmatrix} 0 & -kGA \\ 0 & 0 \end{bmatrix}$$

$$\boldsymbol{K}_{11} = \begin{bmatrix} -\rho A\omega^2 & 0 \\ 0 & kGA - \rho I\omega^2 \end{bmatrix}, \quad \boldsymbol{g} = \begin{Bmatrix} q \\ m \end{Bmatrix} \tag{3.1.18}$$

The variation of Eq. (3.1.16) yields the differential equation

$$\frac{\mathrm{d}}{\mathrm{d}x}\left(\frac{\partial \mathscr{L}}{\partial \dot{\boldsymbol{q}}}\right) - \frac{\partial \mathscr{L}}{\partial \boldsymbol{q}} = \boldsymbol{0} \tag{3.1.19}$$

which is the **Lagrange equation** with the time coordinate t replaced by the space coordinate x. Substituting Eq. (3.1.17) into the above equation yields

$$\boldsymbol{K}_{22}\ddot{\boldsymbol{q}} + (\boldsymbol{K}_{21} - \boldsymbol{K}_{12})\dot{\boldsymbol{q}} - \boldsymbol{K}_{11}\boldsymbol{q} + \boldsymbol{g} = \boldsymbol{0} \tag{3.1.20}$$

which is the matrix (vector) form of Eq. (3.1.14) where $\boldsymbol{K}_{12} = \boldsymbol{K}_{21}^{\mathrm{T}}$.

The equation above is the fundamental equation and the principle of potential energy of the Lagrange system with a single kind of variables for solving the dynamic problem of Timoshenko beam. For $\omega = 0$, obviously, the equation above degenerates to the fundamental equation and the principle of minimum potential energy of static bending of Timoshenko beam.

As the vector $\boldsymbol{q}$ in the Lagrange equation (3.1.19) for Timoshenko beam is two-dimensional, it is not difficult to solve it directly in the displacement space. In order to better describe to the readers the concepts and physical interpretation of Hamiltonian systems and symplectic mathematics, the derivation and solution of Hamiltonian system for bending of Timoshenko beam is presented in this chapter. Certainly, the approach can be generalized to the n-dimensional cases. This is the main intention of this chapter.

3.2. Derivation of Hamiltonian System

To derive Hamiltonian system[4–8] from Eq. (3.1.20), we first introduce the dual variable of variable $\boldsymbol{q}$ according to Legendre's transformation

$$\boldsymbol{p} = \frac{\partial \mathscr{L}}{\partial \dot{\boldsymbol{q}}} = \boldsymbol{K}_{22}\dot{\boldsymbol{q}} + \boldsymbol{K}_{21}\boldsymbol{q} \tag{3.2.1}$$

By virtue of Eq. (3.2.1), the solution of $\dot{\boldsymbol{q}}$ is

$$\dot{\boldsymbol{q}} = -\boldsymbol{K}_{22}^{-1}\boldsymbol{K}_{21}\boldsymbol{q} + \boldsymbol{K}_{22}^{-1}\boldsymbol{p} \tag{3.2.2}$$

Then we introduce the **Hamiltonian density function**

$$\begin{aligned}\mathscr{H}(\boldsymbol{q},\boldsymbol{p}) &= \boldsymbol{p}^{\mathrm{T}}\dot{\boldsymbol{q}} - \mathscr{L}(\boldsymbol{q},\dot{\boldsymbol{q}}) \\ &= \boldsymbol{p}^{\mathrm{T}}\boldsymbol{A}\boldsymbol{q} - \frac{1}{2}\boldsymbol{q}^{\mathrm{T}}\boldsymbol{B}\boldsymbol{q} + \frac{1}{2}\boldsymbol{p}^{\mathrm{T}}\boldsymbol{D}\boldsymbol{p} + \boldsymbol{h}_q^{\mathrm{T}}\boldsymbol{p} - \boldsymbol{h}_p^{\mathrm{T}}\boldsymbol{q}\end{aligned} \tag{3.2.3}$$

where

$$\begin{aligned}&\boldsymbol{A} = -\boldsymbol{K}_{22}^{-1}\boldsymbol{K}_{21}, \quad \boldsymbol{B} = \boldsymbol{K}_{11} - \boldsymbol{K}_{12}\boldsymbol{K}_{22}^{-1}\boldsymbol{K}_{21} \\ &\boldsymbol{D} = \boldsymbol{K}_{22}^{-1}, \quad \boldsymbol{h}_q = \boldsymbol{0}, \quad \boldsymbol{h}_p = -\boldsymbol{g}\end{aligned} \tag{3.2.4}$$

Hence from Eqs. (1.4.6) and (3.1.19), we obtain the equations of dual Hamiltonian system as

$$\left.\begin{aligned}\dot{\boldsymbol{q}} &= \frac{\partial \mathscr{H}}{\partial \boldsymbol{p}} = \boldsymbol{A}\boldsymbol{q} + \boldsymbol{D}\boldsymbol{p} + \boldsymbol{h}_q \\ \dot{\boldsymbol{p}} &= \frac{\partial \mathscr{L}}{\partial \boldsymbol{q}} = -\frac{\partial \mathscr{H}}{\partial \boldsymbol{q}} = \boldsymbol{B}\boldsymbol{q} - \boldsymbol{A}^{\mathrm{T}}\boldsymbol{p} + \boldsymbol{h}_p\end{aligned}\right\} \tag{3.2.5}$$

Introducing the **full state vector**

$$\boldsymbol{v} = \begin{Bmatrix} \boldsymbol{q} \\ \boldsymbol{p} \end{Bmatrix} \tag{3.2.6}$$

the equations of Hamiltonian dual system (3.2.5) can also be expressed as

$$\dot{\boldsymbol{v}} = \boldsymbol{H}\boldsymbol{v} + \boldsymbol{h} \tag{3.2.7}$$

where

$$\boldsymbol{H} = \begin{bmatrix} \boldsymbol{A} & \boldsymbol{D} \\ \boldsymbol{B} & -\boldsymbol{A}^{\mathrm{T}} \end{bmatrix}, \quad \boldsymbol{h} = \begin{Bmatrix} \boldsymbol{h}_q \\ \boldsymbol{h}_p \end{Bmatrix} \tag{3.2.8}$$

As $\boldsymbol{K}_{22}$ and $\boldsymbol{K}_{11}$ are both symmetric matrices, $\boldsymbol{B}$ and $\boldsymbol{D}$ are also symmetric matrices. It is then not difficult to prove that matrix $\boldsymbol{H}$ satisfies the condition of Eq. (1.3.14) and, therefore, it is a Hamiltonian matrix.

Obviously, the Hamiltonian density function can also be expressed in terms of Hamiltonian matrix as

$$\mathscr{H}(\boldsymbol{q}, \boldsymbol{p}) = -\frac{1}{2}\boldsymbol{v}^{\mathrm{T}}(\boldsymbol{J}\boldsymbol{H})\boldsymbol{v} + \boldsymbol{h}^{\mathrm{T}}\boldsymbol{J}\boldsymbol{v} \tag{3.2.9}$$

Both systems of Eqs. (3.2.5) and (3.2.7) are all systems of first-order equations. The time differential of the full state vector is only present on the left-hand-side of the equations while $\boldsymbol{q}$ and $\boldsymbol{p}$ only present on the right-hand-side. Hence, the original system of n second-order differential equations (3.1.20) can be transformed into a system of $2n$ first-order differential equations (3.2.5). Thus we complete the transformation from Lagrange system to Hamiltonian system.

The Hamiltonian density function is a quadratic form of $\boldsymbol{q}$ and $\boldsymbol{p}$ as expressed in Eq. (3.2.3). It becomes a quadratic homogeneous form if there is no external force. This is a special characteristic of a linear system. For a nonlinear system, $\mathscr{H}(\boldsymbol{q}, \boldsymbol{p})$ is a general function of $\boldsymbol{q}$ and $\boldsymbol{p}$ without differential of x.

The Hamiltonian density function $\mathscr{H}(\boldsymbol{q}, \boldsymbol{p})$ is also known as the **mixed energy density**. The corresponding Lagrange density function $\mathscr{L}(\boldsymbol{q}, \dot{\boldsymbol{q}})$ is also the total potential energy density. It is noted that the deformation energy function in accordance with the variational principle of elasticity discussed earlier employs either strain (displacement related), i.e. $v_{\varepsilon}(\boldsymbol{\varepsilon})$, or stress, i.e. $v_{\mathrm{c}}(\boldsymbol{\sigma})$, as the independent variables. Such expressions are not the expressions of mixed energy, but rather the strain energy density or the strain complementary energy density. The expression of Hamiltonian function employs displacement and its dual variable as the independent variables. It is neither the strain energy nor the strain complementary energy, but rather the mixed energy density. The mixed energy density is indefinite, it can be either positive or negative. However, the strain energy density and the strain complementary energy density are both positive definite.

It may be a little contradictory to the convention in the discussion above. Similarly, in many discussions on dynamics systems, the Lagrange function in a form of the difference of kinetic energy and potential energy is also rather puzzling. Nevertheless, the corresponding Hamiltonian function constructed as the sum of kinetic energy and potential energy is comparatively easier to be comprehended. Similar situation can be observed if a Hamiltonian system for dynamics is derived by modeling the spatial x-coordinate as the temporal t-coordinate. Here, contrary to the case of dynamics, confusion is due to the Lagrange function as the total potential energy density while the Hamiltonian function as the mixed energy.

The theory of Hamiltonian system has long been introduced in analytical dynamics in classes because it has been regarded as a subject concerned with dynamics only. But now it is clear that the theory is also applicable to the spatial x-coordinate and is not limited to dynamics. Many mathematicians have already asserted that the theory of Hamiltonian system is a set of mathematically constructed system[9–11]. The functions can be separated from their specific physical meaning when discussed. Much information for applied mechanics can be acquired much from the theory.

The theory of Hamiltonian system is general and it is not limited to linear systems. The many approaches used currently to solve linear elasticity problems embody the content of linear Hamiltonian system. This is an extraordinary point in this book. It should be emphasized that the Hamiltonian system is also applicable to nonlinear elasticity systems although only linear systems are discussed in this book.

Finally, the physical meaning of the dual variable $\boldsymbol{p}$ is discussed. From Eq. (3.2.1), we have

$$\begin{aligned}\boldsymbol{p} &= \boldsymbol{K}_{22}\dot{\boldsymbol{q}} + \boldsymbol{K}_{21}\boldsymbol{q} \\ &= \left\{ kGA(\dot{w} - \theta), \quad EI\dot{\theta} \right\}^{\mathrm{T}} = \left\{ F_{\mathrm{s}}, \quad M \right\}^{\mathrm{T}}\end{aligned} \tag{3.2.10}$$

as the internal force. Referring to Fig. 3.2, F_{s} is consistent with w and M is consistent with θ on the plane with x-axis as its normal. While on the plane with x-axis opposite its normal, the internal force and displacement on the plane are in opposite directions. Hence the full state vector of Timoshenko beam is

$$\boldsymbol{v} = \{w, \quad \theta; \quad F_{\mathrm{s}}, \quad M\}^{\mathrm{T}} \tag{3.2.11}$$

The expressions of Hamiltonian matrix and mixed energy density are

$$\boldsymbol{H} = \begin{bmatrix} 0 & 1 & (kGA)^{-1} & 0 \\ 0 & 0 & 0 & (EI)^{-1} \\ -\rho\omega^2 A & 0 & 0 & 0 \\ 0 & -\rho\omega^2 I & -1 & 0 \end{bmatrix} \tag{3.2.12}$$

$$\begin{aligned} \mathscr{H}(w, \theta, F_{\mathrm{s}}, M) = F_{\mathrm{s}}\theta + \frac{1}{2}\rho\omega^2(Aw^2 + I\theta^2) \\ + \frac{1}{2}\frac{F_{\mathrm{s}}^2}{kGA} + \frac{1}{2}\frac{M^2}{EI} + qw + m\theta \end{aligned} \tag{3.2.13}$$

3.3. The Method of Separation of Variables

There are generally two kinds of solution methodology for the system of dual equations (3.2.7): direct integration and separation of variables. Direct integration is rather difficult for the two-point boundary-value problem here. When the dimension n is not too high, the precise integration method is applicable. As elasticity problems concern continuous bodies, there are equivalently infinite unknown variables. For a continuous body or a case of high dimension n, the method of separation of variables is more appropriate. The latter is, actually, also applicable to cases of low dimension n.

According to the theory of ordinary differential equation, it is necessary to solve the linear homogeneous system of equations in advance before solving the linear inhomogeneous system of Eqs. (3.2.7). That is, we first solve

$$\dot{\boldsymbol{v}} = \boldsymbol{H}\boldsymbol{v} \tag{3.3.1}$$

where $\boldsymbol{v}$ is a $2n$-dimensional vector and $\boldsymbol{H}$ is a Hamiltonian matrix. The **method of separation of variables** requires to seek a solution in the form of

$$\boldsymbol{v}(x) = \xi(x)\boldsymbol{\psi} \tag{3.3.2}$$

where $\xi(x)$ is a function of x and it is independent of any components of vector $\boldsymbol{\psi}$; and $\boldsymbol{\psi}$ is a $2n$-dimensional vector which is independent of x, or

$$\boldsymbol{\psi} = \{\,\psi_1, \quad \psi_2, \quad \cdots, \quad \psi_{2n}\,\}^{\mathrm{T}} \tag{3.3.3}$$

which is a function of some transverse quantities. Substituting Eq. (3.3.2) into Eq. (3.3.1) yields

$$\frac{(\boldsymbol{H}\boldsymbol{\psi})_i}{\psi_i} = \frac{\dot{\xi}(x)}{\xi(x)} \quad (i = 1, 2, \ldots, 2n) \tag{3.3.4}$$

The left-hand-side of this expression is independent of x, and the right-hand-side is independent of subscript i. Hence, the quality can only be equal to a constant which is denoted as μ. Then we have

$$\boldsymbol{H}\boldsymbol{\psi} = \mu\boldsymbol{\psi} \tag{3.3.5}$$

and

$$\xi(x) = \mathrm{e}^{\mu x} \tag{3.3.6}$$

Equation (3.3.5) is the eigenvalue problem of Hamiltonian matrix.

Characteristics of the eigenvalue problem of Hamiltonian matrix has been presented in Sec. 1.3.

If μ is an eigenvalue of a Hamiltonian matrix, then $-\mu$ must also be an eigenvalue. Hence the $2n$ eigenvalues of the $2n$-dimensional Hamiltonian matrix $\boldsymbol{H}$ can be divided into two sets as follows

$$(\alpha)\,\mu_i, \quad \mathrm{Re}(\mu_i) < 0 \quad \text{or} \quad \mathrm{Re}(\mu_i) = 0 \wedge \mathrm{Im}(\mu_i) < 0 \quad (i = 1, 2, \ldots, n) \tag{3.3.7a}$$

$$(\beta)\,\mu_{n+i} = -\mu_i \tag{3.3.7b}$$

The eigenvalues in the (α)-set can be arranged in an ascending order according to the value of $\mathrm{Re}(\mu_i)$.

The eigenvectors of Hamiltonian matrix are mutually adjoint symplectic orthogonal. Let $\boldsymbol{\psi}_i$ and $\boldsymbol{\psi}_j$ be the eigenvectors corresponding to the eigenvalues μ_i and μ_j, respectively. They satisfy

$$\langle \boldsymbol{\psi}_i, \boldsymbol{\psi}_j \rangle = \boldsymbol{\psi}_i^{\mathrm{T}} \boldsymbol{J} \boldsymbol{\psi}_j = 0 \quad \text{for } \mu_i + \mu_j \neq 0 \tag{3.3.8}$$

The eigenvector which is symplectic adjoint with $\boldsymbol{\psi}_i$ must be the eigenvector of eigenvalue $-\mu_i$ (or the Jordan form eigenvector).

As the Hamiltonian matrix $\boldsymbol{H}$ is not a symmetric matrix, it is possible to have repeated eigenvalues and also Jordan form eigenvectors. If $\boldsymbol{\psi}^{(0)}$ is the basic eigenvector of repeated eigenvalues μ, we can obtain each of the Jordan form eigenvectors $\boldsymbol{\psi}^{(1)}, \ldots, \boldsymbol{\psi}^{(k)}$ according to Eq. (1.1.19) from the following equations

$$\left.\begin{aligned} \boldsymbol{H}\boldsymbol{\psi}^{(1)} &= \mu\boldsymbol{\psi}^{(1)} + \boldsymbol{\psi}^{(0)} \\ \boldsymbol{H}\boldsymbol{\psi}^{(2)} &= \mu\boldsymbol{\psi}^{(2)} + \boldsymbol{\psi}^{(1)} \\ \cdots&\cdots\cdots \\ \boldsymbol{H}\boldsymbol{\psi}^{(k)} &= \mu\boldsymbol{\psi}^{(k)} + \boldsymbol{\psi}^{(k-1)} \end{aligned}\right\} \tag{3.3.9}$$

For the basic eigenvector $\boldsymbol{\psi}^{(0)}$, the solution of the associated original problem (3.3.1) is

$$\boldsymbol{v}^{(0)} = \mathrm{e}^{\mu x}\boldsymbol{\psi}^{(0)} \tag{3.3.10}$$

The Jordan form eigenvectors $\boldsymbol{\psi}^{(1)}, \ldots, \boldsymbol{\psi}^{(k)}$ cannot be used directly to obtain the solution of homogenous equation (3.3.1) in the form of Eq. (3.3.10), but they can be used to construct the solution of the original equation (3.3.1), as

$$\left.\begin{aligned} \boldsymbol{v}^{(1)} &= \mathrm{e}^{\mu x}[\boldsymbol{\psi}^{(1)} + x\boldsymbol{\psi}^{(0)}] \\ \boldsymbol{v}^{(2)} &= \mathrm{e}^{\mu x}\left[\boldsymbol{\psi}^{(2)} + x\boldsymbol{\psi}^{(1)} + \frac{1}{2}x^2\boldsymbol{\psi}^{(0)}\right] \\ \cdots&\cdots\cdots \\ \boldsymbol{v}^{(k)} &= \mathrm{e}^{\mu x}\left[\boldsymbol{\psi}^{(k)} + x\boldsymbol{\psi}^{(k-1)} + \cdots + \frac{1}{k!}x^k\boldsymbol{\psi}^{(0)}\right] \end{aligned}\right\} \tag{3.3.11}$$

It should be emphasized here that the eigenvalues $\mu = 0$ is a special case not included in Eq. (3.3.7). As a result, the expressions (3.3.7) are not strictly correct. The case with $\mu = 0$ is very common in the study of elastic statics, and the Jordan form usually exists. As the dual eigenvectors mix with the Jordan form solutions, the development of theory becomes rather inconvenient to a certain extent. We should seek the subspace of eigen-solutions associated with zero eigenvalue in advance, and then reduce the dimension of Hamiltonian matrix such that the zero eigenvalue is eliminated. Hence, the grouping in Eq. (3.3.7) applies.

The eigen-solutions of zero eigenvalue form an important portion of the solutions of elasticity. They frequently appear in the following chapters.

3.4. Reciprocal Theorem for Work and Adjoint Symplectic Orthogonality

The property of adjoint symplectic orthogonality between the eigen-solutions of Hamiltonian matrix has been introduced in Sec. 1.3. Symplectic orthogonality is a mathematical term. It is the key for symplectic subspace and solution by expansion of transverse quantities of eigenvectors. In this section, the adjoint symplectic orthogonality is demonstrated via the reciprocal theorem for work with physical interpretation for better understanding[12].

After obtaining the eigen-solutions $(\mu_i, \boldsymbol{\psi}_i)$, $(\mu_j, \boldsymbol{\psi}_j)$ for the homogenous equations (3.3.1) of a mechanical system by means of the method of separation of variables, the solution of the original equation (3.3.1) can be formed by

$$\boldsymbol{v}_i = \mathrm{e}^{\mu_i x} \boldsymbol{\psi}_i, \quad \boldsymbol{v}_j = \mathrm{e}^{\mu_j x} \boldsymbol{\psi}_j \tag{3.4.1}$$

Equation (3.3.1) is derived from a conservative system and, therefore, the reciprocal theorem for work is applicable.

For a segment of beam $a \leq x \leq b$, denote the generalized displacements and the generalized internal forces on cross section $x = a$ corresponding to the solution $\boldsymbol{v}_i$ are, respectively, $\boldsymbol{q}_{ai}$ and $\boldsymbol{p}_{ai}$, while they are $\boldsymbol{q}_{bi}$ and $\boldsymbol{p}_{bi}$ on cross section $x = b$. Similarly, the generalized displacement and the generalized internal force corresponding to the solution $\boldsymbol{v}_j$ are $\boldsymbol{q}_{aj}$, $\boldsymbol{p}_{aj}$ and $\boldsymbol{q}_{bj}, \boldsymbol{p}_{bj}$. It is clear that the work done by the generalized internal force of solution $\boldsymbol{v}_i$ with respect to the generalized displacement of solution $\boldsymbol{v}_j$ on this section of beam is (noted that the generalized internal force is in opposite direction to the generalized displacement at $x = a$).

$$\boldsymbol{p}_{bi}^{\mathrm{T}} \boldsymbol{q}_{bj} - \boldsymbol{p}_{ai}^{\mathrm{T}} \boldsymbol{q}_{aj} = \left[\mathrm{e}^{(\mu_i + \mu_j) b} - \mathrm{e}^{(\mu_i + \mu_j) a}\right] (\hat{\boldsymbol{p}}_i^{\mathrm{T}} \hat{\boldsymbol{q}}_j) \tag{3.4.2}$$

where $\hat{\boldsymbol{p}}, \hat{\boldsymbol{q}}$ etc. are respectively the values of related physical quantities, which are independent of x, in eigenvector $\boldsymbol{\psi}$ after separating the variables of Eq. (3.4.1).

On the other hand, the work done by the generalized internal force $\boldsymbol{p}_j$ of solution $\boldsymbol{v}_j$ in the generalized displacement $\boldsymbol{q}_i$ of solution $\boldsymbol{v}_i$ is

$$\boldsymbol{p}_{bj}^{\mathrm{T}} \boldsymbol{q}_{bi} - \boldsymbol{p}_{aj}^{\mathrm{T}} \boldsymbol{q}_{ai} = \left[\mathrm{e}^{(\mu_i + \mu_j) b} - \mathrm{e}^{(\mu_i + \mu_j) a}\right] (\hat{\boldsymbol{p}}_j^{\mathrm{T}} \hat{\boldsymbol{q}}_i) \tag{3.4.3}$$

These two works are equal according to the reciprocal theorem. Then,

$$\left[\mathrm{e}^{(\mu_i + \mu_j) b} - \mathrm{e}^{(\mu_i + \mu_j) a}\right] \left(\hat{\boldsymbol{p}}_j^{\mathrm{T}} \hat{\boldsymbol{q}}_i - \hat{\boldsymbol{p}}_i^{\mathrm{T}} \hat{\boldsymbol{q}}_j\right) = 0 \tag{3.4.4}$$

Thus, $\boldsymbol{\psi}_i$ and $\boldsymbol{\psi}_j$ are symplectic orthogonal when $(\mu_i + \mu_j) \neq 0$, i.e.

$$\langle \boldsymbol{\psi}_i, \boldsymbol{\psi}_j \rangle \equiv (\boldsymbol{\psi}_i, \boldsymbol{J}\boldsymbol{\psi}_j) \equiv \boldsymbol{\psi}_i^{\mathrm{T}} \boldsymbol{J} \boldsymbol{\psi}_j = \hat{\boldsymbol{p}}_j^{\mathrm{T}} \hat{\boldsymbol{q}}_i - \hat{\boldsymbol{p}}_i^{\mathrm{T}} \hat{\boldsymbol{q}}_j = 0 \tag{3.4.5}$$

Therefore, there is specific physical meaning to state that **symplectic orthogonality** is equivalent to **the reciprocal theorem for work**.

The derivation above is referred on the basic eigenvectors. For a repeated eigenvalue, it is possible to have Jordan form eigenvectors. In this respect, the property of symplectic orthogonality can be proved in a similarly manner. Let μ_i be a repeated eigenvalue with multiplicity $m_i + 1$ and the corresponding basic eigenvectors and Jordan form eigenvectors are $\boldsymbol{\psi}_i^{(0)}, \ldots, \boldsymbol{\psi}_i^{(m_i)}$. Hence the solution formed by the first-order Jordan form eigenvector $\boldsymbol{\psi}_i^{(1)}$ is

$$\boldsymbol{v}_i^{(1)} = \mathrm{e}^{\mu_i x}(\boldsymbol{\psi}_i^{(1)} + x\boldsymbol{\psi}_i^{(0)}) \tag{3.4.6}$$

Obviously, the work done by the generalized internal force of solution $\boldsymbol{v}_i^{(1)}$ with respect to the generalized displacement of solution $\boldsymbol{v}_j$ is

$$W_1 = \left[\mathrm{e}^{(\mu_i+\mu_j)b} - \mathrm{e}^{(\mu_i+\mu_j)a}\right]\hat{\boldsymbol{p}}_i^{(1)\mathrm{T}} \hat{\boldsymbol{q}}_j + \left[b\mathrm{e}^{(\mu_i+\mu_j)b} - a\mathrm{e}^{(\mu_i+\mu_j)a}\right]\hat{\boldsymbol{p}}_i^{(0)\mathrm{T}} \hat{\boldsymbol{q}}_j \tag{3.4.7}$$

On the other hand, the work done by the generalized internal force of solution $\boldsymbol{v}_j$ with respect to the generalized displacement of solution $\boldsymbol{v}_i^{(1)}$ is

$$W_2 = [\mathrm{e}^{(\mu_i+\mu_j)b} - \mathrm{e}^{(\mu_i+\mu_j)a}]\hat{\boldsymbol{p}}_j^{\mathrm{T}} \hat{\boldsymbol{q}}_i^{(1)} + [b\mathrm{e}^{(\mu_i+\mu_j)b} - a\mathrm{e}^{(\mu_i+\mu_j)a}]\hat{\boldsymbol{p}}_j^{\mathrm{T}} \hat{\boldsymbol{q}}_i^{(0)} \tag{3.4.8}$$

According to the reciprocal theorem for work, $W_1 = W_2$. Then

$$\begin{aligned} &[\mathrm{e}^{(\mu_i+\mu_j)b} - \mathrm{e}^{(\mu_i+\mu_j)a}](\hat{\boldsymbol{p}}_j^{\mathrm{T}} \hat{\boldsymbol{q}}_i^{(1)} - \hat{\boldsymbol{p}}_i^{(1)\mathrm{T}} \hat{\boldsymbol{q}}_j) \\ &\quad + [b\mathrm{e}^{(\mu_i+\mu_j)b} - a\mathrm{e}^{(\mu_i+\mu_j)a}](\hat{\boldsymbol{p}}_j^{\mathrm{T}} \hat{\boldsymbol{q}}_i^{(0)} - \hat{\boldsymbol{p}}_i^{(0)\mathrm{T}} \hat{\boldsymbol{q}}_j) = 0 \end{aligned} \tag{3.4.9}$$

From Eq. (3.4.5), $\boldsymbol{\psi}_i^{(0)}$ and $\boldsymbol{\psi}_j$ are symplectic orthogonal when $\mu_i + \mu_j \neq 0$. Hence

$$\langle \boldsymbol{\psi}_i^{(1)}, \boldsymbol{\psi}_j \rangle \equiv (\boldsymbol{\psi}_i^{(1)}, \boldsymbol{J}\boldsymbol{\psi}_j) \equiv \boldsymbol{\psi}_i^{(1)\mathrm{T}} \boldsymbol{J}\boldsymbol{\psi}_j = 0 \quad \text{when } \mu_i + \mu_j \neq 0 \tag{3.4.10}$$

Thus, $\boldsymbol{\psi}_i^{(1)}$ and $\boldsymbol{\psi}_j$ must be also symplectic orthogonal. In the same way, $\boldsymbol{\psi}_i^{(2)}, \ldots, \boldsymbol{\psi}_i^{(m_i)}$ and $\boldsymbol{\psi}_j$ are also symplectic orthogonal by repeatedly applying the reciprocal theorem for work. When μ_j is a repeated eigenvalue with multiplicity $m_j + 1$, we can prove that the eigenvectors $\boldsymbol{\psi}_i^{(s)} (s = 0, 1, \ldots, m_i)$ corresponding to μ_i are all symplectic orthogonal to the eigenvectors $\boldsymbol{\psi}_j^{(t)}(t = 0, 1, \ldots m_j)$ corresponding to μ_j by virtue of the reciprocal theorem for work in a similar way.

$$\langle \boldsymbol{\psi}_i^{(\mathrm{s})}, \boldsymbol{\psi}_j^{(t)} \rangle = \boldsymbol{\psi}_i^{(\mathrm{s})\mathrm{T}} \boldsymbol{J} \boldsymbol{\psi}_j^{(t)} = 0 \quad \text{when } \mu_i + \mu_j \neq 0 \tag{3.4.11}$$

The eigenvectors corresponding to non-symplectic adjoint eigenvalues are symplectic orthogonal and this property is shown above by applying the reciprocal theorem for work. Now we discuss the adjoint symplectic orthogonality between the eigenvectors corresponding to eigenvalues which are symplectic adjoint. Let μ_j be an eigenvalue which is symplectic adjoint with the nonzero repeated eigenvalue μ_i with multiplicity $m+1 (\mu_j = -\mu_i)$ then its multiplicity must also be $m+1$.

First, the work done by the generalized internal force of solution $\boldsymbol{v}_j^{(0)}$ with respect to the generalized displacement of solution $\boldsymbol{v}_i^{(m)}$ is

$$\begin{aligned} W_1 = {} & (b-a)(\hat{\boldsymbol{p}}_j^{(0)\mathrm{T}} \hat{\boldsymbol{q}}_i^{(m-1)}) + \frac{1}{2}(b^2 - a^2)(\hat{\boldsymbol{p}}_j^{(0)\mathrm{T}} \hat{\boldsymbol{q}}_i^{(m-2)}) \\ & + \cdots + \frac{1}{m!}(b^m - a^m)(\hat{\boldsymbol{p}}_j^{(0)\mathrm{T}} \hat{\boldsymbol{q}}_i^{(0)}) \end{aligned} \tag{3.4.12}$$

On the other hand, the work done by the generalized internal force of solution $\boldsymbol{v}_i^{(m)}$ with respect to the generalized displacement of solution $\boldsymbol{v}_j^{(0)}$ is

$$\begin{aligned} W_2 = {} & (b-a)(\hat{\boldsymbol{p}}_i^{(m-1)\mathrm{T}} \hat{\boldsymbol{q}}_j^{(0)}) + \frac{1}{2}(b^2 - a^2)(\hat{\boldsymbol{p}}_i^{(m-2)\mathrm{T}} \hat{\boldsymbol{q}}_j^{(0)}) \\ & + \cdots + \frac{1}{m!}(b^m - a^m)(\hat{\boldsymbol{p}}_i^{(0)\mathrm{T}} \hat{\boldsymbol{q}}_j^{(0)}) \end{aligned} \tag{3.4.13}$$

According to the reciprocal theorem for work, $W_1 = W_2$. Hence,

$$(b-a)(\boldsymbol{\psi}_i^{(m-1)\mathrm{T}} \boldsymbol{J} \boldsymbol{\psi}_j^{(0)}) + \cdots + \frac{1}{m!}(b^m - a^m)(\boldsymbol{\psi}_i^{(0)\mathrm{T}} \boldsymbol{J} \boldsymbol{\psi}_j^{(0)}) = 0 \tag{3.4.14}$$

Since Eq. (3.4.14) is true for any arbitrary a, b, therefore $\boldsymbol{\psi}_j^{(0)}$ is symplectic orthogonal to all $\boldsymbol{\psi}_i^{(0)}, \ldots, \boldsymbol{\psi}_i^{(m-1)}$. Furthermore, $\boldsymbol{\psi}_j^{(0)}$ must be symplectic adjoint with $\boldsymbol{\psi}_i^{(m)}$. Otherwise, $\boldsymbol{\psi}_j^{(0)}$ is symplectic orthogonal to any vector, then $\boldsymbol{\psi}_j^{(0)}$ is a zero vector, thus it leads to a contradictory conclusion. In the same way, $\boldsymbol{\psi}_i^{(0)}$ must be mutually symplectic adjoint with $\boldsymbol{\psi}_j^{(m)}$.

Next, we prove that it is possible to have $\boldsymbol{\psi}_j^{(1)}$ adjoint symplectic orthogonal with $\boldsymbol{\psi}_i^{(s)} (s = 0, \ldots, m)$. Applying the reciprocal theorem to

$\boldsymbol{v}_j^{(1)}$ and $\boldsymbol{v}_i^{(m-1)}$, and making use of the symplectic orthogonality between $\boldsymbol{\psi}_j^{(0)}$ and $\boldsymbol{\psi}_i^{(0)}, \ldots, \boldsymbol{\psi}_i^{(m-1)}$, we can prove that $\boldsymbol{\psi}_j^{(1)}$ is symplectic orthogonal with any of $\boldsymbol{\psi}_i^{(0)}, \ldots, \boldsymbol{\psi}_i^{(m-2)}$ in a similar way as discussed above. Applying the reciprocal theorem to $\boldsymbol{v}_j^{(1)}$ and $\boldsymbol{v}_i^{(m)}$, we derive

$$\boldsymbol{\psi}_j^{(0)\mathrm{T}} \boldsymbol{J} \boldsymbol{\psi}_i^{(m)} + \boldsymbol{\psi}_j^{(1)^{\mathrm{T}}} \boldsymbol{J} \boldsymbol{\psi}_i^{(m-1)} = 0 \tag{3.4.15}$$

As $\boldsymbol{\psi}_j^{(0)}$ and $\boldsymbol{\psi}_i^{(m)}$ are symplectic adjoint, $\boldsymbol{\psi}_j^{(1)}$ must be also symplectic adjoint with $\boldsymbol{\psi}_i^{(m-1)}$. Then, we discuss the symplectic orthogonality between $\boldsymbol{\psi}_j^{(1)}$ and $\boldsymbol{\psi}_i^{(m)}$. Obviously, the Jordan form eigenvector $\boldsymbol{\psi}_j^{(1)}$ can be superposed on any basic eigenvectors $C\boldsymbol{\psi}_j^{(0)}$. Equation (3.3.9) remains valid if $\boldsymbol{\psi}_j^{(k)}$ is substituted by $\boldsymbol{\psi}_j^{(k)} + C\boldsymbol{\psi}_j^{(k-1)} (k = 1, 2, \ldots, m)$, and the adjoint symplectic orthonormality proved above remains unchanged. As $\boldsymbol{\psi}_j^{(0)}$ is symplectic adjoint with $\boldsymbol{\psi}_i^{(m)}$, it is always possible to have $\boldsymbol{\psi}_j^{(1)} + C\boldsymbol{\psi}_j^{(0)}$ symplectic orthogonal to $\boldsymbol{\psi}_i^{(m)}$ by choosing an appropriate constant C and denoting $\boldsymbol{\psi}_j^{(k)} + C\boldsymbol{\psi}_j^{(k-1)}$ as $\boldsymbol{\psi}_j^{(k)} (k = 1, 2, \ldots, m)$. Here, $\boldsymbol{\psi}_j^{(1)}$ is symplectic adjoint with $\boldsymbol{\psi}_i^{(m-1)}$ while it is symplectic orthogonal to the other vectors $\boldsymbol{\psi}_i^{(s)} (s \neq m-1)$. Similarly, it is possible to have $\boldsymbol{\psi}_j^{(2)}, \ldots, \boldsymbol{\psi}_j^{(m)}$ adjoint symplectic orthogonal to $\boldsymbol{\psi}_i^{(s)} (s = 0, 1, \ldots, m)$ through proper selection. Hence, there exists an adjoint symplectic orthonormal vector set for eigenvalues which are mutually symplectic adjoint, as

$$\langle \boldsymbol{\psi}_i^{(s)}, \boldsymbol{\psi}_j^{(t)} \rangle \equiv \boldsymbol{\psi}_i^{(s)} \boldsymbol{J} \boldsymbol{\psi}_j^{(t)} = \begin{cases} \neq 0; & \text{for } s + t = m \\ = 0; & \text{for } s + t \neq m \end{cases} \tag{3.4.16}$$

The proof for Jordan form above is confined to the condition $\mu_i \neq 0$. The case of zero eigenvalue should then be considered. The eigen-solutions of a zero eigenvalue are usually very important in structural statics and elastic statics, particularly in Saint–Venant problems, etc[13]. Because a zero eigenvalue has itself the symplectic adjoint eigenvalue, the eigenvector corresponding to this zero eigenvalue is symplectic orthogonal to all the other eigenvectors corresponding to the nonzero eigenvalues. Applying the reciprocal theorem for work, we can prove that the eigenvectors corresponding to zero eigenvalue can themselves be adjoint symplectic orthonormal. The proof is more complicated than the case of nonzero eigenvalues and it omitted here. Interested readers are referred this section and Theorem 1.17 in Sec. 1.3 for establishing the proof.

Having established the adjoint symplectic orthogonality of eigenvectors, the problem can be normalized into a standard form in order to facilitate its application for eigen-solution expansion.

3.5. Solution for Non-Homogeneous Equations

The sections above discuss solutions of homogeneous equations. In this section, we discuss the solution of non-homogeneous equation (3.2.7).

First, according to the theorem of expansion of eigenvectors, the full state vector is

$$\boldsymbol{v}(x) = \sum_{i=1}^{n} [a_i(x)\boldsymbol{\psi}_i + b_i(x)\boldsymbol{\psi}_{n+i}] \tag{3.5.1}$$

where $\boldsymbol{\psi}_i(i = 1, 2, \ldots, 2n)$ are the eigenvectors of eigenvalues μ_i associated with the homogeneous equation. It is assumed here that the eigenvectors have been normalized with respect to adjoint symplectic orthogonality. Hence,

$$\left.\begin{aligned} \boldsymbol{\psi}_i^{\mathrm{T}}\boldsymbol{J}\boldsymbol{\psi}_j &= \boldsymbol{\psi}_{n+i}^{\mathrm{T}}\boldsymbol{J}\boldsymbol{\psi}_{n+j} = 0 \\ \boldsymbol{\psi}_i^{\mathrm{T}}\boldsymbol{J}\boldsymbol{\psi}_{n+j} &= \begin{cases} 1, & \text{for } j = i \\ 0, & \text{for } j \neq i \end{cases} \end{aligned}\right\} \quad (i, j = 1, \ldots, n) \tag{3.5.2}$$

At the same time, the known external force vector function $\boldsymbol{h}(x)$ is expanded in eigenvectors as

$$\boldsymbol{h}(x) = \sum_{i=1}^{n} [c_i(x)\boldsymbol{\psi}_i + d_i(x)\boldsymbol{\psi}_{n+i}] \tag{3.5.3}$$

where c_i and d_i are known functions given by

$$c_i(x) = -\boldsymbol{\psi}_{n+i}^{\mathrm{T}}\boldsymbol{J}\boldsymbol{h}(x), d_i(x) = \boldsymbol{\psi}_i^{\mathrm{T}}\boldsymbol{J}\boldsymbol{h}(x) \tag{3.5.4}$$

which is determined by the adjoint symplectic orthonormalization relation (3.5.2).

Next, substituting Eqs. (3.5.1) and (3.5.3) into Eq. (3.2.7) and using the adjoint symplectic orthonormalization relation (3.5.2) yield

$$\dot{a}_i = \mu_i a_i + c_i, \quad \dot{b}_i = -\mu_i b_i + d_i \qquad (1 \leq i \leq n) \tag{3.5.5}$$

Hence the original problem is decoupled into $2n$ first-order nonhomogeneous differential equations with one unknown. There is a standard method for solving first-order differential equations with one unknown and general solutions are

$$\left.\begin{aligned} a_i(x) &= A_i \mathrm{e}^{\mu_i x} + \int_0^x c_i(\xi)\mathrm{e}^{\mu_i(x-\xi)}\mathrm{d}\xi \\ b_i(x) &= B_i \mathrm{e}^{-\mu_i x} + \int_0^x d_i(\xi)\mathrm{e}^{-\mu_i(x-\xi)}\mathrm{d}\xi \end{aligned}\right\} \tag{3.5.6}$$

Hence the general solution of original problem is

$$\boldsymbol{v}(x) = \sum_{i=1}^{n} [A_i \mathrm{e}^{\mu_i x}\boldsymbol{\psi}_i + B_i \mathrm{e}^{-\mu_i x}\boldsymbol{\psi}_{n+i}] + \overline{\boldsymbol{v}}(x) \tag{3.5.7}$$

where $\overline{\boldsymbol{v}}(x)$ is a particular solution corresponding to the nonhomogeneous term $\boldsymbol{h}(x)$ formed by Eq. (3.5.6) and $A_i, B_i \quad (i = 1, 2, \ldots, n)$ are unknown constants to be determined by imposing the boundary conditions at both ends.

Certainly Eq. (3.5.7) can also be written as

$$\boldsymbol{v}(x) = \sum_{i=1}^{n} [A_i \boldsymbol{v}_i(x) + B_i \boldsymbol{v}_{n+i}(x)] + \overline{\boldsymbol{v}}(x) \tag{3.5.8}$$

where $\boldsymbol{v}_i, \boldsymbol{v}_{n+i} (i = 1, 2, \ldots, n)$ are the solutions formed by Eq. (3.4.1) and the solutions of the homogeneous equation (3.3.1) corresponding to the eigenvectors $\boldsymbol{\psi}_i, \boldsymbol{\psi}_{n+i}$.

3.6. Two-Point Boundary Conditions

The equations in the study of elasticity are elliptic and conditions on the boundary are necessary. For a system with a single coordinate, it becomes the boundary conditions at two ends. This is the so-called two-point boundary value problem in mathematics.

There are $2n$ unknown constants of integration in $2n$ first-order differential equations and there should be a total of $2n$ conditions for such two-point boundary value problem with n conditions in general at each end. It is reasonable or otherwise there will be numerical instablility due to the presence of exponents. For instance, the boundary conditions of Timoshenko beam usually are

$$\begin{aligned} \text{Free end:} &\quad F_{\mathrm{s}} = 0, \quad M = 0 \\ \text{Simply supported end:} &\quad w = 0, \quad M = 0 \\ \text{Fixed end:} &\quad w = 0, \quad \theta = 0 \end{aligned} \tag{3.6.1}$$

There are two conditions at each end and these are ideal boundary conditions. There is a variety of other combinations of boundary conditions which are omitted here.

There are various solution methods for two-point boundary value problems. The method of expansion of eigenvectors is an effective approach, particularly in two- and three-dimensional problems in elasticity.

Since there are only n boundary conditions at each end, it is rather difficult to solve directly at each end. In general, we combine the conditions at both ends to establish a set of algebra equations for the solutions. There are various ways to establish the equations, but we should observe the symmetry of equations which represents the reciprocal theorems. A variational method is provided here[5,13]. For the boundary conditions in Eq. (3.6.1), either the force or the displacement is known. In variational principle, this corresponds to the boundary Γ_σ for specified forces or the boundary Γ_u for specific displacements. For the ends at $x = 0$ and $x = l$, the respective boundary conditions are denoted as $\Gamma_{\sigma 0}, \Gamma_{\sigma l}$ and Γ_{u0}, Γ_{ul}. In fact, these ends are only discrete points, for example, $\Gamma_{uf} + \Gamma_{\sigma f}$ are n conditions. Hence we can express the boundary conditions in a most general form, as

$$[\boldsymbol{q} = \overline{\boldsymbol{q}}_0]_{\Gamma_{u0}}; \quad [\boldsymbol{p} = \overline{\boldsymbol{p}}_0]_{\Gamma_{\sigma 0}} \quad \text{at } x = 0 \tag{3.6.2a}$$

$$[\boldsymbol{q} = \overline{\boldsymbol{q}}_l]_{\Gamma_{ul}}; \quad [\boldsymbol{p} = \overline{\boldsymbol{p}}_l]_{\Gamma_{\sigma l}} \quad \text{at } x = l \tag{3.6.2b}$$

It states that a portion of the conditions is designated for specified forces while the remaining portion is designated for specified displacements for n conditions at the end. The expressions in Eq. (3.6.2) are not only applicable to the three different boundary conditions (3.6.1) for a Timoshenko beam, but also to other more general two-point boundary value problems.

With the boundary conditions stated, it is now possible to formulate the Hamiltonian **mixed energy variational principle** which is equivalent to the dual equations (3.2.5) and boundary conditions (3.6.2).

$$\delta\left\{\int_0^l [\boldsymbol{p}^{\mathrm{T}}\dot{\boldsymbol{q}} - \mathscr{H}(\boldsymbol{q},\boldsymbol{p})]\mathrm{d}x - [\boldsymbol{p}^{\mathrm{T}}(\boldsymbol{q} - \overline{\boldsymbol{q}}_l)]_{\Gamma_{ul}} - [\overline{\boldsymbol{p}}_l^{\mathrm{T}}\boldsymbol{q}]_{\Gamma_{\sigma l}} + [\boldsymbol{p}^{\mathrm{T}}(\boldsymbol{q} - \overline{\boldsymbol{q}}_0)]_{\Gamma_{u0}} + [\overline{\boldsymbol{p}}_0^{\mathrm{T}}\boldsymbol{q}]_{\Gamma_{\sigma 0}}\right\} = 0 \tag{3.6.3}$$

Expanding the variational expression (3.6.3) yields

$$\int_0^l \left[(\delta \boldsymbol{p}^{\mathrm{T}}) \left(\dot{\boldsymbol{q}} - \frac{\partial \mathscr{H}}{\partial \boldsymbol{p}} \right) - (\delta \boldsymbol{q}^{\mathrm{T}}) \left(\dot{\boldsymbol{p}} + \frac{\partial \mathscr{H}}{\partial \boldsymbol{q}} \right) \right] \mathrm{d}x$$
$$+ [\boldsymbol{p}^{\mathrm{T}} \delta \boldsymbol{q}]_{x=0}^{l} - [(\delta \boldsymbol{p}^{\mathrm{T}})(\boldsymbol{q} - \overline{\boldsymbol{q}}_l)]_{\Gamma_{ul}} - [\overline{\boldsymbol{p}}_l^{\mathrm{T}} \delta \boldsymbol{q}]_{\Gamma_{\sigma l}}$$
$$+ [(\delta \boldsymbol{p}^{\mathrm{T}})(\boldsymbol{q} - \overline{\boldsymbol{q}}_0)]_{\Gamma_{u0}} + [\overline{\boldsymbol{p}}_0^{\mathrm{T}} \delta \boldsymbol{q}]_{\Gamma_{\sigma 0}} = 0 \tag{3.6.4}$$

Because $\delta \boldsymbol{p}$ and $\delta \boldsymbol{q}$ in the region are arbitrary, we have the dual equations (3.2.5). At the two ends there exists

$$[(\delta \boldsymbol{q})^{\mathrm{T}} (\boldsymbol{p} - \overline{\boldsymbol{p}}_l)]_{\Gamma_{\sigma l}} - [(\delta \boldsymbol{p}^{\mathrm{T}})(\boldsymbol{q} - \overline{\boldsymbol{q}}_l)]_{\Gamma_{ul}}$$
$$- [(\delta \boldsymbol{q})^{\mathrm{T}} (\boldsymbol{p} - \overline{\boldsymbol{p}}_0)]_{\Gamma_{\sigma 0}} + [(\delta \boldsymbol{p}^{\mathrm{T}})(\boldsymbol{q} - \overline{\boldsymbol{q}}_0)]_{\Gamma_{u0}} = 0 \tag{3.6.5}$$

Further, since $\delta \boldsymbol{p}$ and $\delta \boldsymbol{q}$ are arbitrary on Γ_u for specified displacements and on Γ_σ for specified forces, respectively, we obtain the boundary conditions (3.6.2).

In accordance with the solution method by expansion of eigenvectors discussed earlier, the solution of the original problem can be expressed as

$$\boldsymbol{v}(x) = \sum_{i=1}^{n} [A_i \boldsymbol{v}_i(x) + B_i \boldsymbol{v}_{n+i}(x)] = \sum_{i=1}^{n} [A_i \mathrm{e}^{\mu_i x} \boldsymbol{\psi}_i + B_i \mathrm{e}^{-\mu_i x} \boldsymbol{\psi}_{n+i}] \tag{3.6.6}$$

It should be stated here that the factors of particular solution $\overline{\boldsymbol{v}}(x)$ associated with inhomogeneous term have been eliminated by using the principle of superposition. Accordingly, the values of each specified displacement and specified force $\overline{\boldsymbol{q}}_0, \overline{\boldsymbol{q}}_l, \overline{\boldsymbol{p}}_0, \overline{\boldsymbol{p}}_l$ in Eqs. (3.6.2) and (3.6.3) should be regarded as the known values after the elimination of particular solution.

As expansion of eigenvectors is applied to Eq. (3.6.6), all differential equations in the region are satisfied. Further substitution of Eq. (3.6.6) into the variational principle (3.6.3) yields only the remaining variational equation (3.6.5) at the ends.

As an application of the variational equation (3.6.5), we first discuss a simple example of a semi-infinite beam. Let $l \to \infty$ and assume the eigenvalues in Eq. (3.6.6) be arranged according to the rule in Eq. (3.3.7). As $\exp(-\mu_i l) \to \infty$, we have $B_i = 0 (i = 1, 2, \ldots, n)$. Then only n boundary conditions at $x = 0$ remain in the variational equation (3.6.5) for determining the n constants $A_i (i = 1, 2, \ldots, n)$.

Substituting the expression at $x = 0$

$$\boldsymbol{v}(0) = \sum_{i=1}^{n} [A_i \boldsymbol{\psi}_i] = \sum_{i=1}^{n} \left[A_i \begin{Bmatrix} \boldsymbol{q}_i \\ \boldsymbol{p}_i \end{Bmatrix} \right] \tag{3.6.7}$$

into Eq. (3.6.5), we obtain

$$\begin{aligned} & - \left[\left(\sum_{i=0}^{n} \delta A_i \boldsymbol{q}_i \right)^{\mathrm{T}} \left(\sum_{j=0}^{n} A_j \boldsymbol{p}_j - \overline{\boldsymbol{p}}_0 \right) \right]_{\Gamma_{\sigma 0}} \\ & + \left[\left(\sum_{i=0}^{n} \delta A_i \boldsymbol{p}_i \right)^{\mathrm{T}} \left(\sum_{j=0}^{n} A_j \boldsymbol{q}_j - \overline{\boldsymbol{q}}_0 \right) \right]_{\Gamma_{u 0}} = 0 \end{aligned} \tag{3.6.8}$$

Rearranging Eq. (3.6.8), we have

$$\begin{aligned} \sum_{i=0}^{n} \Bigg[& \left(\sum_{j=0}^{n} A_j \boldsymbol{p}_i^{\mathrm{T}} \boldsymbol{q}_j - \boldsymbol{p}_i^{\mathrm{T}} \overline{\boldsymbol{q}}_0 \right)_{\Gamma_{u0}} \\ & - \left(\sum_{j=0}^{n} A_j \boldsymbol{q}_i^{\mathrm{T}} \boldsymbol{p}_j - \boldsymbol{q}_i^{\mathrm{T}} \overline{\boldsymbol{p}}_0 \right)_{\Gamma_{\sigma 0}} \Bigg] \delta A_i = 0 \end{aligned} \tag{3.6.9}$$

Since $\delta A_i (i = 1, 2, \ldots, n)$ is arbitrary, a set of algebraic equations for solving the n constants

$$\begin{bmatrix} c_{11} & c_{12} & \cdots & c_{1n} \\ c_{21} & c_{22} & \cdots & c_{2n} \\ \vdots & \vdots & \ddots & \vdots \\ c_{n1} & c_{n2} & \cdots & c_{nn} \end{bmatrix} \begin{Bmatrix} A_1 \\ A_2 \\ \vdots \\ A_n \end{Bmatrix} = \begin{Bmatrix} d_1 \\ d_2 \\ \vdots \\ d_n \end{Bmatrix} \tag{3.6.10}$$

can be established where

$$\left. \begin{aligned} c_{ij} &= (\boldsymbol{p}_i^{\mathrm{T}} \boldsymbol{q}_j)_{\Gamma_{u0}} - (\boldsymbol{q}_i^{\mathrm{T}} \boldsymbol{p}_j)_{\Gamma_{\sigma 0}}; \quad (i, j = 1, 2, \ldots, n) \\ d_i &= (\boldsymbol{p}_i^{\mathrm{T}} \overline{\boldsymbol{q}}_0)_{\Gamma_{u0}} - (\boldsymbol{q}_i^{\mathrm{T}} \overline{\boldsymbol{p}}_0)_{\Gamma_{\sigma 0}}; \quad (i = 1, 2, \ldots, n) \end{aligned} \right\} \tag{3.6.11}$$

Solving Eq. (3.6.10) yields the values of A_i and the solution of the original problem is then obtained.

Furthermore, we can prove by using symplectic orthogonality that

$$\begin{aligned} c_{ij} - c_{ji} &= (\boldsymbol{p}_i^{\mathrm{T}} \boldsymbol{q}_j)_{\Gamma_{u0}} - (\boldsymbol{q}_i^{\mathrm{T}} \boldsymbol{p}_j)_{\Gamma_{\sigma 0}} - (\boldsymbol{p}_j^{\mathrm{T}} \boldsymbol{q}_i)_{\Gamma_{u0}} + (\boldsymbol{q}_j^{\mathrm{T}} \boldsymbol{p}_i)_{\Gamma_{\sigma 0}} \\ &= (\boldsymbol{p}_i^{\mathrm{T}} \boldsymbol{q}_j)_{\Gamma_{u0} + \Gamma_{\sigma 0}} - (\boldsymbol{q}_i^{\mathrm{T}} \boldsymbol{p}_j)_{\Gamma_{\sigma 0} + \Gamma_{u0}} = \boldsymbol{\psi}_j^{\mathrm{T}} \boldsymbol{J} \boldsymbol{\psi}_i = 0 \end{aligned} \tag{3.6.12}$$

Therefore, the coefficient matrix of the set of Eqs. (3.6.10) must be a symmetric matrix. The significance of symplectic orthogonality is again observed here.

Obviously, if the boundary conditions at $x = 0$ are all specified displacements, the set of Eqs. (3.6.10) for determining the coefficients should be

$$\left.\begin{aligned} c_{ij} &= \boldsymbol{p}_i^{\mathrm{T}}\boldsymbol{q}_j; \quad (i,j = 1,2,\ldots,n) \\ d_i &= \boldsymbol{p}_i^{\mathrm{T}}\overline{\boldsymbol{q}}_0; \quad\ (i = 1,2,\ldots,n) \end{aligned}\right\} \tag{3.6.13}$$

If the boundary conditions at $x = 0$ are all specified forces, the set of Eqs. (3.6.10) for determining the coefficients, after uniformly modifying the signs, should be

$$\left.\begin{aligned} c_{ij} &= \boldsymbol{p}_j^{\mathrm{T}}\boldsymbol{q}_i; \quad (i,j = 1,2,\ldots,n) \\ d_i &= \overline{\boldsymbol{p}}_0^{\mathrm{T}}\boldsymbol{q}_i; \quad\ (i = 1,2,\ldots,n) \end{aligned}\right\} \tag{3.6.14}$$

Finally, we discuss the boundary conditions of an elastic support, i.e. there are elastic supports attached to the boundaries $\Gamma_{\sigma 0}, \Gamma_{\sigma l}$ for specified forces in addition to the boundary Γ_{u0}, Γ_{ul} for specified displacements. The stiffness matrices of elastic supports are denoted as $\boldsymbol{R}_0$ and $\boldsymbol{R}_l$, respectively. Here, the corresponding Hamiltonian variational principle can be established by adding deformation energy of elastic supports to Eq. (3.6.3), as

$$\begin{aligned} \delta\Bigg\{ \int_0^l & [\boldsymbol{p}^{\mathrm{T}}\dot{\boldsymbol{q}} - \mathscr{H}(\boldsymbol{q},\boldsymbol{p})]\mathrm{d}x - [\boldsymbol{p}^{\mathrm{T}}(\boldsymbol{q} - \overline{\boldsymbol{q}}_l)]_{\Gamma_{ul}} \\ & - [\overline{\boldsymbol{p}}_l^{\mathrm{T}}\boldsymbol{q}]_{\Gamma_{\sigma l}} + [\boldsymbol{p}^{\mathrm{T}}(\boldsymbol{q} - \overline{\boldsymbol{q}}_0)]_{\Gamma_{u0}} + [\overline{\boldsymbol{p}}_0^{\mathrm{T}}\boldsymbol{q}]_{\Gamma_{\sigma 0}} \\ & + \left[\frac{1}{2}\boldsymbol{q}^{\mathrm{T}}\boldsymbol{R}_l\boldsymbol{q}\right]_{\Gamma_{\sigma l}} + \left[\frac{1}{2}\boldsymbol{q}^{\mathrm{T}}\boldsymbol{R}_0\boldsymbol{q}\right]_{\Gamma_{\sigma 0}} \Bigg\} = 0 \end{aligned} \tag{3.6.15}$$

If the solution method of expansion of eigenvectors in Eq. (3.6.6) is applied, the variational equation (3.6.15) becomes

$$\begin{aligned} & [(\delta\boldsymbol{q})^{\mathrm{T}}(\boldsymbol{p} - \overline{\boldsymbol{p}}_l + \boldsymbol{R}_l\boldsymbol{q})]_{\Gamma_{\sigma l}} - [(\delta\boldsymbol{p}^{\mathrm{T}})(\boldsymbol{q} - \overline{\boldsymbol{q}}_l)]_{\Gamma_{ul}} \\ & - [(\delta\boldsymbol{q})^{\mathrm{T}}(\boldsymbol{p} - \overline{\boldsymbol{p}}_0 - \boldsymbol{R}_0\boldsymbol{q})]_{\Gamma_{\sigma 0}} + [(\delta\boldsymbol{p}^{\mathrm{T}})(\boldsymbol{q} - \overline{\boldsymbol{q}}_0)]_{\Gamma_{u0}} = 0 \end{aligned} \tag{3.6.16}$$

Hence for the semi-infinite beam with elastic supports, the coefficients $c_{ij}(i,j = 1,2,\ldots,n)$ of Eq. (3.6.10) should be rewritten as

$$c_{ij} = (\boldsymbol{p}_i^{\mathrm{T}}\boldsymbol{q}_j)_{\Gamma_{u0}} - (\boldsymbol{q}_i^{\mathrm{T}}\boldsymbol{p}_j - \boldsymbol{q}_i^{\mathrm{T}}\boldsymbol{R}_0\boldsymbol{q}_j)_{\Gamma_{\sigma 0}} \tag{3.6.17}$$

It should be noted and considered that the variational equation above is derived under the condition of real eigenvectors $\boldsymbol{\psi}$. If complex eigenvectors exist, they should be transformed into real canonical equation in advance.

For real Hamiltonian matrix $\boldsymbol{H}$, the eigenvalue μ_i and its corresponding eigenvector $\boldsymbol{\psi}_i$ (or $\boldsymbol{v}_i$) are both either real or complex simultaneously. Furthermore, if μ_i and $\boldsymbol{\psi}_i$ (or $\boldsymbol{v}_i$) are the complex eigen-solutions of Hamiltonian matrix $\boldsymbol{H}$, then the complex conjugates of μ_i and $\boldsymbol{\psi}_i$ (or $\boldsymbol{v}_i$), i.e. μ_i^* and $\boldsymbol{\psi}_i^*$ (or $\boldsymbol{v}_i^*$), are also the complex eigen-solutions. It should be clear that symplectic adjoint and complex conjugate are two different concepts.

For concise expressions, the complex conjugate eigen-solutions are assumed to have been arranged in an appropriate order. Assuming that the eigenvalues $\mu_k (1 \le k < n)$ in expansion (3.6.6) are complex and their complex conjugate eigenvalues are μ_{k+1}, hence

$$\mu_{k+1} = \mu_k^*, \quad \boldsymbol{v}_{k+1} = \boldsymbol{v}_k^* \tag{3.6.18}$$

As the left-hand-side of solution (3.6.6) is real, the unknown constants on the right-hand-side of Eq. (3.6.6) must be in complex conjugate pairs

$$A_{k+1} = A_k^* \tag{3.6.19}$$

Hence, we have

$$\begin{aligned} A_k \boldsymbol{v}_k + A_{k+1} \boldsymbol{v}_{k+1} &= 2\,\mathrm{Re}(A_k)\mathrm{Re}(\boldsymbol{v}_k) - 2\,\mathrm{Im}(A_k)\mathrm{Im}(\boldsymbol{v}_k) \\ &= A'_k \boldsymbol{v}'_k + A'_{k+1} \boldsymbol{v}'_{k+1} \end{aligned} \tag{3.6.20}$$

In this way, solution of the original problem can be expressed in the form of Eq. (3.6.6). For complex eigen-solutions, the expansion of vectors should be regarded as the real part $\mathrm{Re}(\boldsymbol{v}_k)$ or the imaginary part $\mathrm{Im}(\boldsymbol{v}_k)$ of the complex eigenvector $\boldsymbol{v}_k$. Because the symplectic adjoint eigenvalues of μ_k and μ_{k+1}, i.e. $-\mu_k$ and $-\mu_{k+1}$, are a pair of complex conjugate eigenvalues, the terms related to B_i can be treated in a similar way. Thus, we have hitherto accomplished the transformation of canonical equations from complex form to real form. Subsequently, the unknown constants can be determined in the same way as discussed above.

3.7. Static Analysis of Timoshenko Beam

The methods introduced in the last few previous sections can be applied to solve various problems of Hamiltonian system. In this section we begin to discuss the bending of Timoshenko beam in detail[8]. At a first instance, we consider the special case $\omega = 0$, i.e. the static bending of Timoshenko beam. Here, the Hamiltonian matrix (3.2.12) degenerates to

$$\boldsymbol{H} = \begin{bmatrix} 0 & 1 & (kGA)^{-1} & 0 \\ 0 & 0 & 0 & (EI)^{-1} \\ 0 & 0 & 0 & 0 \\ 0 & 0 & -1 & 0 \end{bmatrix} \tag{3.7.1}$$

After separating the variables of homogeneous equation (3.3.1), we obtain the eigenvalue equation (3.3.5), as

$$(\boldsymbol{H} - \mu \boldsymbol{I})\boldsymbol{\psi} = \boldsymbol{0} \tag{3.7.2}$$

For nontrivial solution of Eq. (3.7.2), the determinant of its coefficient vanishes

$$\det(\boldsymbol{H} - \mu \boldsymbol{I}) = \mu^4 = 0 \tag{3.7.3}$$

Hence, the solution of eigenvalue equation (3.7.2) is a repeated zero eigenvalue with quadruple multiplicity.

For solving the eigenvector, first we solve

$$\boldsymbol{H}\boldsymbol{\psi} = \boldsymbol{0} \tag{3.7.4}$$

to obtain the basic eigenvector

$$\boldsymbol{\psi}_0^{(0)} = \{1, \quad 0, \quad 0, \quad 0\}^{\mathrm{T}} \tag{3.7.5}$$

The eigenvector

$$\boldsymbol{v}_0^{(0)} = \boldsymbol{\psi}_0^{(0)} \tag{3.7.6}$$

is the solution of the original problem (3.3.1) which implies **rigid body translation** of beam. As there is only one chain for the eigenvalue equation (3.7.4), a Jordan form eigenvector exists.

From Eq. (3.3.9), the following equation

$$\boldsymbol{H}\boldsymbol{\psi}_0^{(1)} = \boldsymbol{\psi}_0^{(0)} \tag{3.7.7}$$

should be solved for the first-order Jordan form eigenvector. The solution is

$$\boldsymbol{\psi}_0^{(1)} = \{0, \quad 1, \quad 0, \quad 0\}^{\mathrm{T}} \tag{3.7.8}$$

According to Eq. (3.3.11), the solution formed by $\boldsymbol{\psi}_0^{(1)}$ for the original problem (3.3.1) is

$$\boldsymbol{v}_0^{(1)} = \boldsymbol{\psi}_0^{(1)} + x\boldsymbol{\psi}_0^{(0)} = \{x, \quad 1, \quad 0, \quad 0\}^{\mathrm{T}} \tag{3.7.9}$$

which implies **rigid body rotation** of beam. Through direct examination, the basic eigenvectors $\boldsymbol{\psi}_0^{(0)}$ and $\boldsymbol{\psi}_0^{(1)}$ are symplectic orthogonal.

$$\langle \boldsymbol{\psi}_0^{(1)}, \boldsymbol{\psi}_0^{(0)} \rangle \equiv (\boldsymbol{\psi}_0^{(1)}, \boldsymbol{J}\boldsymbol{\psi}_0^{(0)}) \equiv \boldsymbol{\psi}_0^{(1)\mathrm{T}} \boldsymbol{J}\boldsymbol{\psi}_0^{(0)} = 0 \tag{3.7.10}$$

Hence a second-order Jordan form eigenvector exists.

Similarly, we solve the equation

$$\boldsymbol{H}\boldsymbol{\psi}_0^{(2)} = \boldsymbol{\psi}_0^{(1)} \tag{3.7.11}$$

and obtain the second-order Jordan form eigenvector as

$$\boldsymbol{\psi}_0^{(2)} = \{c, \quad 0, \quad 0, \quad EI\}^{\mathrm{T}} \tag{3.7.12}$$

The solution formed by $\boldsymbol{\psi}_0^{(2)}$ for the original problem (3.3.1) is

$$\begin{aligned} \boldsymbol{v}_0^{(2)} &= \boldsymbol{\psi}_0^{(2)} + x\boldsymbol{\psi}_0^{(1)} + \frac{1}{2}x^2\boldsymbol{\psi}_0^{(0)} \\ &= \left\{\frac{1}{2}x^2 + c, \quad x, \quad 0, \quad EI\right\}^{\mathrm{T}} \end{aligned} \tag{3.7.13}$$

which implies **pure bending** of beam. Through direct examination, the basic eigenvectors $\boldsymbol{\psi}_0^{(0)}$ and $\boldsymbol{\psi}_0^{(2)}$ are again symplectic orthogonal.

$$\langle \boldsymbol{\psi}_0^{(2)}, \boldsymbol{\psi}_0^{(0)} \rangle \equiv (\boldsymbol{\psi}_0^{(2)}, \boldsymbol{J}\boldsymbol{\psi}_0^{(0)}) \equiv \boldsymbol{\psi}_0^{(2)\mathrm{T}} J\boldsymbol{\psi}_0^{(0)} = 0 \tag{3.7.14}$$

Hence a third-order Jordan form eigenvector exists.

Similarly, we solve the equation

$$\boldsymbol{H}\boldsymbol{\psi}_0^{(3)} = \boldsymbol{\psi}_0^{(2)} \tag{3.7.15}$$

and obtain the third-order Jordan form eigenvector as

$$\boldsymbol{\psi}_0^{(3)} = \left\{0, \quad \frac{EI}{kGA} + c, \quad -EI, \quad 0\right\}^{\mathrm{T}} \tag{3.7.16}$$

The solution formed by $\boldsymbol{\psi}_0^{(3)}$ for the original problem is

$$\begin{aligned} \boldsymbol{v}_0^{(3)} &= \boldsymbol{\psi}_0^{(3)} + x\boldsymbol{\psi}_0^{(2)} + \frac{1}{2}x^2\boldsymbol{\psi}_0^{(1)} + \frac{1}{6}x^3\boldsymbol{\psi}_0^{(0)} \\ &= \left\{\frac{1}{6}x^3 + cx, \quad \frac{1}{2}x^2 + \frac{EI}{kGA} + c; \quad -EI, EIx\right\}^{\mathrm{T}} \end{aligned} \tag{3.7.17}$$

which implies **constant shear force bending** of beam. By direct examination, the basic eigenvectors $\boldsymbol{\psi}_0^{(0)}$ and $\boldsymbol{\psi}_0^{(3)}$ are symplectic adjoint.

$$\langle \boldsymbol{\psi}_0^{(3)}, \boldsymbol{\psi}_0^{(0)} \rangle = \boldsymbol{\psi}_0^{(3)\mathrm{T}} \boldsymbol{J} \boldsymbol{\psi}_0^{(0)} = EI \neq 0 \tag{3.7.18}$$

Hence, a fourth-order Jordan form eigenvector does not exist and thus the Jordan chain is terminated. There is a total of four eigenvectors for this Jordan chain and it is completely consistent with the solutions that the zero eigenvalue is a quadruple root of eigenvalue equation (3.7.2).

Besides knowing $\boldsymbol{\psi}_0^{(3)}$ and $\boldsymbol{\psi}_0^{(0)}$ are mutually symplectic adjoint, it is also easy to examine directly that $\boldsymbol{\psi}_0^{(1)}$ and $\boldsymbol{\psi}_0^{(2)}$ are mutually symplectic adjoint, or

$$\langle \boldsymbol{\psi}_0^{(1)}, \boldsymbol{\psi}_0^{(2)} \rangle = \langle \boldsymbol{\psi}_0^{(3)}, \boldsymbol{\psi}_0^{(0)} \rangle = EI \neq 0 \tag{3.7.19}$$

By assigning

$$c = -\frac{EI}{2kGA} \tag{3.7.20}$$

symplectic orthogonality of $\boldsymbol{\psi}_0^{(2)}$ and $\boldsymbol{\psi}_0^{(3)}$ can be verified. Symplectic orthogonality of the other eigenvectors are satisfied automatically. Hence, $\boldsymbol{\psi}_0^{(0)}, \boldsymbol{\psi}_0^{(1)}, \boldsymbol{\psi}_0^{(3)}$ and $\boldsymbol{\psi}_0^{(2)}$ form a set of adjoint symplectic orthonormal basis. Thus we have established all basic eigenvectors and Jordan form eigenvectors for the static bending of Timoshenko beam.

These physical interpretation of these solutions is explicit. The symplectic adjoint solutions of rigid translation and rigid rotation are, respectively, the solutions of constant shear force bending and pure bending. The Jordan form solutions with zero eigenvalue are typical. These solutions with specific physical interpretation are also common in plane and three-dimensional Saint–Venant problems in elastic statics.

3.8. Wave Propagation Analysis of Timoshenko Beam

In Sec. 3.2, the derivation of wave propagation problem of Timoshenko beam into Hamiltonian system with a set of dual equations (3.2.7) has been established where the Hamiltonian matrix $\boldsymbol{H}$ and the Hamiltonian density function $\mathscr{H}$ are expressed in Eqs. (3.2.12) and (3.2.13), respectively.

We discuss the eigenvalue problem of Hamiltonian matrix $\boldsymbol{H}$ first[8]. Expanding the determinant as follows

$$\det(\boldsymbol{H} - \mu \boldsymbol{I}) = 0 \tag{3.8.1}$$

the characteristic polynomial is

$$\mu^4 + \mu^2 \rho\omega^2 \left(\frac{1}{E} + \frac{1}{kG}\right) + \frac{\rho^2\omega^4}{kGE} - \frac{\rho\omega^2 A}{EI} = 0 \tag{3.8.2}$$

It shows that if μ is an eigenvalue, then $-\mu$ must be also an eigenvalue. This is consistent with a Hamiltonian matrix. Equation (3.8.2) is a quadratic equation in unknown μ^2 with discriminant

$$\left[\rho\omega^2 \left(\frac{1}{E} + \frac{1}{kG}\right)\right]^2 - 4\left[\frac{\rho^2\omega^4}{kGE} - \frac{\rho\omega^2 A}{EI}\right]$$

$$= \rho^2\omega^4 \left(\frac{1}{E} - \frac{1}{kG}\right)^2 + \frac{4\rho\omega^2 A}{EI} > 0 \tag{3.8.3}$$

Hence, μ^2 must be a real root. Furthermore, there are three different cases for μ^2: (1) there are two negative roots, (2) there are a positive root and a negative root, (3) there are a negative root and a zero root. The separation is according to

$$\frac{\rho^2\omega_{\text{cr}}^4}{kGE} - \frac{\rho\omega_{\text{cr}}^2 A}{EI} = 0, \quad \text{i.e. } \omega_{\text{cr}}^2 = \frac{kGA}{\rho I} \tag{3.8.4}$$

(1) For $\omega^2 > \omega_{\text{cr}}^2$, there are two negative roots for μ^2. The eigenvalues can be denoted as

$$\mu_1 \mathrm{i}, \quad \mu_2 \mathrm{i}; \quad -\mu_1 \mathrm{i}, \quad -\mu_2 \mathrm{i} \tag{3.8.5}$$

Hence solution of the full state vector is

$$\boldsymbol{v}_1(x) = \mathrm{e}^{\mathrm{i}\mu_1 x}\boldsymbol{\psi}_1; \quad \boldsymbol{v}_2(x) = \mathrm{e}^{\mathrm{i}\mu_2 x}\boldsymbol{\psi}_2,$$

$$\boldsymbol{v}_3(x) = \mathrm{e}^{-\mathrm{i}\mu_1 x}\boldsymbol{\psi}_3, \quad \boldsymbol{v}_4(x) = \mathrm{e}^{-\mathrm{i}\mu_2 x}\boldsymbol{\psi}_4 \tag{3.8.6}$$

and solution of the problem (3.1.11) before separating the time variable is

$$\left.\begin{aligned} \tilde{\boldsymbol{v}}_1(x,t) &= \mathrm{e}^{\mathrm{i}(\mu_1 x + \omega t)}\boldsymbol{\psi}_1 \\ \tilde{\boldsymbol{v}}_2(x,t) &= \mathrm{e}^{\mathrm{i}(\mu_2 x + \omega t)}\boldsymbol{\psi}_2 \\ \tilde{\boldsymbol{v}}_3(x,t) &= \mathrm{e}^{\mathrm{i}(-\mu_1 x + \omega t)}\boldsymbol{\psi}_3 \\ \tilde{\boldsymbol{v}}_4(x,t) &= \mathrm{e}^{\mathrm{i}(-\mu_2 x + \omega t)}\boldsymbol{\psi}_4 \end{aligned}\right\} \tag{3.8.7}$$

where $\tilde{\boldsymbol{v}} = \{\tilde{w}, \tilde{\theta}, \tilde{F}_{\text{s}}, \tilde{M}\}^{\text{T}}$. Obviously, it indicates propagation of two pairs of waves with velocities ω/μ_1 and ω/μ_2, and traveling along $-x$ and $+x$, respectively.

(2) For $\omega^2 < \omega_{cr}^2$, there are a positive root and a negative root for μ^2. The negative root indicates propagation of a pair of waves along the positive and negative directions of x. The positive root yields

$$\boldsymbol{v}_1(x) = \mathrm{e}^{-\mu_1 x}\boldsymbol{\psi}_1, \quad \boldsymbol{v}_3(x) = \mathrm{e}^{\mu_1 x}\boldsymbol{\psi}_3 \tag{3.8.8}$$

which indicates local vibration capable of creating resonance. Such phenomenon is important and it will be discussed in detail in the next section.

(3) For $\omega^2 = \omega_{cr}^2$, there is a zero root for μ^2, i.e. there are two zero roots for μ and there will be Jordan form solutions. First, from $\boldsymbol{H}\boldsymbol{\psi} = \boldsymbol{0}$, or

$$\left.\begin{array}{lllll} 0 & +\theta & +F_s/kGA & +0 & =0 \\ 0 & +0 & +0 & +M/EI & =0 \\ -kGA^2w/I & +0 & +0 & +0 & =0 \\ 0 & -kGA\theta & -F_s & +0 & =0 \end{array}\right\} \tag{3.8.9}$$

the basic eigenvector is

$$\boldsymbol{\psi}_{cr}^{(0)} = \{0, \quad -1, \quad kGA, \quad 0\}^{\mathrm{T}} \tag{3.8.10}$$

Next, the Jordan form eigenvector is solved from $\boldsymbol{H}\boldsymbol{\psi}_{cr}^{(1)} = \boldsymbol{\psi}_{cr}^{(0)}$, or

$$\left.\begin{array}{lllll} 0 & +\theta & +F_s/kGA & +0 & =0 \\ 0 & +0 & +0 & +M/EI & =-1 \\ -kGA^2w/I & +0 & +0 & +0 & =kGA \\ 0 & -kGA\theta & -F_s & +0 & =0 \end{array}\right\} \tag{3.8.11}$$

The solution is

$$\boldsymbol{\psi}_{cr}^{(1)} = \{-I/A, \quad 0, \quad 0, \quad -EI\}^{\mathrm{T}} \tag{3.8.12}$$

Indeed the sub-order eigenvector can be superposed by an arbitrary $a\boldsymbol{\psi}_{cr}^{(0)}$.

By virtue of eigenvector $\boldsymbol{\psi}_{cr}^{(0)}$ and according to Eq. (3.3.10), the solution of homogeneous equation (3.3.1) is

$$\boldsymbol{v}_{cr}^{(0)} = \boldsymbol{\psi}_{cr}^{(0)}, \quad \tilde{\boldsymbol{v}}_{cr}^{(0)}(x,t) = \mathrm{e}^{\mathrm{i}\omega_{cr}t}\boldsymbol{\psi}_{cr}^{(0)} \tag{3.8.13}$$

However, the sub-order eigenvector itself is not a solution. It follows the form of Eq. (3.3.11) to construct the solution of (3.3.1), as

$$\boldsymbol{v}_{cr}^{(1)} = [\boldsymbol{\psi}_{cr}^{(1)} + x\boldsymbol{\psi}_{cr}^{(0)}], \quad \tilde{\boldsymbol{v}}_{cr}^{(1)} = \mathrm{e}^{\mathrm{i}\omega_{cr}t}\boldsymbol{v}_{cr}^{(1)} \tag{3.8.14}$$

It is easy to verify directly that $\psi_{\mathrm{cr}}^{(0)}$ and $\psi_{\mathrm{cr}}^{(1)}$ are mutually symplectic adjoint.

For ω_{cr}^2, there is still a negative root $\mu^2 = -A(1 + kG/E)/I$ for μ^2. We have

$$\mu_{3,4} = \pm\,\mathrm{i}\sqrt{\frac{A}{I}\left(1 + \frac{kG}{E}\right)} \tag{3.8.15}$$

The corresponding eigenvectors satisfy

$$\left.\begin{array}{lllll} 0 & +\theta & +F_{\mathrm{s}}/kGA & +0 & = \mu w \\ 0 & +0 & +0 & +M/EI & = \mu\theta \\ -kGA^2 w/I & +0 & +0 & +0 & = \mu F_{\mathrm{s}} \\ 0 & -kGA\theta & -F_{\mathrm{s}} & +0 & = \mu M \end{array}\right\} \tag{3.8.16}$$

and the eigenvectors are

$$\psi_i = \left\{-\frac{1}{kGA}, \frac{1}{\mu_i EI}, \frac{A}{\mu_i I}, 1\right\}^{\mathrm{T}} \quad (i = 3, 4) \tag{3.8.17}$$

These two vectors are also mutually symplectic adjoint. In addition to the two pairs of symplectic adjoint vectors above, it can be verified directly that the other vectors are symplectic orthogonal.

From the point of view of wave propagation, the phase velocity approaches infinity because $\omega_{\mathrm{cr}}^2 \neq 0$. This is a conclusion with respect to a monochromatic wave. As wave propagation represents a process of energy transmission, it is more meaningful to observe its group velocity.

3.9. Wave Induced Resonance

For wave propagation of Timoshenko beam, $\omega^2 < \omega_{\mathrm{cr}}^2$ yields a positive root and a negative root for μ^2. Suppose that $\mu_1^2 < 0, \mu_2^2 > 0$, then

$$\mu_1 = -\mathrm{i}\frac{\pi}{l}, \quad \mu_3 = \mathrm{i}\frac{\pi}{l}; \tag{3.9.1}$$

$$\mu_2 = -\lambda, \quad \mu_4 = \lambda \tag{3.9.2}$$

where

$$l^2 = \frac{2\pi^2}{\rho\omega^2\left(\frac{1}{E}+\frac{1}{kG}\right)+\sqrt{4\rho\omega^2\frac{A}{EI}+\rho^2\omega^4\left(\frac{1}{E}-\frac{1}{kG}\right)^2}} \tag{3.9.3}$$

$$\lambda^2 = \sqrt{\rho\omega^2\frac{A}{EI}+\frac{1}{4}\rho^2\omega^4\left(\frac{1}{E}-\frac{1}{kG}\right)^2}-\frac{1}{2}\rho\omega^2\left(\frac{1}{E}+\frac{1}{kG}\right) \tag{3.9.4}$$

The following equations

$$\left.\begin{array}{ccccc} -\mu w & +\theta & +F_s/kGA & +0 & =0 \\ 0 & -\mu\theta & +0 & +M/EI & =0 \\ -\rho\omega^2 Aw & +0 & -\mu F_s & +0 & =0 \\ 0 & -\rho\omega^2 I\theta & -F_s & -\mu M & =0 \end{array}\right\} \tag{3.9.5}$$

should be solve for the eigenvectors corresponding to the eigenvalues above. As μ has been chosen such that the determinant of coefficient matrix vanishes, hence

$$\left.\begin{array}{l} w = \dfrac{1}{A}\left(\dfrac{1}{E}+\dfrac{\mu^2}{\rho\omega^2}\right)M, \quad \theta = \dfrac{1}{\mu EI}M; \\ F_s = -\mu\left(1+\dfrac{\rho\omega^2}{\mu^2 E}\right)M \end{array}\right\} \tag{3.9.6}$$

The eigenvectors corresponding to the eigenvalues in Eq. (3.9.1) are

$$\boldsymbol{q}_1 = \left\{\begin{array}{c} \dfrac{1}{A}\left(\dfrac{1}{E}-\dfrac{\pi^2}{\rho\omega^2 l^2}\right) \\ \mathrm{i}\dfrac{l}{\pi EI} \end{array}\right\}, \quad \boldsymbol{p}_1 = \left\{\begin{array}{c} \mathrm{i}\dfrac{\pi}{l}\left(1-\dfrac{\rho\omega^2 l^2}{\pi^2 E}\right) \\ 1 \end{array}\right\} \tag{3.9.7a}$$

$$\boldsymbol{q}_3 = \left\{\begin{array}{c} \dfrac{1}{A}\left(\dfrac{1}{E}-\dfrac{\pi^2}{\rho\omega^2 l^2}\right) \\ -\mathrm{i}\dfrac{l}{\pi EI} \end{array}\right\}, \quad \boldsymbol{p}_3 = \left\{\begin{array}{c} -\mathrm{i}\dfrac{\pi}{l}\left(1-\dfrac{\rho\omega^2 l^2}{\pi^2 E}\right) \\ 1 \end{array}\right\} \tag{3.9.7b}$$

These are two complex eigenvectors, and $\boldsymbol{q}_1, \boldsymbol{p}_1$ and $\boldsymbol{q}_3, \boldsymbol{p}_3$ are respectively complex conjugate vectors.

The eigenvectors corresponding to the eigenvalues in Eq. (3.9.2) are

$$\boldsymbol{q}_2 = \begin{Bmatrix} \dfrac{1}{A}\left(\dfrac{\lambda^2}{\rho\omega^2} + \dfrac{1}{E}\right) \\ -\dfrac{1}{\lambda EI} \end{Bmatrix}, \quad \boldsymbol{p}_2 = \begin{Bmatrix} \lambda\left(1 + \dfrac{\rho\omega^2}{\lambda^2 E}\right) \\ 1 \end{Bmatrix} \tag{3.9.8a}$$

$$\boldsymbol{q}_4 = \begin{Bmatrix} \dfrac{1}{A}\left(\dfrac{\lambda^2}{\rho\omega^2} + \dfrac{1}{E}\right) \\ \dfrac{1}{\lambda EI} \end{Bmatrix}, \quad \boldsymbol{p}_4 = \begin{Bmatrix} -\lambda\left(1 + \dfrac{\rho\omega^2}{\lambda^2 E}\right) \\ 1 \end{Bmatrix} \tag{3.9.8b}$$

These are real vectors. Although there are complex vectors $\boldsymbol{q}_1, \boldsymbol{p}_1$ and $\boldsymbol{q}_3, \boldsymbol{p}_3$, it is clear through direct examination that $\boldsymbol{q}_1, \boldsymbol{p}_1$ (or $\boldsymbol{q}_3, \boldsymbol{p}_3$) and $\boldsymbol{q}_2, \boldsymbol{p}_2$ (or $\boldsymbol{q}_4, \boldsymbol{p}_4$) are symplectic orthogonal.

The physical interpretation of eigen-solutions is given here. For solutions of μ_2 and μ_4, the real vectors indicate local vibration as shown by their mode shapes. These modes do not exist in region with two infinite ends. In a region $x \geq 0$ with one infinite end, the eigen-solution of $(\mu_2, \boldsymbol{\psi}_2)$ exists and it approaches zero at $x \to \infty$ because of the presence of factor $\mathrm{e}^{-\lambda x}$. For solutions of μ_1 and μ_3, the eigenvalues are purely imaginary. Hence the solutions of the original problem are

$$\tilde{\boldsymbol{v}}_1 = \mathrm{e}^{\mathrm{i}(\omega t - \pi x/l)}\boldsymbol{\psi}_1, \quad \tilde{\boldsymbol{v}}_3 = \mathrm{e}^{\mathrm{i}(\omega t + \pi x/l)}\boldsymbol{\psi}_3 \tag{3.9.9}$$

Obviously, these are traveling waves along the x-direction where $\tilde{\boldsymbol{v}}_1$ travels along the positive x-direction while $\tilde{\boldsymbol{v}}_3$ travels along the negative x-direction.

Having analyzed the properties of solutions, we present the boundary value problem. Consider a semi-infinite region $x \geq 0$ with an incidence wave at infinity and an attached mass at $x = 0$. With no external excitation in the entire region, the response of structure is analyzed here.

Let the dynamic stiffness matrix (D'Alembert's principle) due to the attached mass at the end $x = 0$ be

$$\boldsymbol{R} = -\omega^2 \boldsymbol{R}_0 \quad (\boldsymbol{R}_0^{\mathrm{T}} = \boldsymbol{R}_0) \tag{3.9.10}$$

then the solution according to the analysis above is

$$\tilde{\boldsymbol{v}} = \mathrm{e}^{\mathrm{i}\omega t}[A_1 \mathrm{e}^{-\mathrm{i}\pi x/l}\boldsymbol{\psi}_1 + A_2 \mathrm{e}^{-\lambda x}\boldsymbol{\psi}_2 + \mathrm{e}^{\mathrm{i}\pi x/l}\boldsymbol{\psi}_3] \tag{3.9.11}$$

At boundary $x = 0$, we have

$$\boldsymbol{p} = A_1\boldsymbol{p}_1 + A_2\boldsymbol{p}_2 + \boldsymbol{p}_3, \quad \boldsymbol{q} = A_1\boldsymbol{q}_1 + A_2\boldsymbol{q}_2 + \boldsymbol{q}_3 \tag{3.9.12}$$

in which factor $e^{i\omega t}$ has been eliminated because it has no effect on the variational expression (3.6.16).

Although of $\boldsymbol{q}$ and $\boldsymbol{p}$ in Eq. (3.9.12) with undetermined constants A_1 and A_2 have been obtained, the expressions are complex. Only real solutions are required here. As $\boldsymbol{p}_1$ and $\boldsymbol{p}_3$, and $\boldsymbol{q}_1$ and $\boldsymbol{q}_3$ are two pairs of mutually complex conjugate vector functions, a real solution exists only if $A_1 = 1$ and A_2 is real. Hence, at $x = 0$

$$\boldsymbol{q} = \begin{Bmatrix} \dfrac{1}{A}\left[2\left(\dfrac{1}{E} - \dfrac{\pi^2}{\rho\omega^2 l^2}\right) + A_2\left(\dfrac{\lambda^2}{\rho\omega^2} + \dfrac{1}{E}\right)\right] \\ -\dfrac{1}{\lambda EI}A_2 \end{Bmatrix} \tag{3.9.13a}$$

$$\boldsymbol{p} = \begin{Bmatrix} A_2\lambda\left(1 + \dfrac{\rho\omega^2}{\lambda^2 E}\right) \\ A_2 + 2 \end{Bmatrix} \tag{3.9.13b}$$

with only one undetermined real constant A_2. Substituting the expression above into Eq. (3.6.16) and noticing $\overline{\boldsymbol{p}}_0 = \boldsymbol{0}$, we have

$$A_2\left\{\left[\frac{\lambda}{A}\left(\frac{2}{E} + \frac{\rho\omega^2}{\lambda^2 E^2} + \frac{\lambda^2}{\rho\omega^2}\right) - \frac{1}{\lambda EI}\right]\right.$$

$$\left. - \left[\frac{1}{A}\left(\frac{\lambda^2}{\rho\omega^2} + \frac{1}{E}\right);\quad -\frac{1}{\lambda EI}\right]\boldsymbol{R}\begin{bmatrix} \dfrac{1}{A}\left(\dfrac{\lambda^2}{\rho\omega^2} + \dfrac{1}{E}\right) \\ -\dfrac{1}{\lambda EI} \end{bmatrix}\right\}$$

$$= \frac{2}{\lambda EI} + \left[\frac{2}{A}\left(\frac{1}{E} - \frac{\pi^2}{\rho\omega^2 l^2}\right);\quad 0\right]\boldsymbol{R}\begin{bmatrix} \dfrac{1}{A}\left(\dfrac{\lambda^2}{\rho\omega^2} + \dfrac{1}{E}\right) \\ -\dfrac{1}{\lambda EI} \end{bmatrix} \tag{3.9.14}$$

The undetermined constant A_2 can then be solved using this equation.

It should be noted that resonance will occur if the coefficient of A_2 on the left-hand-side of the expression above approaches zero. The matrix $\boldsymbol{R}$

at the end represents the boundary conditions at $x = 0$. Resonance may occur under various boundary conditions. Such resonance occurs as a result of the propagation and reflection of wave and it is called **wave induced resonance**[14] which should not be overlooked.

There are two essential requirements for the existence of wave induced resonance, i.e. the coexistence of propagating wave and non-propagating wave, and the existence of boundary constraints or partial nonhomogeneity. With respect to the former, for any given ω, there are many more non-propagating waves in an elastic medium than the propagating waves. For the latter, boundary conditions or partial nonhomogeneity do always exist. Propagating wave is specifically indicated by the presence of factor $e^{i(\omega t - kx)}$ while $e^{\mu x + i\omega t}$ shows the presence of non-propagating wave. Attention should be paid to these points when solutions are derived[15,16].

References

1. S. Timoshenko, *Strength of Materials*, 3rd edn., New York: Van Nostrand, 1955–1956.
2. Haichang Hu, *Variational Principle of Elasticity and Its Application*, Beijing: Science Press, 1981 (in Chinese).
3. Zudao Luo and Sijian Li, *Mechanics of Anisotropic Materials*, Shanghai: Shanghai Jiaotong University Press, 1994 (in Chinese).
4. Wanxie Zhong and Xiangxiang Zhong, On the transverse eigenvector solution of the elliptic partial differential equation in the prismatic domain, *Journal on Numerical Methods and Computer Applications*, 1992, 13(2): 107–118 (in Chinese).
5. Wanxie Zhong and Zaishi Yang, Partial differential equations and Hamiltonian system, in: F.Y. Cheng and Zizhi Fu, eds., *Computational Mechanics in Structural Engineering*, London: Elsevier, 1992: 32–48.
6. Wanxie Zhong and F.W. Williams, On the localization of the vibration mode of sub-structural chain-type structure, *Proceedings of the Institution of Mechanical Engineers*, Part C, 1991, 205(C4): 281–288.
7. Wanxie Zhong, Huajiang Ouyang and Zicheng Deng, *Computational Structural Mechanics and Optimal Control*, Dalian: Dalian University of Technology Press, 1993 (in Chinese).
8. Wanxie Zhong, *A New Systematic Methodology for Theory of Elasticity*, Dalian University of Technology Press, 1995 (in Chinese).
9. E.T. Whittaker, *A Treatise on the Analytical Dynamics*, 4th edn., Combridge: Cambridge University Press, 1960.
10. V.I. Arnold, *Mathemaical Methods of Classical Mechanics*, New York: Springer-Verlag, 1978.

11. Kang Feng and Mengzhao Qin, *Hamilton Methodology for Hamiltonian Dynamical Systems*, Progress in Natural Science, 1991, 1(2): 102–112 (in Chinese).
12. Wanxie Zhong, On the reciprocal theorem and adjoint symplectic orthogonal relation, *ACTA Mechanica Sinica*, 1992, 24(4): 432–437 (in Chinese).
13. Wanxie Zhong, Plane elasticity problem in strip domain and Hamiltonian system, *Journal of Dalian University of Technology*, 1991, 31(4): 373–384 (in Chinese).
14. Wanxie Zhong and F.W. Williams, On the direct solution of wave propagation for repetitive structures, *Journal of Sound and Vibration*, 1995, 181(3): 485–501.
15. K.F. Graff, Wave motion in elastic solids, Oxford: Clarendon Press, 1975.
16. J.D. Achenbach, *Wave Propagation in Elastic Solids*, New York: North-Holland, 1973.

Chapter 4

Plane Elasticity in Rectangular Coordinates

In this chapter we introduce in detail the Hamiltonian system of plane elasticity in rectangular coordinates. Based on the method of separation of variables and the solution of eigenvalue problems in the transverse direction, we establish a complete system of symplectic solution for plane problems. Accordingly we can obtain directly the solutions for problems in a plane rectangular domain. Emphasis will be given on the eigen-solutions of zero eigenvalue and the analytical solutions of Saint–Venant problems.

4.1. The Fundamental Equations of Plane Elasticity

Strictly speaking, every elastic body is a space body and, therefore, the governing three-dimensional partial differential equations should be solved. However, many problems in engineering practices can be represented by plane elasticity models after simplification and treatment by mechanics. They are usually divided into plane stress and plane strain problems.

Research in **plane stress problems** deals with thin plates with constant thickness. In such problems, the dimension of body in one coordinate direction (for instance, the y-axis) is much smaller than dimensions in the other two coordinate directions. We further assume that the body forces and lateral surface forces are parallel to the plane of plate and distributed uniformly across the thickness. Additionally the geometric constraints of plate do not vary across the thickness likewise.

As the plate is very thin and the external forces are constant across the thickness, it is reasonable to approximately assume constant stress across the thickness. In addition, if there is no force acting on both faces of the plate, we have within the plate

$$\sigma_y = \tau_{xy} = \tau_{yz} = 0 \tag{4.1.1}$$

and the other stress components σ_x, σ_z and τ_{xz} are all functions of x, z only. From the stress-strain relations (2.3.13), we obtain

$$\gamma_{xy} = \gamma_{yz} = 0; \quad \varepsilon_y = -\nu(\sigma_x + \sigma_z)/E \tag{4.1.2}$$

and the simplified stress-strain relations for plane stress problems are

$$\varepsilon_x = \frac{1}{E}[\sigma_x - \nu\sigma_z]; \quad \varepsilon_z = \frac{1}{E}[\sigma_z - \nu\sigma_x]; \quad \gamma_{xz} = \frac{2(1+\nu)}{E}\tau_{xz} \tag{4.1.3}$$

The expressions above can be rearranged as

$$\left.\begin{aligned} \sigma_x &= \frac{E}{1-\nu^2}(\varepsilon_x + \nu\varepsilon_z) \\ \sigma_z &= \frac{E}{1-\nu^2}(\varepsilon_z + \nu\varepsilon_x) \\ \tau_{xz} &= \frac{E}{2(1+\nu)}\gamma_{xz} \end{aligned}\right\} \tag{4.1.4}$$

The displacements w, u are also functions of x, z only. Their relations with strains are

$$\varepsilon_z = \frac{\partial w}{\partial z}; \quad \varepsilon_x = \frac{\partial u}{\partial x}; \quad \gamma_{xz} = \frac{\partial u}{\partial z} + \frac{\partial w}{\partial x} \tag{4.1.5}$$

and the equations of equilibrium (2.1.10) can be simplified to

$$\frac{\partial \sigma_x}{\partial x} + \frac{\partial \tau_{xz}}{\partial z} + F_x = 0 \tag{4.1.6a}$$

$$\frac{\partial \tau_{xz}}{\partial x} + \frac{\partial \sigma_z}{\partial z} + F_z = 0 \tag{4.1.6b}$$

In addition, there are the corresponding boundary conditions. For example, the boundary conditions on the boundary Γ_σ for specified tractions are

$$\left.\begin{aligned} F_{nx} &= \sigma_x l + \tau_{xz} n = \overline{F}_{nx} \\ F_{nz} &= \tau_{xz} l + \sigma_z n = \overline{F}_{nz} \end{aligned}\right\} \quad \text{on } \Gamma_\sigma \tag{4.1.7}$$

where l, n are direction cosines of the exterior normal vector $\boldsymbol{n}$ of the boundary surface with the positive directions of x and z, respectively. Likewise, the boundary conditions on the boundary Γ_u for specified displacements are

$$u = \overline{u}, \quad w = \overline{w} \quad \text{on } \Gamma_u \tag{4.1.8}$$

Equations (4.1.3) to (4.1.8) constitute the complete differential equations and boundary conditions of plane stress problems.

Contrary to plane stress problems, research in **plane strain problems** deals with long cylindrical solids (cylindrical solids of infinite length strictly speaking). In such problems, the dimension of body in one coordinate direction (for instance, the y-direction) is much larger than dimensions in the other two coordinate directions. We further assume that the body forces and lateral surface forces are parallel to the cross section of the cylinder and distributed uniformly along the y-direction. Additionally the geometric constraints of cylinder do not vary along the y-direction likewise.

As any cross section can be considered as a symmetry plane, the displacement of any point is parallel to the xz-plane, i.e.

$$w = w(x, z), \quad u = u(x, z), \quad v = 0 \tag{4.1.9}$$

From the geometric equation (2.2.2), we have

$$\varepsilon_y = \gamma_{xy} = \gamma_{yz} = 0 \tag{4.1.10}$$

According to the stress-strain relation (2.3.13), we obtain

$$\tau_{xy} = \tau_{yz} = 0, \quad \sigma_y = \nu(\sigma_x + \sigma_z) \tag{4.1.11}$$

and the simplified stress-strain relations for plane strain problems are

$$\left.\begin{aligned} \varepsilon_x &= \frac{1-\nu^2}{E}\left(\sigma_x - \frac{\nu}{1-\nu}\sigma_z\right) \\ \varepsilon_z &= \frac{1-\nu^2}{E}\left(\sigma_z - \frac{\nu}{1-\nu}\sigma_x\right) \\ \gamma_{xz} &= \frac{2(1+\nu)}{E}\tau_{xz} \end{aligned}\right\} \tag{4.1.12}$$

The other fundamental equations, such as the strain-displacement geometric relations, the equations of equilibrium and boundary conditions, are identical with those of plane stress problems, as in Eqs. (4.1.5) to (4.1.8).

Incidentally, if we denote

$$E_1 = \frac{E}{1-\nu^2}, \quad \nu_1 = \frac{\nu}{1-\nu} \tag{4.1.13}$$

then Eq. (4.1.12) can be rewritten as

$$\varepsilon_x = \frac{1}{E_1}[\sigma_x - \nu_1\sigma_z]; \quad \varepsilon_z = \frac{1}{E_1}[\sigma_z - \nu_1\sigma_x]; \quad \gamma_{xz} = \frac{2(1+\nu_1)}{E_1}\tau_{xz} \tag{4.1.14}$$

which is absolutely identical with Eq. (4.1.3).

For the eight unknown physical quantities $u, w, \varepsilon_x, \varepsilon_z, \gamma_{xz}$ and $\sigma_x, \sigma_z, \tau_{xz}$ the fundamental equations and boundary conditions of plane strain problems and those of plane stress problems are identical. The only difference lies in the interpretation equation (4.1.13) of elastic constants. Henceforth we do not distinguish these two classes of problems which are referred as plane elasticity problems. The fundamental equations and boundary conditions to be satisfied are from Eqs. (4.1.3) to (4.1.8).

There has been extensive and profound research in plane elasticity[1,2]. Most of the solution methods are based on the Airy stress function of biharmonic equation. In this approach, Eq. (4.1.6) can be transformed to a homogeneous equation

$$\frac{\partial \sigma_x}{\partial x} + \frac{\partial \tau_{xz}}{\partial z} = 0, \qquad \frac{\partial \tau_{xz}}{\partial x} + \frac{\partial \sigma_z}{\partial z} = 0 \tag{4.1.15}$$

by virtue of a special solution. Then, the Airy stress function $\varphi_{\mathrm{f}}(x, z)$ is introduced such that

$$\sigma_x = \frac{\partial^2 \varphi_{\mathrm{f}}}{\partial z^2}, \qquad \sigma_z = \frac{\partial^2 \varphi_{\mathrm{f}}}{\partial x^2}, \qquad \tau_{xz} = -\frac{\partial^2 \varphi_{\mathrm{f}}}{\partial x \partial z} \tag{4.1.16}$$

In this way the homogeneous equation (4.1.15) is satisfied. In order to have the stress expressions satisfy the deformation compatibility equations, the Airy stress function $\varphi_{\mathrm{f}}(x, z)$ should fulfill

$$\nabla^2 \nabla^2 \varphi_{\mathrm{f}} = \frac{\partial^4 \varphi_{\mathrm{f}}}{\partial x^4} + 2\frac{\partial^4 \varphi_{\mathrm{f}}}{\partial x^2 \partial z^2} + \frac{\partial^4 \varphi_{\mathrm{f}}}{\partial z^4} = 0 \tag{4.1.17}$$

The corresponding boundary conditions can be expressed in terms of stress functions accordingly. This is the stress solution methodology with one kind of variable. Other solution methodologies include the trial-and-error polynomials approach in rectangular coordinates; the trial-and-error trigonometric functions approach in polar coordinates[3]; and rational solution approach by complex functions[4]; etc. It is difficult for the trial-and-error solutions to satisfy the boundary conditions which have been commonly dealt with by the Saint–Venant principle. The solution method by complex functions is a significant advancement and it is applicable to structures made of single, homogeneous material and with common boundary geometry. It is not convenient for some special domains such as a rectangle, etc. and also difficult for composite materials.

At this point, there has been detailed discussion on numerous research works in elasticity. A large portion has been devoted to the discussion on static problems of rectangular domain in rectangular coordinates. In this chapter, we continue to introduce the Hamiltonian system method based on the specific, exemplary problem. It is a generalization of the method introduced last chapter on the set of partial differential equation.

4.2. Hamiltonian System in Rectangular Domain[5]

Consider a rectangular domain as shown in Fig. 4.1

$$V: 0 \le z \le l, \quad -h \le x \le h \tag{4.2.1}$$

where l is comparatively larger than h. There are surface forces acting on both side boundary surfaces ($x = \pm h$)

$$\sigma_x = \overline{F}_{x1}(z), \quad \tau_{xz} = \overline{F}_{z1}(z) \quad \text{at } x = -h \tag{4.2.2a}$$

$$\sigma_x = \overline{F}_{x2}(z), \quad \tau_{xz} = \overline{F}_{z2}(z) \quad \text{at } x = h \tag{4.2.2b}$$

and there are body forces F_x, F_z in the domain acting along the x- and z-directions, respectively.

Applying the principle of minimum potential energy to the problem above yields

$$\delta E_{\mathrm{p}} = \delta\left\{\int_0^l \int_{-h}^h (v_\varepsilon - wF_z - uF_x)\mathrm{d}x\mathrm{d}z - \int_0^l [(w\overline{F}_{z2} + u\overline{F}_{x2})_{x=h} - (w\overline{F}_{z1} + u\overline{F}_{x1})_{x=-h}]\mathrm{d}z\right\} = 0 \tag{4.2.3}$$

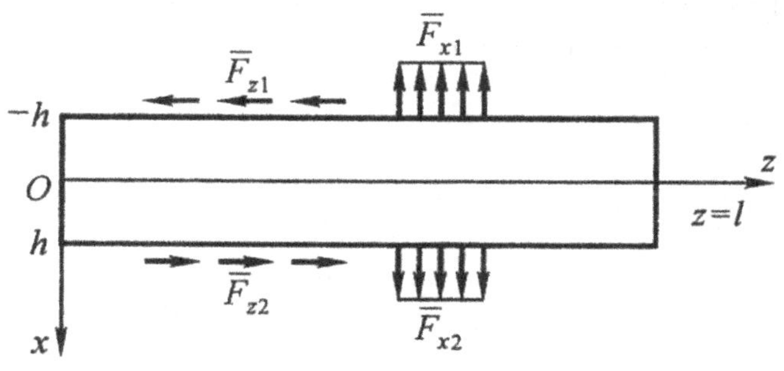

Fig. 4.1. Plane problem in rectangular domain.

where the strain energy density is

$$v_\varepsilon = \frac{E}{2(1-\nu^2)}\left[\left(\frac{\partial u}{\partial x}\right)^2 + \left(\frac{\partial w}{\partial z}\right)^2 + 2\nu\left(\frac{\partial u}{\partial x}\right)\left(\frac{\partial w}{\partial z}\right)\right] + \frac{E}{4(1+\nu)}\left(\frac{\partial u}{\partial z} + \frac{\partial w}{\partial x}\right)^2 \tag{4.2.4}$$

In addition, there are the corresponding boundary conditions at both end ($z = 0$ or l) which will be discussed in due course.

Here, the z-coordinate is treated as the time coordinate and differentiation with respect to z is indicated by a dot, i.e. $(\dot{\ }) = \partial/\partial z$. Then the Lagrange density function is the integrand in Eq. (4.2.3)

$$\mathscr{L}(w, u, \dot{w}, \dot{u}) = v_\varepsilon - wF_z - uF_x \tag{4.2.5}$$

First of all, the dual variables of displacement w, u, i.e. the stresses σ_z, τ_{xz},

$$\left.\begin{aligned} \sigma &= \frac{\partial \mathscr{L}}{\partial \dot{w}} = \frac{E}{1-\nu^2}\left(\dot{w} + \nu\frac{\partial u}{\partial x}\right) \\ \tau &= \frac{\partial \mathscr{L}}{\partial \dot{u}} = \frac{E}{2(1+\nu)}\left(\dot{u} + \frac{\partial w}{\partial x}\right) \end{aligned}\right\} \tag{4.2.6}$$

are introduced by applying Legendre's transformation in accordance with the rule of Hamiltonian system. Hence the original variable $\boldsymbol{q}$ and the dual variable $\boldsymbol{p}$ are

$$\boldsymbol{q} = \{w, u\}^{\mathrm{T}}, \quad \boldsymbol{p} = \{\sigma, \tau\}^{\mathrm{T}} \tag{4.2.7}$$

From Eq. (4.2.6), we obtain

$$\dot{w} = -\nu\frac{\partial u}{\partial x} + \frac{1-\nu^2}{E}\sigma, \qquad \dot{u} = -\frac{\partial w}{\partial x} + \frac{2(1+\nu)}{E}\tau \tag{4.2.8}$$

Next, applying the equations of equilibrium (4.1.6) and eliminating σ_x and $\dot{w}$ by virtue of stress-strain relations (4.1.4) and (4.2.8) yield

$$\dot{\sigma} = -\frac{\partial \tau}{\partial x} - F_z \tag{4.2.9a}$$

$$\begin{aligned} \dot{\tau} &= -\frac{\partial \sigma_x}{\partial x} - F_x \\ &= -\frac{E}{1-\nu^2}\frac{\partial^2 u}{\partial x^2} - \frac{E\nu}{1-\nu^2}\frac{\partial \dot{w}}{\partial x} - F_x \\ &= -E\frac{\partial^2 u}{\partial x^2} - \nu\frac{\partial \sigma}{\partial x} - F_x \end{aligned} \tag{4.2.9b}$$

Then Eqs. (4.2.8) and (4.2.9) form a set of Hamiltonian dual equations

$$\begin{Bmatrix} \dot{w} \\ \dot{u} \\ \dot{\sigma} \\ \dot{\tau} \end{Bmatrix} = \begin{bmatrix} 0 & -\nu\dfrac{\partial}{\partial x} & \dfrac{1-\nu^2}{E} & 0 \\ -\dfrac{\partial}{\partial x} & 0 & 0 & \dfrac{2(1+\nu)}{E} \\ 0 & 0 & 0 & -\dfrac{\partial}{\partial x} \\ 0 & -E\dfrac{\partial^2}{\partial x^2} & -\nu\dfrac{\partial}{\partial x} & 0 \end{bmatrix} \begin{Bmatrix} w \\ u \\ \sigma \\ \tau \end{Bmatrix} + \begin{Bmatrix} 0 \\ 0 \\ -F_z \\ -F_x \end{Bmatrix} \tag{4.2.10}$$

It is also possible to introduce the **Hamiltonian density function**

$$\begin{aligned} \mathscr{H}(w,u,\sigma,\tau) &= \sigma\dot{w} + \tau\dot{u} - \mathscr{L}(w,u,\dot{w},\dot{u}) \\ &= \frac{1-\nu^2}{2E}\sigma^2 + \frac{1+\nu}{E}\tau^2 - \nu\sigma\frac{\partial u}{\partial x} \\ &\quad - \tau\frac{\partial w}{\partial x} - \frac{1}{2}E\left(\frac{\partial u}{\partial x}\right)^2 + wF_z + uF_x \end{aligned} \tag{4.2.11}$$

to obtain the Hamiltonian dual equations directly, i.e. the matrix form of Eq. (4.2.10), as

$$\left.\begin{aligned} \dot{\boldsymbol{q}} &= \frac{\partial \mathscr{H}}{\partial \boldsymbol{p}} = \boldsymbol{A}\boldsymbol{q} + \boldsymbol{D}\boldsymbol{p} \\ \dot{\boldsymbol{p}} &= -\frac{\partial \mathscr{H}}{\partial \boldsymbol{q}} = \boldsymbol{B}\boldsymbol{q} - \boldsymbol{A}^{\mathrm{T}}\boldsymbol{p} - \boldsymbol{X} \end{aligned}\right\} \tag{4.2.12}$$

where

$$\boldsymbol{A} = \begin{bmatrix} 0 & -\nu\dfrac{\partial}{\partial x} \\ -\dfrac{\partial}{\partial x} & 0 \end{bmatrix}, \qquad \boldsymbol{B} = \begin{bmatrix} 0 & 0 \\ 0 & -E\dfrac{\partial^2}{\partial x^2} \end{bmatrix}$$

$$\boldsymbol{D} = \begin{bmatrix} \dfrac{1-\nu^2}{E} & 0 \\ 0 & \dfrac{2(1+\nu)}{E} \end{bmatrix}, \quad \boldsymbol{A}^{\mathrm{T}} = \begin{bmatrix} 0 & \dfrac{\partial}{\partial x} \\ \nu\dfrac{\partial}{\partial x} & 0 \end{bmatrix} \quad \boldsymbol{X} = \begin{bmatrix} F_z \\ F_x \end{bmatrix} \tag{4.2.13}$$

The transpose of an operator matrix here should be noted as not just simple transposition but it refers to the adjoint operator matrix. With respect to differentiation, it includes integration by parts, as

$$\int_{-h}^{h} \boldsymbol{p}^{\mathrm{T}} \boldsymbol{A}\boldsymbol{q} \mathrm{d}x = \int_{-h}^{h} \boldsymbol{q}^{\mathrm{T}} \boldsymbol{A}^{\mathrm{T}} \boldsymbol{p} \mathrm{d}x - [\nu\sigma u + \tau w]_{-h}^{h} \tag{4.2.14}$$

There is no difference within the domain $-h < x < h$. The remaining boundary term and the boundary terms of $\boldsymbol{B}$ make up the boundary conditions of the problem.

With Hamiltonian density function, the Hamiltonian variational principle (or mixed energy variational principle) can be stated as

$$\delta\left\{\int_0^l \int_{-h}^{h} [\boldsymbol{p}^{\mathrm{T}}\dot{\boldsymbol{q}} - \mathscr{H}(\boldsymbol{q},\boldsymbol{p})]\mathrm{d}x\mathrm{d}z - \int_0^l [(\overline{\boldsymbol{X}}_2^{\mathrm{T}}\boldsymbol{q})_{x=h} - (\overline{\boldsymbol{X}}_1^{\mathrm{T}}\boldsymbol{q})_{x=-h}]\mathrm{d}z\right\} = 0 \tag{4.2.15}$$

where

$$\overline{\boldsymbol{X}}_1 = \{\overline{F}_{z1}, \quad \overline{F}_{x1}\}^{\mathrm{T}}, \qquad \overline{\boldsymbol{X}}_2 = \{\overline{F}_{z2}, \quad \overline{F}_{x2}\}^{\mathrm{T}} \tag{4.2.16}$$

The variation of Eq. (4.2.15) yields the set of dual equations (4.2.12) in the domain and the boundary conditions (4.2.2) on both side boundary surfaces ($x = \pm h$). Actually, the Hamiltonian variational principle (4.2.15) can be also derived from the Hellinger–Reissner variational principle with two kinds of variables. For rectangular plane problems, the Hellinger–Reissner variational principle yields

$$\delta\left\{\int_0^l \int_{-h}^{h} \left[\sigma_x \frac{\partial u}{\partial x} + \sigma_z \frac{\partial w}{\partial z} + \tau_{xz}\left(\frac{\partial u}{\partial z} + \frac{\partial w}{\partial x}\right) - v_{\mathrm{c}} - F_x u - F_z w\right] \mathrm{d}x\mathrm{d}z \right.$$
$$\left. - \int_0^l [(w\overline{F}_{z2} + u\overline{F}_{x2})_{x=h} - (w\overline{F}_{z1} + u\overline{F}_{x1})_{x=-h}]\mathrm{d}z\right\} = 0 \tag{4.2.17}$$

where

$$v_{\mathrm{c}} = \frac{1}{2E}(\sigma_x^2 + \sigma_z^2 - 2\nu\sigma_x\sigma_z) + \frac{1+\nu}{E}\tau_{xz}^2 \tag{4.2.18}$$

Because the z-direction is assigned longitudinal and the x-direction transverse, the transverse stress σ_x should be eliminated. The variation of

Eq. (4.2.17) with respect to σ_x yields

$$\sigma_x = E\left(\frac{\partial u}{\partial x}\right) + \nu\sigma_z \tag{4.2.19}$$

Substituting the expression above into Eq. (4.2.17) and eliminating σ_x yield the Hamiltonian variational principle (4.2.15).

Subsequently we introduce the **full state vector** $\boldsymbol{v}$ and the operator matrix $\boldsymbol{H}$, as

$$\boldsymbol{v} = \begin{Bmatrix} \boldsymbol{q} \\ \boldsymbol{p} \end{Bmatrix} = \begin{Bmatrix} w \\ u \\ \sigma \\ \tau \end{Bmatrix}, \qquad \boldsymbol{H} = \begin{bmatrix} \boldsymbol{A} & \boldsymbol{D} \\ \boldsymbol{B} & -\boldsymbol{A}^{\mathrm{T}} \end{bmatrix} \tag{4.2.20}$$

Then the Hamiltonian dual equations (4.2.12) can be rewritten as

$$\dot{\boldsymbol{v}} = \boldsymbol{H}\boldsymbol{v} + \boldsymbol{h}; \quad \boldsymbol{h} = \{\boldsymbol{0}^{\mathrm{T}}, \quad -\boldsymbol{X}^{\mathrm{T}}\}^{\mathrm{T}} \tag{4.2.21}$$

In order to discuss the properties of operator matrix $\boldsymbol{H}$, we introduce unit symplectic matrix

$$\boldsymbol{J} = \begin{bmatrix} \boldsymbol{0} & \boldsymbol{I}_2 \\ -\boldsymbol{I}_2 & \boldsymbol{0} \end{bmatrix} \tag{4.2.22}$$

and denote

$$\langle \boldsymbol{v}_1, \boldsymbol{v}_2 \rangle \stackrel{\mathrm{def}}{=} \int_{-h}^{h} \boldsymbol{v}_1^{\mathrm{T}} \boldsymbol{J} \boldsymbol{v}_2 \mathrm{d}x = \int_{-h}^{h} (w_1\sigma_2 + u_1\tau_2 - \sigma_1 w_2 - \tau_1 u_2)\mathrm{d}x \tag{4.2.23}$$

It is clear that Eq. (4.2.23) satisfies the four conditions of symplectic inner product as expressed in Eq. (1.3.2). Hence, the full state vectors $\boldsymbol{v}$ form a symplectic space according to the definition of symplectic inner product (4.2.23).

For the purpose of examining the properties of operator which is independent of load, we should consider the homogeneous linear differential equation

$$\dot{\boldsymbol{v}} = \boldsymbol{H}\boldsymbol{v} \tag{4.2.24}$$

and the homogeneous boundary conditions

$$E\frac{\partial u}{\partial x} + \nu\sigma = 0, \quad \tau = 0 \text{ at } x = \pm h \tag{4.2.25}$$

By integration by parts, it is easy to prove that

$$
\begin{aligned}
\langle \boldsymbol{v}_1, \boldsymbol{H}\boldsymbol{v}_2 \rangle &= \int_{-h}^{h} \boldsymbol{v}_1^{\mathrm{T}} \boldsymbol{J}\boldsymbol{H}\boldsymbol{v}_2 \,\mathrm{d}x \\
&= \int_{-h}^{h} \left[\nu\sigma_1 \frac{\partial u_2}{\partial x} + \tau_1 \frac{\partial w_2}{\partial x} - E u_1 \frac{\partial^2 u_2}{\partial x^2} - \nu u_1 \frac{\partial \sigma_2}{\partial x} \right. \\
&\qquad \left. - w_1 \frac{\partial \tau_2}{\partial x} - \frac{1-\nu^2}{E}\sigma_1\sigma_2 - \frac{2(1+\nu)}{E}\tau_1\tau_2 \right] \mathrm{d}x \\
&= \int_{-h}^{h} \left[-\nu u_2 \frac{\partial \sigma_1}{\partial x} - w_2 \frac{\partial \tau_1}{\partial x} - E u_2 \frac{\partial^2 u_1}{\partial x^2} + \nu\sigma_2 \frac{\partial u_1}{\partial x} \right. \\
&\qquad \left. + \tau_2 \frac{\partial w_1}{\partial x} - \frac{1-\nu^2}{E}\sigma_2\sigma_1 - \frac{2(1+\nu)}{E}\tau_2\tau_1 \right] \mathrm{d}x \\
&\quad + \left[u_2 \left(E\frac{\partial u_1}{\partial x} + \nu\sigma_1 \right) + w_2\tau_1 - u_1 \left(E\frac{\partial u_2}{\partial x} + \nu\sigma_2 \right) - w_1\tau \right]_{-h}^{h} \\
&= \langle \boldsymbol{v}_2, \boldsymbol{H}\boldsymbol{v}_1 \rangle + \left[u_2 \left(E\frac{\partial u_1}{\partial x} + \nu\sigma_1 \right) + w_2\tau_1 \right]_{-h}^{h} \\
&\quad - \left[u_1 \left(E\frac{\partial u_2}{\partial x} + \nu\sigma_2 \right) + w_1\tau_2 \right]_{-h}^{h}
\end{aligned} \tag{4.2.26}
$$

As a result, if $\boldsymbol{v}_1$,$\boldsymbol{v}_2$ are continuously differentiable full state vectors satisfying the boundary conditions (4.2.25), there exists an identity

$$
\langle \boldsymbol{v}_1, \boldsymbol{H}\boldsymbol{v}_2 \rangle = \langle \boldsymbol{v}_2, \boldsymbol{H}\boldsymbol{v}_1 \rangle \tag{4.2.27}
$$

Hence the operator matrix $\boldsymbol{H}$ is a Hamiltonian transformation (operator matrix) in a symplectic space.

It is observed from Eq. (4.2.26) that Eq. (4.2.27) remains an identity as long as $\boldsymbol{v}_1, \boldsymbol{v}_2$ fulfill the boundary conditions even for clamped and simply supported boundaries.

4.3. Separation of Variables and Transverse Eigen-Problems[6]

The common method of separation of variables is not applicable to Lame equations. However, after transforming the equations into Hamiltonian dual equations, it is natural to apply the method of separation of variables to the resulted homogeneous equations (4.2.24). Let

$$\boldsymbol{v}(z,x)=\xi(z)\boldsymbol{\psi}(x) \tag{4.3.1}$$

and substitute it into Eq. (4.2.24), we obtain

$$\xi(z)=\mathrm{e}^{\mu z} \tag{4.3.2}$$

and the eigenvalue equation

$$\boldsymbol{H}\boldsymbol{\psi}(x)=\mu\boldsymbol{\psi}(x) \tag{4.3.3}$$

where μ is the unknown eigenvalue and $\boldsymbol{\psi}(x)$ is the eigenvector which is to satisfy the homogeneous boundary conditions (4.2.25). Here, the longitudinal z-coordinate has been separated and, therefore, it is an eigenvalue problem on the transverse cross section. As it is a continuous body, there are infinite dimensions in the transverse direction.

It has been proved in the section above that $\boldsymbol{H}$ is a Hamiltonian operator matrix. This eigenvalue problem has special properties as that of eigenvalue problem of Hamiltonian matrix of finite dimensions discussed in Chapters 1 and 3, i.e.

(1) If μ is an eigenvalue of a Hamiltonian operator matrix $\boldsymbol{H}$, $-\mu$ is also an eigenvalue.
As the Hamiltonian eigenvalue problem discussed here has infinite dimensions, there are infinite eigenvalues which can be divided into two sets

$$(\alpha)\ \ \mu_i,\quad \mathrm{Re}(\mu_i)<0\ \text{ or }\ \mathrm{Re}(\mu_i)=0\wedge\mathrm{Im}(\mu_i)<0\quad(i=1,2,\ldots) \tag{4.3.4a}$$

$$(\beta)\ \ \mu_{-i}=-\mu_i \tag{4.3.4b}$$

The eigenvalues in the (α)-set are arranged in an ascending order according to the magnitude $|\mu_i|$.

(2) The eigenvectors of Hamiltonian operator matrix are mutually adjoint symplectic orthogonal. Let $\boldsymbol{\psi}_i$ and $\boldsymbol{\psi}_j$ be the eigenvectors corresponding to the eigenvalues μ_i and μ_j, respectively, then for $\mu_i+\mu_j\neq 0$, they are symplectic orthogonal

$$\langle\boldsymbol{\psi}_i,\boldsymbol{\psi}_j\rangle=\int_{-h}^{h}\boldsymbol{\psi}_i^{\mathrm{T}}\boldsymbol{J}\boldsymbol{\psi}_j\mathrm{d}x=0 \tag{4.3.5}$$

The eigenvector symplectic adjoint with $\boldsymbol{\psi}_i$ must be the eigenvector (or the Jordan form eigenvector) corresponding to eigenvalue $-\mu_i$.

With the adjoint symplectic orthogonal relation, the full state vector $\boldsymbol{v}$ on any transverse cross section can always be expanded in terms of eigen-solutions,

$$\boldsymbol{v} = \sum_{i=1}^{\infty} (a_i \boldsymbol{\psi}_i + b_i \boldsymbol{\psi}_{-i}) \tag{4.3.6}$$

where a_i and b_i are undetermined coefficients while $\boldsymbol{\psi}_i$ and $\boldsymbol{\psi}_{-i}$ are eigen-vectors satisfying the following adjoint symplectic orthonormal relations

$$\left.\begin{aligned} \langle \boldsymbol{\psi}_i, \boldsymbol{\psi}_j \rangle &= \langle \boldsymbol{\psi}_{-i}, \boldsymbol{\psi}_{-j} \rangle = 0 \\ \langle \boldsymbol{\psi}_i, \boldsymbol{\psi}_{-j} \rangle &= \begin{cases} 1 & (i = j) \\ 0 & (i \neq j) \end{cases} \end{aligned}\right\} \quad (i, j = 1, 2, \ldots) \tag{4.3.7}$$

It should be noted that the validity of expansion of eigen-solutions depends on the completeness of these eigen-solutions in full state vector space. Such completeness problem also exists in the Sturm–Liouville problem. It can be first transformed into an integral equation with a symmetrical kernel and then the completeness will be proven by the Hibert–Schmidt theory. There is a complete set of theory in functional analysis on the spectrum analysis of symmetric operators. The completeness theorem constitutes an important part of these subjects.

The operator in hand is a Hamiltonian operator which is not self-adjoint. Therefore, the situation is more complicated. Although its eigenvalue spectrum has certain characteristics, e.g. Eq. (4.3.4), the eigenvalues are not necessarily real and Jordan forms are also possible. The kernel of integral equation corresponding to the Hamiltonian operator is also of Hamiltonian type, i.e. it is an eigenvelue problem of **Hamiltonian integral equation**. There is a lot of subjects, for examples, adjoint symplectic orthonormal relation of the eigenvectors, completeness problem similar to the Hibert–Schmidt theory, degenerated kernel integral equation, etc. which are not explored yet. Consequently, the theory of spectrum analysis for Hamiltonian operators as well as its completeness theorem should be available in functional analysis. In summary, it is a very vast field of research and a large amount of work is to be completed.

In this book, we deal with the Hamiltonian solution of elasticity. Therefore, we do not discuss these problems further.

4.4. Eigen-Solutions of Zero Eigenvalue

The zero eigenvalue is a very special eigenvalue in the Hamiltonian eigenvalue problems. It is not included in the expression (4.3.4). The eigen-solution of this eigenvalue contains particular significance in elasticity.

Repeated zero eigenvalue exists for elastic problems in a rectangular domain because the boundaries at both sides ($x = \pm h$) are free. To obtain the eigen-solution of zero eigenvalue, we solve the following differential equation

$$\boldsymbol{H}\boldsymbol{\psi}(x) = \mathbf{0} \tag{4.4.1}$$

Expanding the equation yields

$$\left.\begin{array}{lllll} 0 & -\nu\dfrac{\mathrm{d}u}{\mathrm{d}x} & +\dfrac{1-\nu^2}{E}\sigma & +0 & =0 \\ -\dfrac{\mathrm{d}w}{\mathrm{d}x} & +0 & +0 & +\dfrac{2(1+\nu)}{E}\tau & =0 \\ 0 & +0 & +0 & -\dfrac{\mathrm{d}\tau}{\mathrm{d}x} & =0 \\ 0 & -E\dfrac{\mathrm{d}^2u}{\mathrm{d}x^2} & -\nu\dfrac{\mathrm{d}\sigma}{\mathrm{d}x} & +0 & =0 \end{array}\right\} \tag{4.4.2}$$

Certainly, the eigen-solution should fulfill the homogeneous boundary conditions (4.2.25) at both sides ($x = \pm h$).

Equation (4.4.2) and boundary conditions (4.2.25) shows that the solution can be decoupled into two sets. The first set of equations with respect to w, τ consists of the second and the third equations of Eq. (4.4.2) and the second equation of boundary conditions (4.2.25), while the second set with respect to u, σ consists of the first and the fourth equations of Eq. (4.4.2) and the first equation of boundary conditions (4.2.25). Solve the former yields

$$w = c_1, \quad \tau = 0 \tag{4.4.3}$$

while solving the latter yields

$$u = c_2, \quad \sigma = 0 \tag{4.4.4}$$

where c_1 and c_2 are arbitrary constants. Therefore, the linearly independent basic eigen-solutions are

$$\boldsymbol{\psi}_{0f}^{(0)} = \{1, \ \ 0; \ \ 0, \ \ 0\}^{\mathrm{T}} \tag{4.4.5}$$

$$\boldsymbol{\psi}_{0s}^{(0)} = \{0, \ \ 1; \ \ 0, \ \ 0\}^{\mathrm{T}} \tag{4.4.6}$$

Here, there are two chains denoted by the subscript f and s, respectively. From Eqs. (4.3.1) and (4.3.2), these eigenvectors are the solutions of the original equations (4.2.24) with boundary conditions (4.2.25),

$$\boldsymbol{v}_{0f}^{(0)} = \boldsymbol{\psi}_{0f}^{(0)}, \quad \boldsymbol{v}_{0s}^{(0)} = \boldsymbol{\psi}_{0s}^{(0)} \tag{4.4.7}$$

These two solutions are physically interpreted, respectively, as **rigid body translations** along the z- and x-direction.

Next, we solve for the Jordan form eigen-solutions of zero eigenvalue. We consider the following equation

$$\boldsymbol{H}\boldsymbol{\psi}_{0}^{(i)} = \boldsymbol{\psi}_{0}^{(i-1)} \tag{4.4.8}$$

where superscript i, $i-1$ denote the ith, $(i-1)$th order Jordan form (or basic) eigen-solution.

To obtain the first-order Jordan form eigen-solution on chain one, we first solve the following equations with homogeneous boundary conditions Eq. (4.2.25),

$$\boldsymbol{H}\boldsymbol{\psi}_{0f}^{(1)} = \boldsymbol{\psi}_{0f}^{(0)} \tag{4.4.9}$$

and we obtain

$$\boldsymbol{\psi}_{0f}^{(1)} = \{0, \quad -\nu x; \quad E, \quad 0\}^{\mathrm{T}} \tag{4.4.10}$$

Here the first-order Jordan form eigenvector $\boldsymbol{\psi}_{0f}^{(1)}$ is no longer the solution of the original equations (4.2.24) with homogeneous boundary conditions (4.2.25). However, the solutions can be constructed similar to expression (3.3.11), as

$$\boldsymbol{v}_{0f}^{(1)} = \boldsymbol{\psi}_{0f}^{(1)} + z\boldsymbol{\psi}_{0f}^{(0)} \tag{4.4.11}$$

The components of displacement and stress are

$$w = z, \quad u = -\nu x, \quad \sigma = E, \quad \tau = 0 \tag{4.4.12}$$

This solution is physically interpreted as uniform tension along the axial direction.

Similarly, the first-order Jordan form eigen-solution on chain two, can be obtained by first solving the following equations with homogeneous boundary conditions Eq. (4.2.25)

$$\boldsymbol{H}\boldsymbol{\psi}_{0s}^{(1)} = \boldsymbol{\psi}_{0s}^{(0)} \tag{4.4.13}$$

and we obtain

$$\boldsymbol{\psi}_{0s}^{(1)} = \{-x, \quad 0; \quad 0, \quad 0\}^{\mathrm{T}} \tag{4.4.14}$$

Likewise, the first-order Jordan form eigenvector $\boldsymbol{\psi}_{0s}^{(1)}$ is no longer the solution of the original equations (4.2.24) with homogeneous boundary condition (4.2.25). The solution of the original equation thus is composed of

$$\boldsymbol{v}_{0s}^{(1)} = \boldsymbol{\psi}_{0s}^{(1)} + z\boldsymbol{\psi}_{0s}^{(0)} \tag{4.4.15}$$

The components of displacement and stress are

$$w = -x, \quad u = z, \quad \sigma = 0, \quad \tau = 0 \tag{4.4.16}$$

This solution is physically interpreted as in-plane **rigid body rotation**.

Having obtained the first-order Jordan form eigen-solutions, we can now solve the second-order Jordan form eigen-solutions. On chain one, we need to solve the following equation

$$\boldsymbol{H}\boldsymbol{\psi}_{0f}^{(2)} = \boldsymbol{\psi}_{0f}^{(1)} \tag{4.4.17}$$

Solving the third expression of Eq. (4.4.17) yields $\tau = -Ex + c$ where c is an arbitrary constant. As this expression cannot simultaneously satisfy the homogeneous boundary conditions $\tau = 0$ at $x = \pm h$, hence there exists no solution and this chain of Jordan form eigen-solution is terminated.

Consider the other chain of Jordan form; the second-order Jordan form solution can be obtained by solve the following equation

$$\boldsymbol{H}\boldsymbol{\psi}_{0s}^{(2)} = \boldsymbol{\psi}_{0s}^{(1)} \tag{4.4.18}$$

Obviously, we can first obtain $w = \tau = 0$. Then from the first expression and the integration of the fourth expression of Eq. (4.4.18), we get

$$u_{0s}^{(2)} = \frac{1}{2}\nu x^2 + c_3 x + c_4, \qquad \sigma_{0s}^{(2)} = -Ex + \frac{E\nu}{1-\nu^2}c_3 \tag{4.4.19}$$

Substituting the expression above into the first equation of boundary conditions (4.2.25), we find out that the boundary conditions are satisfied for $c_3 = 0$ while c_4 is an arbitrary constant. Hence

$$\boldsymbol{\psi}_{0s}^{(2)} = \left\{0, \quad \frac{1}{2}\nu x^2 + c_4; \quad -Ex, \quad 0\right\}^{\mathrm{T}} \tag{4.4.20}$$

and the solution of original equation thus constructed is

$$\boldsymbol{v}_{0s}^{(2)} = \boldsymbol{\psi}_{0s}^{(2)} + z\boldsymbol{\psi}_{0s}^{(1)} + \frac{1}{2}z^2\boldsymbol{\psi}_{0s}^{(0)} \tag{4.4.21}$$

in which the components of displacement and stress are

$$w = -xz, \quad u = \frac{1}{2}(z^2 + \nu x^2) + c_4, \quad \sigma = -Ex, \quad \tau = 0 \tag{4.4.22}$$

This solution is physically interpreted as **pure bending**.

It should be noted that the presence of c_4 is equivalent to superposing a basic eigen-solution to the Jordan form eigen-solution. Similar to the discussion in Sec. 1.3, the eigen-solutions become symplectic orthogonal if an appropriate value for c_4 is assigned.

Subsequently, we solve the third-order Jordan form eigen-solution. The corresponding equation is

$$\boldsymbol{H}\boldsymbol{\psi}_{0s}^{(3)} = \boldsymbol{\psi}_{0s}^{(2)} \tag{4.4.23}$$

Obviously, $u = \sigma = 0$ is obtained. Integrating the third expression of Eq. (4.4.23) and substituting the boundary condition $\tau = 0$ at $x = \pm h$, we obtain

$$\tau_{0s}^{(3)} = \frac{1}{2}E(x^2 - h^2) \tag{4.4.24}$$

Further substituting into the second expression of Eq. (4.4.23) and integrating yield

$$w_{0s}^{(3)} = -(1+\nu)h^2 x - c_4 x + \frac{1}{6}(2+\nu)x^3 \tag{4.4.25}$$

Hence

$$\boldsymbol{\psi}_{0s}^{(3)} = \begin{Bmatrix} -(1+\nu)h^2 x - c_4 x + \frac{1}{6}(2+\nu)x^3 \\ 0 \\ 0 \\ \frac{1}{2}E(x^2 - h^2) \end{Bmatrix} \tag{4.4.26}$$

and the solution of original equation thus constructed is

$$\boldsymbol{v}_{0s}^{(3)} = \boldsymbol{\psi}_{0s}^{(3)} + z\boldsymbol{\psi}_{0s}^{(2)} + \frac{1}{2}z^2\boldsymbol{\psi}_{0s}^{(1)} + \frac{1}{6}z^3\boldsymbol{\psi}_{0s}^{(0)} \tag{4.4.27}$$

in which the components of displacement and stress are

$$\left.\begin{aligned} w &= -(1+\nu)h^2 x - c_4 x + \frac{1}{6}(2+\nu)x^3 - \frac{1}{2}xz^2 \\ u &= \frac{1}{2}\nu x^2 z + c_4 z + \frac{1}{6}z^3 \\ \sigma &= -Exz \\ \tau &= \frac{1}{2}E(x^2 - h^2) \end{aligned}\right\} \tag{4.4.28}$$

This solution is physically interpreted as **constant shear force bending**.

Finally, we examine whether a higher-order Jordan form eigen-solution exists. The corresponding equation is

$$\boldsymbol{H}\boldsymbol{\psi} = \boldsymbol{\psi}_{0s}^{(3)} \tag{4.4.29}$$

Integrating the fourth expression

$$-E\frac{\mathrm{d}^2 u}{\mathrm{d}x^2} - \nu\frac{\mathrm{d}\sigma}{\mathrm{d}x} = \frac{1}{2}E(x^2 - h^2) \tag{4.4.30}$$

from $x = -h$ to $x = h$ yields

$$-\left[E\frac{\mathrm{d}u}{\mathrm{d}x} + \nu\sigma\right]_{x=-h}^{x=h} = -\frac{2}{3}Eh^3 \tag{4.4.31}$$

According to the homogeneous boundary condition (4.2.25), the left-hand-side of the expression above vanishes. Therefore, there exists no solution and the chain of Jordan form eigen-solution is terminated.

At this point, we have obtained all eigen-solutions of zero eigenvalue and discussed their specific physical interpretation. Obviously, the solutions $\boldsymbol{v}_{0f}^{(0)}$ and $\boldsymbol{v}_{0f}^{(1)}$ on chain one are related to symmetric deformation states with respect to $x = 0$ while the solutions $\boldsymbol{v}_{0s}^{(0)}, \boldsymbol{v}_{0s}^{(1)}, \boldsymbol{v}_{0s}^{(2)}$ and $\boldsymbol{v}_{0s}^{(3)}$ on chain two are related to antisymmetric deformation states.

The different order eigenvectors corresponding to a zero eigenvalue are mutually adjoint symplectic orthogonal. As the eigenvectors $\boldsymbol{\psi}_{0f}^{(0)}$ and $\boldsymbol{\psi}_{0f}^{(1)}$ on chain one are the symmetric deformation while the eigenvectors $\boldsymbol{\psi}_{0s}^{(0)}, \boldsymbol{\psi}_{0s}^{(1)}, \boldsymbol{\psi}_{0s}^{(2)}$ and $\boldsymbol{\psi}_{0s}^{(3)}$ on chain two are antisymmetric deformation, the eigenvectors of chain one and chain two must be mutually symplectic orthogonal.

There are only two vector functions on the Jordan form chain of symmetric deformation $\boldsymbol{\psi}_{0f}^{(0)}$ and $\boldsymbol{\psi}_{0f}^{(1)}$ which are adjoint but not symplectic orthogonal. In respect, it can be verified that

$$\langle \boldsymbol{\psi}_{0f}^{(0)}, \boldsymbol{\psi}_{0f}^{(1)} \rangle = \int_{-h}^{h} \boldsymbol{\psi}_{0f}^{(0)\mathrm{T}} \boldsymbol{J} \boldsymbol{\psi}_{0f}^{(1)} \mathrm{d}x = \int_{-h}^{h} E \mathrm{d}x = 2Eh \neq 0 \tag{4.4.32}$$

We then continue to consider the adjoint symplectic orthonormality on Jordan form chain two of antisymmetric deformation $\boldsymbol{\psi}_{0s}^{(0)}, \boldsymbol{\psi}_{0s}^{(1)}, \boldsymbol{\psi}_{0s}^{(2)}$ and $\boldsymbol{\psi}_{0s}^{(3)}$. Through direct verification or by applying the proof similar to Theorem 1.17, we can prove that $\boldsymbol{\psi}_{0s}^{(0)}$ is symplectic orthogonal to $\boldsymbol{\psi}_{0s}^{(1)}$ and

$\psi_{0s}^{(2)}$ while $\psi_{0s}^{(0)}$ and $\psi_{0s}^{(3)}$ are symplectic adjoint

$$\langle \psi_{0s}^{(0)}, \psi_{0s}^{(3)} \rangle = \int_{-h}^{h} \left[\frac{1}{2} E(x^2 - h^2) \right] dx = -\frac{2}{3} E h^3 \neq 0 \tag{4.4.33}$$

Through direct verification or following the conclusion of Eq. (1.3.20), we also know that $\psi_{0s}^{(1)}$ and $\psi_{0s}^{(2)}$ are symplectic adjoint

$$\langle \psi_{0s}^{(1)}, \psi_{0s}^{(2)} \rangle = -\langle \psi_{0s}^{(0)}, \psi_{0s}^{(3)} \rangle = \frac{2}{3} E h^3 \neq 0 \tag{4.4.34}$$

while $\psi_{0s}^{(1)}$ and $\psi_{0s}^{(3)}$ are symplectic orthogonal.

Finally we consider the symplectic orthogonal relation of $\psi_{0s}^{(2)}$ and $\psi_{0s}^{(3)}$ which can be satisfied through an appropriate choice of c_4. From

$$\langle \psi_{0s}^{(2)}, \psi_{0s}^{(3)} \rangle = \int_{-h}^{h} \left\{ Ex \left[\frac{1}{6}(2+\nu)x^3 - (1+\nu)h^2 x - c_4 x \right] + \frac{1}{2} E(x^2 - h^2) \left(\frac{1}{2}\nu x^2 + c_4 \right) \right\} dx = 0 \tag{4.4.35}$$

we have

$$c_4 = -\left(\frac{2}{5} + \frac{\nu}{2} \right) h^2 \tag{4.4.36}$$

The adjoint symplectic orthonormality of eigenvectors corresponding to zero eigenvalue has been established so far, i.e. an adjoint symplectic orthonormal vector set has been constituted. Similar to Chapter 3, the solutions of axial translation, transverse translation and rigid body rotation are adjoint to, respectively, the solutions of tension, constant shear force bending and pure bending.

The six eigen-solutions of zero eigenvalue are the basic solutions of two-dimensional Saint–Venant problem. These solutions span a complete symplectic subspace of zero eigenvalue. There are three independent plane rigid-body translations and a total of six independent rigid-body translations at both end surfaces of beam. Thus, the six independent solutions of Saint–Venant problem are matched.

It is noted that there is no solution for the set of Eqs. (4.4.17) or (4.4.29) with homogeneous boundary conditions (4.2.25). It only indicates termination of the corresponding Jordan form chain of eigen-solutions, i.e. there exists no further eigen-solution. This method can be further applied to solve for the inhomogeneous particular solutions for cases with uniformly

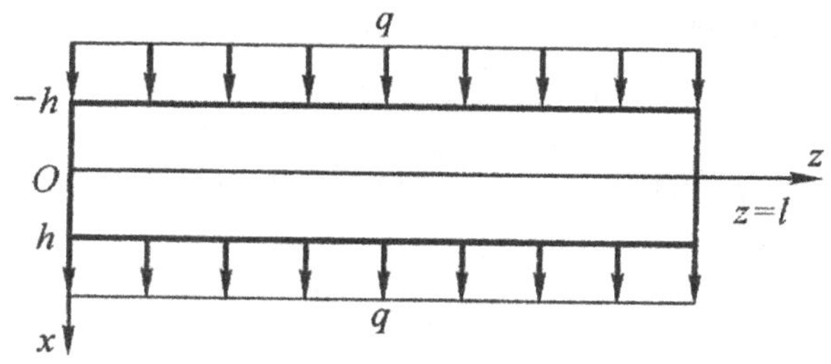

Fig. 4.2. Plane rectangular domain with uniform load.

distributed load. By using the Jordan form on chain one in Eq. (4.4.17), we can obtain the inhomogeneous particular solution along the z-direction for uniformly distributed load. Similarly, by using the Jordan form on chain two in Eq. (4.4.29), we can obtain the inhomogeneous particular solution along the x-direction for uniformly distributed load.

As an example, for a rectangular domain with uniformly distributed force q acting on surfaces $x = \pm h$ as illustrated in Fig. 4.2, we can obtain an inhomogeneous particular solution by using the Jordan form solution.

Apparently, the boundary conditions on both side boundary surfaces ($x = \pm h$) are

$$(\sigma_x =) E\frac{\partial u}{\partial x} + \nu\sigma = -q, \quad \tau_{xz} = 0 \quad \text{at } x = -h \tag{4.4.37a}$$

$$(\sigma_x =) E\frac{\partial u}{\partial x} + \nu\sigma = q, \quad \tau_{xz} = 0 \quad \text{at } x = h \tag{4.4.37b}$$

For Jordan form particular solution, we solve

$$\boldsymbol{H}\tilde{\boldsymbol{\psi}} = k\boldsymbol{\psi}_{0s}^{(3)} \tag{4.4.38}$$

Integrating the fourth equation

$$-E\frac{\mathrm{d}^2\tilde{u}}{\mathrm{d}x^2} - \nu\frac{\mathrm{d}\tilde{\sigma}}{\mathrm{d}x} = \frac{1}{2}kE(x^2 - h^2) \tag{4.4.39}$$

yields

$$E\frac{\mathrm{d}\tilde{u}}{\mathrm{d}x} + \nu\tilde{\sigma} = \frac{1}{6}kE(3h^2x - x^3) + c \tag{4.4.40}$$

and further substituting the boundary conditions (4.4.37) yields

$$k = \frac{3q}{Eh^3}, \quad c = 0 \tag{4.4.41}$$

And therefore we have

$$(\tilde{\sigma}_x =)E\frac{\mathrm{d}\tilde{u}}{\mathrm{d}x} + \nu\tilde{\sigma} = \frac{q}{2h^3}(3h^2x - x^3) \tag{4.4.42}$$

Making use of Eq. (4.4.42) and the first equation of Eq. (4.4.38) simultaneous, we obtain

$$\tilde{\sigma} = \frac{q}{h^3}x^3 - \frac{9q}{5h}x \tag{4.4.43}$$

and

$$\frac{\mathrm{d}\tilde{u}}{\mathrm{d}x} = -\frac{q(1+2\nu)}{2Eh^3}x^3 + \frac{3q(5+6\nu)}{10Eh}x \tag{4.4.44}$$

From to the second and third equations of Eq. (4.4.38) and the boundary conditions (4.4.37), we derive

$$\tilde{\tau} = \frac{\mathrm{d}\tilde{w}}{\mathrm{d}x} = 0 \tag{4.4.45}$$

Without the loss of generality, we establish a particular solution of Eq. (4.4.38) as

$$\tilde{\boldsymbol{\psi}} = \left\{\begin{array}{l} \tilde{w} = 0 \\ \tilde{u} = -\dfrac{q(1+2\nu)}{8Eh^3}x^4 + \dfrac{3q(5+6\nu)}{20Eh}x^2 \\ \tilde{\sigma} = \dfrac{q}{h^3}x^3 - \dfrac{9q}{5h}x \\ \tilde{\tau} = 0 \end{array}\right\} \tag{4.4.46}$$

Obviously $\tilde{\boldsymbol{\psi}}$ is not the particular solution of the original problem in Eq. (4.2.21). However, the particular solution of the original problem can be constructed as

$$\tilde{\boldsymbol{v}} = \tilde{\boldsymbol{\psi}} + k\left(z\boldsymbol{\psi}_{0s}^{(3)} + \frac{1}{2}z^2\boldsymbol{\psi}_{0s}^{(2)} + \frac{1}{6}z^3\boldsymbol{\psi}_{0s}^{(1)} + \frac{1}{24}z^4\boldsymbol{\psi}_{0s}^{(0)}\right) \tag{4.4.47}$$

The corresponding stress field is

$$\left.\begin{array}{l} \sigma = \dfrac{q}{h^3}x^3 - \dfrac{9q}{5h}x - \dfrac{3q}{2h^3}xz^2 \\ \tau = \dfrac{3q}{2h^3}z(x^2 - h^2) \\ \sigma_x = \dfrac{q}{2h^3}(3h^2x - x^3) \end{array}\right\} \tag{4.4.48}$$

and the corresponding displacement field is

$$\left.\begin{aligned} w &= \frac{q}{10Eh^3}xz[5(2+\nu)x^2 - 3(6+5\nu)h^2 - 5z^2] \\ u &= -\frac{q(1+2\nu)}{8Eh^3}x^4 + \frac{3q(5+6\nu)}{20Eh}x^2 \\ &\quad + \frac{3\nu q}{4Eh^3}x^2z^2 - \frac{3q(4+5\nu)}{20Eh}z^2 + \frac{q}{8Eh^3}z^4 \end{aligned}\right\} \tag{4.4.49}$$

Having solved a particular solution, we may discuss the solution of homogeneous equation based on the superposition principle.

Further problems with different forms of distributed load such as linearly distributed load, parabolic distributed load, etc. can be solved in a same way.

4.5. Solutions of Saint–Venant Problems for Rectangular Beam

The Saint–Venant principle is particularly emphasized in the semi-inverse method in elasticity. It states local influences of a system of forces in **self-equilibrium**. Equivalently, the local effects decay drastically with respect to distance. It is a special characteristic of an exponential function which is inherent in the eigen-solutions of nonzero eigenvalues.

Exponential functions do not appear in the solutions of zero eigenvalues and, therefore, the solutions are not sensitive to self-equilibrium system of forces on a certain cross section. The influence of non-equilibrium external load on the cross section propagates to farther region through the solutions. The solution of zero eigenvalue explicitly divides the problem with local and non-local characteristics.

The Saint–Venant principle is applicable to the current problem for $l \gg h$ in a rectangular domain. The influence of self-equilibrium forces at both ends ($z = 0$ or l) is only confined to the vicinity of the region. It is then appropriate to neglect the solutions of nonzero eigenvalues and, therefore, to apply only the solutions of zero eigenvalue in the expansion theorem, as

$$\boldsymbol{v} = a_1\boldsymbol{\psi}_{0f}^{(0)} + a_2\boldsymbol{\psi}_{0f}^{(1)} + a_3\boldsymbol{\psi}_{0s}^{(0)} + a_4\boldsymbol{\psi}_{0s}^{(1)} + a_5\boldsymbol{\psi}_{0s}^{(2)} + a_6\boldsymbol{\psi}_{0s}^{(3)} \tag{4.5.1}$$

or

$$\left.\begin{aligned} w &= a_1 - a_4 x + a_6\left(-\frac{6+5\nu}{10}h^2 x + \frac{2+\nu}{6}x^3\right) \\ u &= -a_2\nu x + a_3 - a_5\left(\frac{4+5\nu}{10}h^2 - \frac{\nu}{2}x^2\right) \\ \sigma &= a_2 E - a_5 E x \\ \tau &= \frac{1}{2}a_6 E(x^2 - h^2) \end{aligned}\right\} \tag{4.5.2}$$

Every term in the expansion of Eq. (4.5.1) has specific physical meaning, likewise each of the undetermined coefficients $a_1 \sim a_6$ has its physical meaning. For instance,

$$a_1 \quad \text{— axial displacement} \tag{4.5.3a}$$

$$a_2 \cdot (2hE) \quad \text{— axial force} \tag{4.5.3b}$$

and these two parameters constitute a set of parameters for symmetric deformation. The set of parameters for antisymmetric deformation includes

$$a_3 \quad \text{— transverse displacement} \tag{4.5.4a}$$

$$a_4 \quad \text{— rotation angle of cross section} \tag{4.5.4b}$$

$$a_5 \cdot (2Eh^3/3) \quad \text{— bending moment (with deformation)} \tag{4.5.4c}$$

$$a_6 \cdot (2Eh^3/3) \quad \text{— shear force (with deformation)} \tag{4.5.4d}$$

The Hamiltonian variational principle (4.2.15) can be rewritten in terms of scalar quantities as

$$\begin{aligned} \delta\Bigg\{ &\int_0^l \int_{-h}^h \Bigg[\sigma\dot{w} + \tau\dot{u} - \frac{1-\nu^2}{2E}\sigma^2 - \frac{1+\nu}{E}\tau^2 \\ &+ \nu\sigma\frac{\partial u}{\partial x} + \tau\frac{\partial w}{\partial x} + \frac{1}{2}E\left(\frac{\partial u}{\partial x}\right)^2 - wF_z - uF_x\Bigg]\mathrm{d}x\mathrm{d}z \\ &- \int_0^l [\overline{F}_{x2}u + \overline{F}_{z2}w]_{x=h}\mathrm{d}z + \int_0^l \left[\overline{F}_{x1}u + \overline{F}_{z1}w\right]_{x=-h}\mathrm{d}z \Bigg\} = 0 \end{aligned} \tag{4.5.5}$$

Hence a_1, a_2 of symmetric deformation and a_3, a_4, a_5, a_6 of antisymmetric deformation are two decoupled sets which can be solved independently.

First, a_1, a_2 of symmetric deformation in the expansion of Eq. (4.5.2) are taken and substituted into Eq. (4.5.5). We have

$$\delta\left\{\int_0^l [(2Eh)a_2\dot{a}_1 - Eha_2^2 - a_1\overline{F}_N - a_2\overline{W}]\,\mathrm{d}z\right\} = 0 \tag{4.5.6}$$

where

$$\overline{F}_N = \int_{-h}^{h} F_z \mathrm{d}x + \overline{F}_{z2} - \overline{F}_{z1} \tag{4.5.7a}$$

$$\overline{W} = -\int_{-h}^{h} F_x \nu x \mathrm{d}x - \nu h(\overline{F}_{x2} + \overline{F}_{x1}) \tag{4.5.7b}$$

Apparently, $\overline{F}_N$ is the axial resultant force while $\overline{W}$ represents the effect of lateral force on the axial tension. The independent variables in the variational equation (4.5.6) are a_1 and a_2 only. The results are

$$\dot{a}_1 = a_2 + \frac{1}{2Eh}\overline{W}, \qquad \dot{a}_2 = -\frac{1}{2Eh}\overline{F}_N \tag{4.5.8}$$

which can be physical interpreted as the axial strain equation and the equilibrium equation, respectively. It is noted here that $\overline{W}$ is resulted from the self-equilibrium external forces on the cross section and also Possion's ratio ν. This term propagates the influence of axial displacement to a farther region and, therefore, attention should be paid when the Saint–Venant principle is applied in such problems.

Next, a_3, a_4, a_5, a_6 of antisymmetric deformation in the expansion of Eq. (4.5.2) are taken and substituted into Eq. (4.5.5). We have

$$\delta\left\{\int_0^l \left[\frac{1}{3}Eh^3(2a_5\dot{a}_4 - 2a_6\dot{a}_3 - a_5^2 + 2a_4a_6) - a_3\overline{F}_S - a_4\overline{M} - a_5\overline{\theta} - a_6\overline{U}\right]\mathrm{d}z \right.$$

$$\left. + \left[\frac{4}{15}(1+\nu)Eh^5 a_5 a_6\right]_{z=0}^{l}\right\} = 0 \tag{4.5.9}$$

where

$$\overline{F}_S = \int_{-h}^{h} F_x \mathrm{d}x + \overline{F}_{x2} - \overline{F}_{x1} \tag{4.5.10a}$$

$$\overline{M} = -\int_{-h}^{h} F_z x \mathrm{d}x - h(\overline{F}_{z2} + \overline{F}_{z1}) \tag{4.5.10b}$$

$$\overline{\theta} = \int_{-h}^{h} F_x \left(-\frac{4+5\nu}{10}h^2 + \frac{\nu}{2}x^2\right)\mathrm{d}x - \frac{2}{5}h^2(\overline{F}_{x2} - \overline{F}_{x1}) \tag{4.5.10c}$$

$$\overline{U} = \int_{-h}^{h} F_z \left(-\frac{6+5\nu}{10}h^2 x + \frac{2+\nu}{6}x^3\right)\mathrm{d}x - h^3(\overline{F}_{z2} + \overline{F}_{z1})\left(\frac{4+5\nu}{15}\right) \tag{4.5.10d}$$

Explanation on these variational equations is required.

First of all, a_3, a_4 of the variables of variation a_3, a_4, a_5, a_6 represent a displacement mode. On the other hand, a_5, a_6 do not represent a pure internal force mode but rather a mixed mode because the corresponding $\boldsymbol{\psi}_{0s}^{(2)}$ and $\boldsymbol{\psi}_{0s}^{(3)}$ not only include distribution of internal forces but also displacement. The displacement portion in this mixed mode includes some correlative deformation. Since the mode is an eigenvector which has been adjoint symplectic orthonormalized, the coefficient of a_6^2 in variational equation (4.5.9) vanishes. However, the term a_4a_6 still exists and it relates a_3, a_6 and a_4, a_5 to form a chain.

It is observed from Eq. (4.5.10) that $\overline{F}_S, \overline{M}$ are respectively the shear force and bending moment on the cross section while $\overline{\theta}, \overline{U}$ represent the effect of the lateral force and axial force on the transverse and axial displacements.

Performing variation on Eq. (4.5.9) yields

$$\dot{a}_3 = a_4 - \frac{3}{2Eh^3}\overline{U} \tag{4.5.11a}$$

$$\dot{a}_4 = a_5 + \frac{3}{2Eh^3}\overline{\theta} \tag{4.5.11b}$$

and the equations of equilibrium

$$\dot{a}_5 = a_6 - \frac{3}{2Eh^3}\overline{M} \tag{4.5.11c}$$

$$\dot{a}_6 = \frac{3}{2Eh^3}\overline{F}_S \tag{4.5.11d}$$

The general solution can be obtained by directly integrating Eq. (4.5.11). In order to obtain the solution of a specific problem, the constants of integration need to be determined by substituting the boundary conditions at both ends $z = 0$ and $z = l$.

The force boundary conditions in the variational principle (4.5.9) are natural boundary conditions, and therefore they can be derived via variation directly as

$$a_5 = a_6 = 0 \quad \text{at } z = 0 \text{ or } l \tag{4.5.12}$$

which can be interpreted physically as zero bending moment and shear force at the ends.

As displacement boundary conditions involve the respective modes, they can better be derived from the natural boundary conditions of variational principle. For example, for fixed end boundaries, Eq. (4.5.5) with added relevant end terms becomes

$$\begin{aligned}\delta\Big\{\int_0^l\int_{-h}^h\Big[&\sigma\dot{w} + \tau\dot{u} - \frac{1-\nu^2}{2E}\sigma^2 - \frac{1+\nu}{E}\tau^2 + \nu\sigma\frac{\partial u}{\partial x} + \tau\frac{\partial w}{\partial x}\\ &+ \frac{1}{2}E\Big(\frac{\partial u}{\partial x}\Big)^2 - wF_z - uF_x\Big]\mathrm{d}x\mathrm{d}z - \int_{-h}^h[\sigma w + \tau u]_{z=0}^{z=l}\mathrm{d}x\\ &- \int_0^l[\overline{F}_{x2}u + \overline{F}_{z2}w]_{x=h}\mathrm{d}z + \int_0^l[\overline{F}_{x1}u + \overline{F}_{z1}w]_{x=-h}\mathrm{d}z\Big\} = 0\end{aligned} \tag{4.5.13}$$

which yields the displacement boundary conditions. The corresponding Eq. (4.5.9) then becomes

$$\begin{aligned}\delta\Big\{\int_0^l\Big[&\frac{1}{3}Eh^3(2a_5\dot{a}_4 + 2a_6\dot{a}_3 - a_5^2 + 2a_4a_6)\\ &- a_3\overline{F}_S - a_4\overline{M} - a_5\overline{\theta} - a_6\overline{U}\Big]\mathrm{d}z\\ &- \frac{2}{3}Eh^3\Big[a_5a_4 - a_6a_3 + \frac{2}{5}(1+\nu)h^2a_5a_6\Big]_{z=0}^{l}\Big\} = 0\end{aligned} \tag{4.5.14}$$

Hence the boundary conditions at the fixed ends are

$$\left.\begin{aligned} a_4 + 0.4(1+\nu)h^2a_6 &= 0\\ a_3 - 0.4(1+\nu)h^2a_5 &= 0\end{aligned}\right\} \quad \text{at } z = 0 \text{ or } l \tag{4.5.15}$$

Although this equation seems somewhat strange, from Eq. (4.5.2)

$$\left.\begin{aligned} w &= -a_4 x + a_6\left(-\frac{6+5\nu}{10}h^2 x + \frac{2+\nu}{6}x^3\right) \\ u &= a_3 - a_5\left(\frac{4+5\nu}{10}h^2 - \frac{\nu}{2}x^2\right) \end{aligned}\right\} \tag{4.5.16}$$

Therefore Eq. (4.5.15) expresses zero equivalent rotation angle and zero equivalent displacement respectively, or

$$\int_{-h}^{h} xw\mathrm{d}x = 0, \qquad \int_{-h}^{h}\left(1-\frac{x^2}{h^2}\right)u\mathrm{d}x = 0 \tag{4.5.17}$$

Having derived the boundary conditions, the solutions of the Saint–Venant problem can be solved easily.

Example 4.1. Let a cantilevered beam be fixed at $z = 0$ and free at $z = l$. Determine the solution if the beam is subjected to unit uniform load on side $x = h$.

Solution. From Eq. (4.5.10), we know

$$\overline{F}_S = 1, \quad \overline{M} = 0; \quad \overline{\theta} = -0.4h^2, \quad \overline{U} = 0 \tag{4.5.18}$$

Integrating Eqs. (4.5.11d) and (4.5.11c), and substituting the free boundary condition

$$a_5 = a_6 = 0 \qquad \text{at } z = l \tag{4.5.19}$$

yield

$$a_6 = \frac{3}{2Eh^3}(z-l), \quad a_5 = \frac{3}{4Eh^3}(z-l)^2 \tag{4.5.20}$$

Further integrating Eqs. (4.5.11b) and (4.5.11a), and substituting the fixed boundary condition

$$\left.\begin{aligned} a_4 + 0.4(1+\nu)h^2 a_6 &= 0 \\ a_3 - 0.4(1+\nu)h^2 a_5 &= 0 \end{aligned}\right\} \quad \text{at } z = 0 \tag{4.5.21}$$

yield

$$a_4 = \frac{1}{4Eh^3}[z^3 - 3z^2 l + 3zl^2] + \frac{3}{5Eh}[l(1+\nu) - z] \tag{4.5.22}$$

and

$$a_3 = \frac{1}{16Eh^3}[z^4 - 4z^3 l + 6z^2 l^2] + \frac{3}{10Eh}[l(1+\nu)(2z+l) - z^2] \tag{4.5.23}$$

Substituting $a_3 \sim a_6$ into Eq. (4.5.2) results in the displacement field and stress field of the cantilevered beam. For example, the deflection of the beam axis is

$$u(0, z) = \frac{1}{16Eh^3}[z^4 - 4z^3 l + 6z^2 l^2] + \frac{3}{10Eh}[l(1+\nu)(2z+l) - z^2] - \frac{3(4+5\nu)}{40Eh}(z-l)^2 \tag{4.5.24}$$

It should be noted that the solution obtained here is not the exact solution of the original problem, but rather the approximate solution of Saint–Venant problem. If the inhomogeneous boundary conditions are satisfied by a particular solution, then the original problem can be transformed to the corresponding one with homogeneous equations and homogeneous boundary conditions. Further we can obtain an analytical solution by means of expansion of eigen-solutions of zero eigenvalue. The latter is the elastic exact solution if the distribution of external forces (a force system in equilibrium) at the free end is the same as that of the solution.

Thus, we have obtained all the Saint–Venant solutions through the method of separation of variables in Hamiltonian dual system. Saint–Venant solutions are eigen-solutions of zero eigenvalue. In conclusion, the completeness of the solutions above have been proved by rational derivation.

The conclusion above can be positively affirmed by applying rational derivation. The traditional semi-inverse solution method requires solutions via trial and error. It is not possible to confirm the number of solutions, whether the search is complete, nor the method to determine the remaining solutions, for example, the solution with local effect. On the contrary, the rational method clearly shows that the solution with local effect is the one corresponding to the eigenvalues with nonzero real part (it can be proved that there exists no pure imaginary eigenvalue in elasticity described in a rectangular coordinate system[7]). This salient will be introduced in the next section.

4.6. Eigen-Solutions of Nonzero Eigenvalues

The eigen-solutions of zero eigenvalue correspond to the solutions of Saint–Venant problems. On the other hand, the solutions of the portion covered by the Saint–Venant principle corresponding to the eigen-solutions with nonzero eigenvalues are very important for satisfying the boundary conditions at two ends ($z = 0$ or l) or when there are sudden changes of external loading within the region.

Expanding the eigenvalue equations (4.3.3) yields

$$\left.\begin{array}{lllll} 0 & -\nu\dfrac{\mathrm{d}u}{\mathrm{d}x} & +\dfrac{1-\nu^2}{E}\sigma & +0 & =\mu w \\ -\dfrac{\mathrm{d}w}{\mathrm{d}x} & +0 & +0 & +\dfrac{2(1+\nu)}{E}\tau & =\mu u \\ 0 & +0 & +0 & -\dfrac{\mathrm{d}\tau}{\mathrm{d}x} & =\mu\sigma \\ 0 & -E\dfrac{\mathrm{d}^2u}{\mathrm{d}x^2} & -\nu\dfrac{\mathrm{d}\sigma}{\mathrm{d}x} & +0 & =\mu\tau \end{array}\right\} \quad (4.6.1)$$

This is a system of ordinary differential equations with respect to x which can be solved by first determining the eigenvalue λ with respect to the x-direction. The corresponding equation is

$$\det\begin{bmatrix} -\mu & -\nu\lambda & (1-\nu^2)/E & 0 \\ -\lambda & -\mu & 0 & 2(1+\nu)/E \\ 0 & 0 & -\mu & -\lambda \\ 0 & -E\lambda^2 & -\nu\lambda & -\mu \end{bmatrix} = 0 \quad (4.6.2)$$

Expanding the determinant yields the eigenvalue equation

$$(\lambda^2+\mu^2)^2 = 0 \quad (4.6.3)$$

The eigenvalues are repeated roots $\lambda = \pm \mathrm{i}\mu$, hence, the general solution is

$$\left.\begin{array}{l} w = A_w\cos(\mu x) + B_w\sin(\mu x) + C_w x\sin(\mu x) + D_w x\cos(\mu x) \\ u = A_u\sin(\mu x) + B_u\cos(\mu x) + C_u x\cos(\mu x) + D_u x\sin(\mu x) \\ \sigma = A_\sigma\cos(\mu x) + B_\sigma\sin(\mu x) + C_\sigma x\sin(\mu x) + D_\sigma x\cos(\mu x) \\ \tau = A_\tau\sin(\mu x) + B_\tau\cos(\mu x) + C_\tau x\cos(\mu x) + D_\tau x\sin(\mu x) \end{array}\right\} \quad (4.6.3')$$

It shows that the partial solutions relevant to A and C are the solutions of symmetric deformation with the z-axis while the partial solutions relevant to B and D are the solutions of antisymmetric deformation with the z-axis.

4.6.1. *Eigen-Solutions of Nonzero Eigenvalues of Symmetric Deformation*

The general solution of symmetric deformation is

$$\left.\begin{aligned} w &= A_w \cos(\mu x) + C_w x \sin(\mu x) \\ u &= A_u \sin(\mu x) + C_u x \cos(\mu x) \\ \sigma &= A_\sigma \cos(\mu x) + C_\sigma x \sin(\mu x) \\ \tau &= A_\tau \sin(\mu x) + C_\tau x \cos(\mu x) \end{aligned}\right\} \tag{4.6.4}$$

where the constants are not all independent. Noting that the expression is true for an arbitrary x, the substitution of Eq. (4.6.4) into Eq. (4.6.1) yields

$$\begin{bmatrix} -\mu & \nu\mu & (1-\nu^2)/E & 0 \\ -\mu & -\mu & 0 & 2(1+\nu)/E \\ 0 & 0 & -\mu & \mu \\ 0 & E\mu^2 & -\nu\mu & -\mu \end{bmatrix} \begin{Bmatrix} C_w \\ C_u \\ C_\sigma \\ C_\tau \end{Bmatrix} = \mathbf{0} \tag{4.6.5}$$

and

$$\begin{bmatrix} -\mu & -\nu\mu & (1-\nu^2)/E & 0 \\ \mu & -\mu & 0 & 2(1+\nu)/E \\ 0 & 0 & -\mu & -\mu \\ 0 & E\mu^2 & \nu\mu & -\mu \end{bmatrix} \begin{Bmatrix} A_w \\ A_u \\ A_\sigma \\ A_\tau \end{Bmatrix} = \begin{Bmatrix} \nu C_u \\ C_w \\ C_\tau \\ \nu C_\sigma - 2E\mu C_u \end{Bmatrix} \tag{4.6.6}$$

As the determinant of coefficient matrix of Eq. (4.6.5) vanishes, the non-trivial solutions are

$$C_w = C_u, \qquad C_\sigma = C_\tau = \frac{E\mu}{1+\nu} C_u \tag{4.6.7}$$

Since Eq. (4.6.6) are compatible, we have

$$\left.\begin{aligned} A_w &= -A_u - \frac{3-\nu}{(1+\nu)\mu} C_u \\ A_\sigma &= -\frac{E\mu}{1+\nu} A_u - \frac{E(3+\nu)}{(1+\nu)^2} C_u \\ A_\tau &= \frac{E\mu}{1+\nu} A_u + \frac{2E}{(1+\nu)^2} C_u \end{aligned}\right\} \tag{4.6.8}$$

Therefore, there are only two independent constants. Although A_u and C_u are chosen here as the independent constants, it is also possible if other constants are chosen. Substituting Eqs. (4.6.4), (4.6.7) and (4.6.8) into the boundary conditions (4.2.25) yields

$$\left.\begin{aligned} A_u\mu\sin(\mu h)+C_u\left[\mu h\cos(\mu h)+\frac{2}{1+\nu}\sin(\mu h)\right]&=0\\ A_u\mu\cos(\mu h)+C_u\left[-\mu h\sin(\mu h)+\frac{1-\nu}{1+\nu}\cos(\mu h)\right]&=0\end{aligned}\right\} \tag{4.6.9}$$

For nontrivial solution to exist, the determinant of coefficient matrix vanishes. Hence, we have

$$2\mu h+\sin(2\mu h)=0 \tag{4.6.10}$$

Obviously, $-\mu$ must be a root if μ is a root which is consistent with the characteristics of Hamiltonian operator matrix. It is also clear that there are no nonzero real roots for Eq. (4.6.10). Because Eq. (4.6.10) is a real equation, the roots are complex conjugates which are denoted as $2\mu h=\pm\alpha\pm \mathrm{i}\beta$ where α and β are real and positive. It is possible to discuss the roots in the first quadrant only, as

$$2\mu h=\alpha+\mathrm{i}\beta \tag{4.6.11}$$

Equation (4.6.10) can be solved by numerical methods. In this respect, Newton's method is able to provide fast convergent numerical solutions but it requires an initial approximate root which can be obtained by an asymptotic method. Because trigonometric functions are periodic, a root for $\beta>0$ exists in the complex domain for each 2π-increment for $2\mu h$. Hence, we denote

$$\alpha=2n\pi+\alpha' \tag{4.6.12}$$

where $0\le\alpha'<2\pi$. For large positive β, Eq. (4.6.10) can be approximated as

$$(\alpha'+2n\pi)+\mathrm{i}\beta-\frac{1}{2\mathrm{i}}\mathrm{e}^{-\mathrm{i}(\alpha'+\mathrm{i}\beta)}\approx 0 \tag{4.6.13}$$

Separating the real and imaginary parts of Eq. (4.6.13) yields

$$\alpha' + 2n\pi + \frac{1}{2}e^{\beta}\sin\alpha' \approx 0 \tag{4.6.14a}$$

$$\beta + \frac{1}{2}e^{\beta}\cos\alpha' \approx 0 \tag{4.6.14b}$$

For large positive β, we have

$$\frac{2\beta}{e^{\beta}} \to 0^{+} \tag{4.6.15}$$

hence we deduce

$$\cos\alpha' \to 0^{-} \tag{4.6.16}$$

In addition, we have $\sin\alpha' < 0$ from Eq. (4.6.14a) and the asymptotic solution is

$$\alpha \to 2n\pi - \frac{\pi}{2} - \varepsilon \tag{4.6.17a}$$

where $n = 1, 2, 3, \ldots$. Substituting Eq. (4.6.17a) into Eq. (4.6.14a), we have approximately

$$\beta \to \ln(2\alpha) \tag{4.6.17b}$$

The expression (4.6.17) can be taken as the initial approximate root in Newton's method which subsequently yields numerically the eigenvalues. Table 4.1 lists the first five eigenvalues.

Only roots in the first quadrant are listed in the table. Each μ_n in reality has its symplectic adjoint eigenvalue $-\mu_n$ and their complex conjugate eigenvalues, i.e. there are totally four eigenvalues in which two belong to the (α)-set while the other two the (β)-set. It is obvious from Eq. (4.6.10) that the nonzero eigenvalues are all single roots.

Having solved the eigenvalues μ_n, the nontrivial solution of Eq. (4.6.9) is

$$A_u = \cos^2(\mu_n h) - \frac{2}{1+\nu}, \quad C_u = \mu_n \tag{4.6.18}$$

Table 4.1. Nonzero eigenvalues for symmetric deformation.

$n =$	1	2	3	4	5
$\mathrm{Re}(\mu_n h)$	$\frac{\pi}{2} + 0.5354$	$\frac{3\pi}{2} + 0.6439$	$\frac{5\pi}{2} + 0.6827$	$\frac{7\pi}{2} + 0.7036$	$\frac{9\pi}{2} + 0.7169$
$\mathrm{Im}(\mu_n h)$	1.1254	1.5516	1.7755	1.9294	2.0469

The other constants are determined from Eqs. (4.6.7) and (4.6.8). Hence, the corresponding eigenvector function is

$$\boldsymbol{\psi}_n = \begin{Bmatrix} w_n \\ u_n \\ \sigma_n \\ \tau_n \end{Bmatrix} = \begin{Bmatrix} -\left[\cos^2(\mu_n h) + \dfrac{1-\nu}{1+\nu}\right]\cos(\mu_n x) + \mu_n x \sin(\mu_n x) \\ \left[\cos^2(\mu_n h) - \dfrac{2}{1+\nu}\right]\sin(\mu_n x) + \mu_n x \cos(\mu_n x) \\ \dfrac{E\mu_n}{1+\nu}[-(1+\cos^2(\mu_n h))\cos(\mu_n x) + \mu_n x \sin(\mu_n x)] \\ \dfrac{E\mu_n}{1+\nu}[\cos^2(\mu_n h)\sin(\mu_n x) + \mu_n x \cos(\mu_n x)] \end{Bmatrix} \tag{4.6.19}$$

As the eigenvalues are complex, the eigen-solutions are also complex. The solution of original problem corresponding to Eq. (4.2.24) is

$$\boldsymbol{v}_n = \mathrm{e}^{\mu_n z}\boldsymbol{\psi}_n \tag{4.6.20}$$

4.6.2. *Eigen-Solutions of Nonzero Eigenvalues of Antisymmetric Deformation*

The general solution of antisymmetric deformation is

$$\left.\begin{aligned} w &= B_w \sin(\mu x) + D_w x \cos(\mu x) \\ u &= B_u \cos(\mu x) + D_u x \sin(\mu x) \\ \sigma &= B_\sigma \sin(\mu x) + D_\sigma x \cos(\mu x) \\ \tau &= B_\tau \cos(\mu x) + D_\tau x \sin(\mu x) \end{aligned}\right\} \tag{4.6.21}$$

where the constants are not all independent. Substituting Eq. (4.6.21) into Eq. (4.6.1) yields the equations

$$\begin{bmatrix} -\mu & -\nu\mu & (1-\nu^2)/E & 0 \\ \mu & -\mu & 0 & 2(1+\nu)/E \\ 0 & 0 & -\mu & -\mu \\ 0 & E\mu^2 & \nu\mu & -\mu \end{bmatrix} \begin{Bmatrix} D_w \\ D_u \\ D_\sigma \\ D_\tau \end{Bmatrix} = \mathbf{0} \tag{4.6.22}$$

and

$$\begin{bmatrix} -\mu & \nu\mu & (1-\nu^2)/E & 0 \\ -\mu & -\mu & 0 & 2(1+\nu)/E \\ 0 & 0 & -\mu & \mu \\ 0 & E\mu^2 & -\nu\mu & -\mu \end{bmatrix} \begin{Bmatrix} B_w \\ B_u \\ B_\sigma \\ B_\tau \end{Bmatrix} = \begin{Bmatrix} \nu D_u \\ D_w \\ D_\tau \\ \nu D_\sigma + 2E\mu D_u \end{Bmatrix} \tag{4.6.23}$$

As these two sets of equations are compatible, hence we obtain

$$\left.\begin{aligned} B_w &= B_u - \frac{3-\nu}{(1+\nu)\mu} D_u; & D_w &= -D_u \\ B_\sigma &= \frac{E\mu}{1+\nu} B_u - \frac{E(3+\nu)}{(1+\nu)^2} D_u; & D_\sigma &= -\frac{E\mu}{1+\nu} D_u \\ B_\tau &= \frac{E\mu}{1+\nu} B_u - \frac{2E}{(1+\nu)^2} D_u; & D_\tau &= \frac{E\mu}{1+\nu} D_u \end{aligned}\right\} \tag{4.6.24}$$

Here B_u and D_u are chosen as the independent constants. Substituting Eqs. (4.6.21) and (4.6.24) into the boundary conditions (4.2.25) yields

$$\left.\begin{aligned} B_u\mu\cos(\mu h) + D_u\left[\mu h\sin(\mu h) - \frac{2}{1+\nu}\cos(\mu h)\right] &= 0 \\ -B_u\mu\sin(\mu h) + D_u\left[\mu h\cos(\mu h) + \frac{1-\nu}{1+\nu}\sin(\mu h)\right] &= 0 \end{aligned}\right\} \tag{4.6.25}$$

Setting the determinant of coefficient matrix to vanish yields

$$2\mu h - \sin(2\mu h) = 0 \tag{4.6.26}$$

Obviously, $-\mu$ must be a root if μ is a root. Similarly, there are no nonzero real roots for Eq. (4.6.26). Denote the roots in the first quadrant as

$$2\mu h = \alpha + \mathrm{i}\beta \tag{4.6.27}$$

where α and β are positive real numbers.

Likewise, Newton's method can be applied to solve Eq. (4.6.26). Similar to the derivation in the previous section, the approximated asymptotic solution is

$$\alpha \to 2n\pi + \frac{\pi}{2} - \varepsilon, \qquad \beta \to \ln(2\alpha) \tag{4.6.28}$$

Table 4.2. Nonzero eigenvalues for antisymmetric deformation.

$n=$	1	2	3	4	5
$\mathrm{Re}(\mu_n h)$	$\pi+0.6072$	$2\pi+0.6668$	$3\pi+0.6954$	$4\pi+0.7109$	$5\pi+0.7219$
$\mathrm{Im}(\mu_n h)$	1.3843	1.6761	1.8584	1.9916	2.0966

where $n=1,2,3,\ldots$, which can then be used as the initial approximate root for Newton's method to solve for the eigenvalues. Table 4.2 lists the first five eigenvalues.

Obviously, there exists a symplectic adjoint eigenvalue $-\mu_n$ for each μ_n as well as their complex conjugate eigenvalues. There are totally four eigenvalues in which two belong to the (α)-set while the other two the (β)-set. From Eq. (4.6.26), the antisymmetric nonzero eigenvalues are all single roots.

The eigenvector function corresponding to the eigenvalue μ_n is

$$\boldsymbol{\psi}_n=\begin{Bmatrix} w_n \\ u_n \\ \sigma_n \\ \tau_n \end{Bmatrix}=\begin{Bmatrix} -\left[\sin^2(\mu_n h)+\dfrac{1-\nu}{1+\nu}\right]\sin(\mu_n x)-\mu_n x\cos(\mu_n x) \\ \left[-\sin^2(\mu_n h)+\dfrac{2}{1+\nu}\right]\cos(\mu_n x)+\mu_n x\sin(\mu_n x) \\ -\dfrac{E\mu_n}{1+\nu}[(1+\sin^2(\mu_n h))\sin(\mu_n x)+\mu_n x\cos(\mu_n x)] \\ \dfrac{E\mu_n}{1+\nu}[-\sin^2(\mu_n h)\cos(\mu_n x)+\mu_n x\sin(\mu_n x)] \end{Bmatrix} \tag{4.6.29}$$

The nonzero eigenvalues and eigen-solutions for antisymmetric deformation are complex. The solution corresponding to the original problem (4.2.24) is

$$\boldsymbol{v}_n=\mathrm{e}^{\mu_n z}\boldsymbol{\psi}_n \tag{4.6.30}$$

Thus, we have obtained all eigen-solutions of nonzero eigenvalues. Except the symplectic adjoint eigenvectors corresponding to the symplectic adjoint eigenvalues, the remaining eigenvectors are symplectic orthogonal including symplectic orthogonality between them and the eigenvectors of zero eigenvalue. Adjoint symplectic orthogonality is a very important characteristic. If normalized, the eigenvectors can be further expanded according to the expansion theorem and this is very helpful for solving the problem.

The eigen-solutions of these nonzero eigenvalues decay with distance depending on the characteristics of eigenvalues. The solutions in (α)-set decay along the positive z-direction while the solutions in (β)-set decay along the negative z-direction. All of these solutions are covered in the Saint–Venant principle.

These eigen-solutions have one common characteristic, i.e. they are symplectic orthogonal to the eigen-solutions of zero eigenvalue. As they are symplectic orthogonal to $\boldsymbol{\psi}_{0f}^{(0)}$, $\boldsymbol{\psi}_{0s}^{(0)}$, $\boldsymbol{\psi}_{0s}^{(1)}$, the distributed forces of these solutions on the cross section are in self-equilibrium, hence the solutions satisfy the requirement of the classical Saint–Venant principle. In short, self-equilibrium of the force system is the key salient point of the classical Saint–Venant principle. However, the satisfaction of this requirement is not sufficient for the eigen-solutions of zero eigenvalue.

The eigen-solutions of homogeneous equation has been discussed above. Subsequently, we consider external forces or two point boundary value problems.

4.7. Solutions of Generalized Plane Problems in Rectangular Domain

The discussion above concerns the solutions of homogeneous equation (4.2.24). With external load, the original equation is Eq. (4.2.21) where the inhomogeneous term $\boldsymbol{h}$ is related to the given external load. While there are various solution methods, it is most effective by means of eigenvectors and the expansion theorems.

Substitute the expansion expression of eigenvector in Eq. (4.3.6) to the full state vector $\boldsymbol{v}$ in Eq. (4.2.21), we obtain the ordinary differential equations with respect of a_i, b_i.

The eigen-solutions of singly repeated eigenvalue μ_i have been introduced in Sec. 3.5 at length. The differential equations have been decoupled into

$$\dot{a}_i = \mu_i a_i + c_i, \qquad \dot{b}_i = -\mu_i b_i + d_i \tag{4.7.1}$$

where

$$c_i = \langle \boldsymbol{h}, \boldsymbol{\psi}_{-i} \rangle, \qquad d_i = -\langle \boldsymbol{h}, \boldsymbol{\psi}_i \rangle \tag{4.7.2}$$

The subscripts i and $-i (i = 1, 2, \ldots)$ denote the corresponding eigenvalues belonging to (α)- and (β)-sets, respectively.

If the eigenvalues $\pm\mu_i$ are repeated eigenvalues, for example, triple roots, then there are correspondingly six unknown functions $a_i^{(0)}, a_i^{(1)}, a_i^{(2)}$ and $b_i^{(0)}, b_i^{(1)}, b_i^{(2)}$. We assume the eigenvectors include Jordan form eigenvectors $\boldsymbol{\psi}_i^{(0)}, \boldsymbol{\psi}_i^{(1)}, \boldsymbol{\psi}_i^{(2)}$ and $\boldsymbol{\psi}_{-i}^{(2)}, \boldsymbol{\psi}_{-i}^{(1)}, \boldsymbol{\psi}_{-i}^{(0)}$ which have been transformed into a set of normal adjoint symplectic orthonormal vectors, and in addition, $\boldsymbol{\psi}_{-i}^{(2)}, \boldsymbol{\psi}_{-i}^{(1)}, \boldsymbol{\psi}_{-i}^{(0)}$ satisfy Eq. (1.3.19′). Then, the equations for solving six functions $a_i^{(0)}, a_i^{(1)}, a_i^{(2)}$ and $b_i^{(0)}, b_i^{(1)}, b_i^{(2)}$ are

$$\left.\begin{aligned} \dot{a}_i^{(0)} &= \mu_i a_i^{(0)} + a_i^{(1)} + c_i^{(0)}, & \dot{b}_i^{(0)} &= -\mu_i b_i^{(0)} + d_i^{(0)} \\ \dot{a}_i^{(1)} &= \mu_i a_i^{(1)} + a_i^{(2)} + c_i^{(1)}, & \dot{b}_i^{(1)} &= -\mu_i b_i^{(1)} - b_i^{(0)} + d_i^{(1)} \\ \dot{a}_i^{(2)} &= \mu_i a_i^{(2)} + c_i^{(2)}, & \dot{b}_i^{(2)} &= -\mu_i b_i^{(2)} - b_i^{(1)} + d_i^{(2)} \end{aligned}\right\} \tag{4.7.3}$$

where

$$c_i^{(j)} = \langle \boldsymbol{h}, \boldsymbol{\psi}_{-i}^{(2-j)} \rangle, \quad d_i^{(j)} = -\langle \boldsymbol{h}, \boldsymbol{\psi}_i^{(2-j)} \rangle \quad (j = 0, 1, 2) \tag{4.7.4}$$

Hence, for Jordan form solutions we need to solve the system of differential equations (4.7.3) one by one.

The solutions a_i, b_i of Eqs. (4.7.1) or (4.7.3) can be substituted into the corresponding boundary conditions to determine the integration constants which then form the solution of the original problem.

The approach above also provides a method to solve the particular solution for the inhomogeneous term $\boldsymbol{h}$. Once the particular solution is obtained, the general solution can then be expressed as the sum of the particular solution and the homogeneous solution according to the superposition principle. In summary, the problem can be treated in advance with the particular solution and then transformed into the solution of homogeneous equation (4.2.24).

The following discussion is restricted to the solution of homogeneous equation (4.2.24). The corresponding boundary conditions should be derived by subtracting the boundary values of the particular solution from the original boundary values.

The end boundary conditions for specified displacements are

$$\left.\begin{aligned} w &= \overline{w}_0(x), \quad u = \overline{u}_0(x) \quad \text{at } z = 0 \\ w &= \overline{w}_l(x), \quad u = \overline{u}_l(x) \quad \text{at } z = l \end{aligned}\right\} \tag{4.7.5}$$

where $\overline{w}_0, \overline{u}_0, \overline{w}_l, \overline{u}_l$ are the specified displacements at the ends. Equation (4.7.5) can be expressed as

$$\left.\begin{aligned} \boldsymbol{q}_0 = \bar{\boldsymbol{q}}_0(x) = \left\{ \overline{w}_0(x), \quad \overline{u}_0(x) \right\}^{\mathrm{T}} \quad \text{at } z = 0 \\ \boldsymbol{q}_l = \bar{\boldsymbol{q}}_l(x) = \left\{ \overline{w}_l(x), \quad \overline{u}_l(x) \right\}^{\mathrm{T}} \quad \text{at } z = l \end{aligned}\right\} \tag{4.7.6}$$

where $\boldsymbol{q}_0, \boldsymbol{q}_l$ denote the values of variable $\boldsymbol{q}$ at $z = 0$ and $z = l$, respectively.

The end boundary conditions for specified forces are

$$\left.\begin{aligned} \sigma = \overline{\sigma}_0(x), \quad \tau = \overline{\tau}_0(x) \quad \text{at } z = 0 \\ \sigma = \overline{\sigma}_l(x), \quad \tau = \overline{\tau}_l(x) \quad \text{at } z = l \end{aligned}\right\} \tag{4.7.7}$$

where $\overline{\sigma}_0, \overline{\tau}_0, \overline{\sigma}_l, \overline{\tau}_l$ are the specified tractions at the ends. Equation (4.7.7) can be expressed as

$$\left.\begin{aligned} \boldsymbol{p}_0 = \overline{\boldsymbol{p}}_0(x) = \left\{ \overline{\sigma}_0(x), \quad \overline{\tau}_0(x) \right\}^{\mathrm{T}} \quad \text{at } z = 0 \\ \boldsymbol{p}_l = \overline{\boldsymbol{p}}_l(x) = \left\{ \overline{\sigma}_l(x), \quad \overline{\tau}_l(x) \right\}^{\mathrm{T}} \quad \text{at } z = l \end{aligned}\right\} \tag{4.7.8}$$

where $\boldsymbol{p}_0, \boldsymbol{p}_l$ denote the values of variable $\boldsymbol{p}$ at $z = 0$ and $z = l$, respectively.

The boundary conditions at both ends ($z = 0$ or l) can also be mixed boundary conditions which are left for the readers to accomplish.

As homogeneous equation (4.2.24) with homogeneous boundary conditions (4.2.25) at both ends ($z = 0$ or l) is considered here, the corresponding Hamiltonian mixed energy variational principle (4.2.15) degenerates to

$$\delta \left\{ \int_0^l \int_{-h}^h [\boldsymbol{p}^{\mathrm{T}} \dot{\boldsymbol{q}} - \mathscr{H}(\boldsymbol{q}, \boldsymbol{p})] \mathrm{d}x \mathrm{d}z + U_e \right\} = 0 \tag{4.7.9}$$

where the Hamiltonian density function is

$$\mathscr{H} = \frac{1 - \nu^2}{2E} \sigma^2 + \frac{1 + \nu}{E} \tau^2 - \nu \sigma \frac{\partial u}{\partial x} - \tau \frac{\partial w}{\partial x} - \frac{1}{2} E \left(\frac{\partial u}{\partial x} \right)^2 \tag{4.7.10}$$

Here the effect U_e at $z = 0$ and $z = l$ has been considered. For boundary conditions (4.7.5) with specified displacements at both ends ($z = 0$ or l), we have

$$U_e = \int_{-h}^h \boldsymbol{p}_0^{\mathrm{T}} (\boldsymbol{q}_0 - \bar{\boldsymbol{q}}_0) \mathrm{d}x - \int_{-h}^h \boldsymbol{p}_l^{\mathrm{T}} (\boldsymbol{q}_l - \bar{\boldsymbol{q}}_l) \mathrm{d}x \tag{4.7.11}$$

For boundary conditions (4.7.7) with specified forces at both ends ($z = 0$ or l), we have

$$U_e = \int_{-h}^{h} \boldsymbol{q}_0^{\mathrm{T}} \overline{\boldsymbol{p}}_0 \, \mathrm{d}x - \int_{-h}^{h} \boldsymbol{q}_l^{\mathrm{T}} \overline{\boldsymbol{p}}_l \, \mathrm{d}x \tag{4.7.12}$$

The solution of homogeneous equations (4.2.24) by the method of separation variables has been discussed in the previous several sections. The analytical expressions of eigen-solutions of zero eigenvalue and of nonzero eigenvalues have also been presented. Based on expansion theorems, the general solution of homogeneous equations (4.2.24) for the plane elasticity in rectangular domain is

$$\boldsymbol{v} = \sum_{i=0}^{1} a_{0f}^{(i)} \boldsymbol{v}_{0f}^{(i)} + \sum_{i=0}^{3} a_{0s}^{(i)} \boldsymbol{v}_{0s}^{(i)} + \sum_{i=1}^{\infty} (\tilde{a}_i \boldsymbol{v}_i + \tilde{b}_i \boldsymbol{v}_{-i}) \tag{4.7.13}$$

where $a_{0f}^{(i)}, a_{0s}^{(i)}, \tilde{a}_i, \tilde{b}_i$ are undetermined constants.

There are complex eigenvectors $\boldsymbol{v}_i, \boldsymbol{v}_{-i}$ for the current problem here because the corresponding nonzero eigenvalues are all complex. The presence of complex operation is, of course, troublesome. The problem is essentially real, complex numbers appear due to the eigen-solutions. When establishing algebraic equation to satisfy the boundary conditions at both ends ($z = 0$ or l), it is better to deal with real numbers. In addition, the extremum condition using the variational principle involves only real numbers. With reference to the derivation of Eq. (3.6.20), Eq. (4.7.13) can be transformed into a real canonical equation as

$$\begin{aligned} \boldsymbol{v} = & \sum_{i=0}^{1} a_{0f}^{(i)} \boldsymbol{v}_{0f}^{(i)} + \sum_{i=0}^{3} a_{0s}^{(i)} \boldsymbol{v}_{0s}^{(i)} \\ & + \sum_{i=1,3,\ldots}^{\infty} [a_i \mathrm{Re}(\boldsymbol{v}_i) + a_{i+1} \mathrm{Im}(\boldsymbol{v}_i) + b_i \mathrm{Re}(\boldsymbol{v}_{-i}) + b_{i+1} \mathrm{Im}(\boldsymbol{v}_{-i})] \end{aligned} \tag{4.7.14}$$

It should be noted that $i = 1, 3, \ldots$ refers to eigen-solutions of eigenvalues $\mathrm{Im}(\mu) > 0$ only in the expansion of Eq. (4.7.14). Thus the transformation from complex form to real canonical equation is completed.

The expression (4.7.14) is real and it satisfy the partial differential equations (4.2.24) and boundary conditions (4.2.25) on both side ($x = \pm h$). The variational equations corresponding to the two point boundary conditions can be obtained by applying the variational principle.

Performing variation on Eqs. (4.7.9) and (4.7.11) with respect to the boundary conditions (4.7.5) for specified displacements at both ends ($z = 0$ or l) yields

$$\int_0^l \int_{-h}^h \left[(\delta \boldsymbol{p}^{\mathrm{T}}) \left(\dot{\boldsymbol{q}} - \frac{\delta \mathscr{H}}{\delta \boldsymbol{p}} \right) - (\delta \boldsymbol{q}^{\mathrm{T}}) \left(\dot{\boldsymbol{p}} + \frac{\delta \mathscr{H}}{\delta \boldsymbol{q}} \right) \right] \mathrm{d}x \mathrm{d}z$$
$$- \int_{-h}^h (\delta \boldsymbol{p}_l^{\mathrm{T}})(\boldsymbol{q}_l - \overline{\boldsymbol{q}}_l) \mathrm{d}x + \int_{-h}^h (\delta \boldsymbol{p}_0^{\mathrm{T}})(\boldsymbol{q}_0 - \overline{\boldsymbol{q}}_0) \mathrm{d}x = 0 \quad (4.7.15)$$

Since $\boldsymbol{q}, \boldsymbol{p}$ are in the form of expanded eigenvectors (4.7.14), the first term of the variational equation vanishes. The remaining variation equations of two point boundary conditions are

$$\int_{-h}^h (\delta \boldsymbol{p}_l^{\mathrm{T}})(\boldsymbol{q}_l - \overline{\boldsymbol{q}}_l) \mathrm{d}x - \int_{-h}^h (\delta \boldsymbol{p}_0^{\mathrm{T}})(\boldsymbol{q}_0 - \overline{\boldsymbol{q}}_0) \mathrm{d}x = 0 \quad (4.7.16)$$

In this way, the boundary conditions (4.7.5) for specified displacements can be expressed as the variational equation (4.7.16). Substituting $z = 0$ and $z = l$ into Eq. (4.7.14), $\boldsymbol{q}_0, \boldsymbol{q}_l$ and $\boldsymbol{p}_0, \boldsymbol{p}_l$ then become functions of undetermined constants $a_{0f}^{(i)}, a_{0s}^{(i)}, a_i, b_i$. These undetermined constants are the parameters of variation and they can be obtained by solving the simultaneous equations derived from variation. These simultaneous equations are the canonical equations. From the solution which satisfies all governing equations except the boundary conditions at two ends ($z = 0$ or l) and the compatibility condition represented by the equations of variation (4.7.16), we know the simultaneous equations obtained are the force canonical break equations.

A very simple example is presented here. Consider a simple tension problem of a semi-infinite strip fixed at $z = 0$, the stress distribution at the fixed end is determined here.

With reference to the problem, there is only the tension stress σ_∞ for $z \to \infty$ and the deformation is symmetric with respect to z-axis. The solution is then formed from Eqs. (4.4.5), (4.4.11) and the eigen-solution of nonzero eigenvalue of symmetric deformation (4.6.20). Only eigen-solutions of (α)-set in Eq. (4.7.14) are adopted, i.e. $\mathrm{Re}(\mu_i) < 0$. Hence the expansion of general solution is

$$\boldsymbol{v} = \left(\frac{\sigma_\infty}{E} \right) \boldsymbol{v}_{0f}^{(1)} + a_0 \boldsymbol{v}_{0f}^{(0)} + \sum_{i=1,3,\ldots}^{\infty} [a_i \mathrm{Re}(\boldsymbol{v}_i) + a_{i+1} \mathrm{Im}(\boldsymbol{v}_i)] \quad (4.7.17)$$

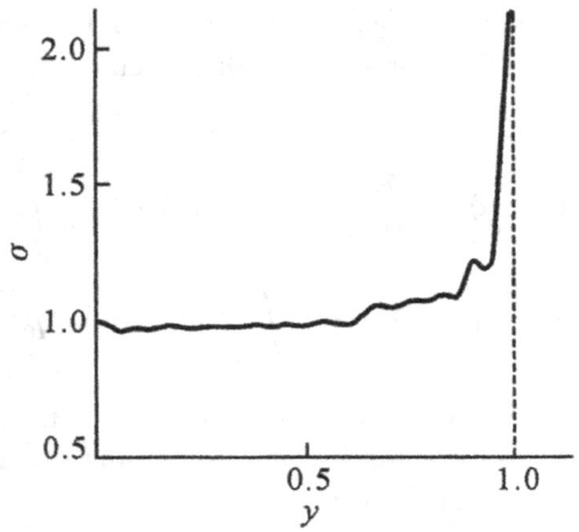

Fig. 4.3. Stress analysis with fixed end.

where $\boldsymbol{v}_{0f}^{(1)}$ indicates the stress at $z \to \infty$, and a_0 the rigid body translation which has no effect on stress. The eigen-solutions of (α)-set decay as z increases, thus consistent with the far-end boundary condition at $z \to \infty$. These terms are those covered by the Saint–Venant principle which has the most significant influence on the eigen-roots closest to the imaginary axis. Substituting Eq. (4.7.17) into the equation of variation (4.7.16) and adopting $i = 1, 3, \ldots, 39$, a total of 20 eigen-solutions of nonzero eigenvalues in the calculation yield σ_z/σ_∞ at the fixed end $z = 0$ as illustrated in Fig. 4.3. The figure shows that there is stress singularity at the edge corner. Fluctuation of stress observed is common when truncated finite terms are adopted in the expansion. Such phenomenon has been observed, for instance, when finite terms in a Fourier series are assumed.

In this chapter, we discuss the problem of plane elasticity in rectangular domain with free side boundary surfaces. Problems with other boundary conditions can be solved in a similar way. In conclusion, the Hamiltonian system approach can be applied to problems with various combinations of boundary conditions.

References

1. Zhilun Xu, *Elasticity*, Beijing: Higher Education Press, 1990 (in Chinese).
2. Changjun Cheng, *Elasticity*, Lanzhou: Lanzhou University Press, 1995 (in Chinese).

3. S.P. Timoshenko and J.N. Goodier, *Theory of Elasticity*, 3rd edn., New York: McGraw-Hill, 1970.
4. N.I. Muschelishvili, *Some Basic Problems of the Mathematical Theory of Elasticity*, Groningen, Netherlands: Noordholf, 1953.
5. Wanxie Zhong, Plane elasticity problem in strip domain and Hamiltonian system, *Journal of Dalian University of Technology*, 1991, 31(4): 373–384 (in Chinese).
6. Wanxie Zhong, Method of separation of variables and Hamiltonian system, *Computational Structural Mechanics and Applications*, 1991, 8(3): 229–240 (in Chinese).
7. Wanxie Zhong, *A New Systematic Methodology for Theory of Elasticity*, Dalian University of Technology Press, 1995 (in Chinese).

Chapter 5

Plane Anisotropic Elasticity Problems

The structural analysis of composite materials is always based on the analysis of anisotropic materials. The solution of anisotropic elasticity problems is more complicated than isotropic problems. Based on symplectic system of isotropic elasticity, the Hamiltonian system is further extended to plane anisotropic elasticity problems in this chapter and thus a complete symplectic solution methodology for these problems is established. A rational analytical solution for Saint–Venant problems is derived by the expansion of eigenvector subspace of zero eigenvalue.

5.1. The Fundamental Equations of Plane Anisotropic Elasticity Problems

Consider in the scope of isotropic elasticity a thin plate with constant thickness and with body forces and forces on side boundary surfaces parallel to the plane of plate. These forces are constant through the thickness of plate. Such problems can be simplified to plane stress problems. On the other hand, for long cylindrical solids with body forces and forces on side boundary surfaces parallel to the cross section of cylinder and all forces are constant in the longitudinal direction, they can be simplified to plane strain problems. For an anisotropic body, the specific geometries and loading conditions above are not the sufficient conditions for simplification to plane elasticity problems. It is necessary for the anisotropic materials to have an elastic symmetric plane which is consistent with the plane of load in order for the problems to be approximately treated as plane elasticity problems.

Consider a thin plate with constant thickness. At any point in the domain, there exists an elastic symmetric plane of material which is parallel to the plane of plate. Let Oxz-plane be the mid-plane of plate, i.e. the dimension along the y-axis is far smaller than the other two dimensions. Assume that the body forces and forces on side boundary surfaces are parallel to the plane and all forces are constant through thickness. Further assume the geometric constraints are also constant through thickness. Such problems can be approximately simplified to plane stress problems. Similar to the plane isotropic elasticity problems, it is required to solve the eight physical quantities u, w, ε_x, ε_z, γ_{xz} and σ_x, σ_z, τ_{xz}. As compared to isotropic problems, the only difference here is the stress-strain relation

$$\begin{Bmatrix} \varepsilon_x \\ \varepsilon_z \\ \gamma_{xz} \end{Bmatrix} = \begin{bmatrix} s_{11} & s_{13} & s_{15} \\ s_{13} & s_{33} & s_{35} \\ s_{15} & s_{35} & s_{55} \end{bmatrix} \begin{Bmatrix} \sigma_x \\ \sigma_z \\ \tau_{xz} \end{Bmatrix} \tag{5.1.1}$$

or in another form

$$\begin{Bmatrix} \sigma_x \\ \sigma_z \\ \tau_{xz} \end{Bmatrix} = \begin{bmatrix} b_{11} & b_{13} & b_{15} \\ b_{13} & b_{33} & b_{35} \\ b_{15} & b_{35} & b_{55} \end{bmatrix} \begin{Bmatrix} \varepsilon_x \\ \varepsilon_z \\ \gamma_{xz} \end{Bmatrix} \tag{5.1.2}$$

where b_{ij} are the **reduced stiffness coefficients** whose determinant is denoted as d in brief. For an elastic body the deformation energy density is positive definite.

For such problems, the strain-displacement relation and equation of equilibrium remain as Eqs. (4.1.5) and (4.1.6), respectively, and the boundary conditions on Γ_σ for specified forces and on Γ_u for specified displacements remain as Eqs. (4.1.7) and (4.1.8), respectively.

Consider a long cylindrical solid with an elastic symmetric plane of material Oxz at every point and the elastic principal axis y is consistent with the axis of the cylinder. In addition, the body forces and forces on the side boundary surfaces are parallel to the cross section. These forces as well as the geometric constraints are constant along the y-axis. Then, such problems can be simplified to plane strain problems. The stress-strain relation of plane strain problems is

$$\begin{Bmatrix} \sigma_x \\ \sigma_z \\ \tau_{xz} \end{Bmatrix} = \begin{bmatrix} c_{11} & c_{13} & c_{15} \\ c_{13} & c_{33} & c_{35} \\ c_{15} & c_{35} & c_{55} \end{bmatrix} \begin{Bmatrix} \varepsilon_x \\ \varepsilon_z \\ \gamma_{xz} \end{Bmatrix} \tag{5.1.3}$$

while the other fundamental equations remain as Eqs. (4.1.5) to (4.1.8). The fundamental equations and boundary conditions of plane strain problems are the same as those of plane stress problems. The only difference lies on the interpretation of elastic constants. Likewise, the two kinds of problems are treated as similar cases and they are generally referred as the plane anisotropic elasticity problems. The fundamental equations and boundary conditions are Eqs. (5.1.1) or (5.1.2) and Eqs. (4.1.5) to (4.1.8), respectively.

5.2. Symplectic Solution Methodology for Anisotropic Elasticity Problems

Consider a rectangular domain as illustrated in Fig. 5.1

$$V: \quad 0 \leq z \leq l, \quad -h \leq x \leq h \tag{5.2.1}$$

where l is relatively larger. The forces on side boundary surfaces ($x = \pm h$) are

$$\sigma_x = \overline{F}_{x1}(z), \quad \tau_{xz} = \overline{F}_{z1}(z) \quad \text{at } x = -h \tag{5.2.2a}$$

$$\sigma_x = \overline{F}_{x2}(z), \quad \tau_{xz} = \overline{F}_{z2}(z) \quad \text{at } x = h \tag{5.2.2b}$$

And there are body forces F_x and F_z acting along the x- and z-directions, respectively, in the domain.

There are the corresponding boundary conditions at the two ends $z = 0, l$. For instance, the boundary conditions for specified displacements are

$$w = \overline{w}, \quad u = \overline{u} \quad \text{at } z = 0 \quad \text{or} \quad l \tag{5.2.3a}$$

while the boundary conditions for specified forces are

$$\sigma_z = \overline{\sigma}_z, \quad \tau_{xz} = \overline{\tau}_{xz} \quad \text{at } z = 0 \quad \text{or} \quad l \tag{5.2.3b}$$

The fundamental equations and boundary conditions of anisotropic elasticity problems mentioned above can be derived from the Hellinger–Reissner

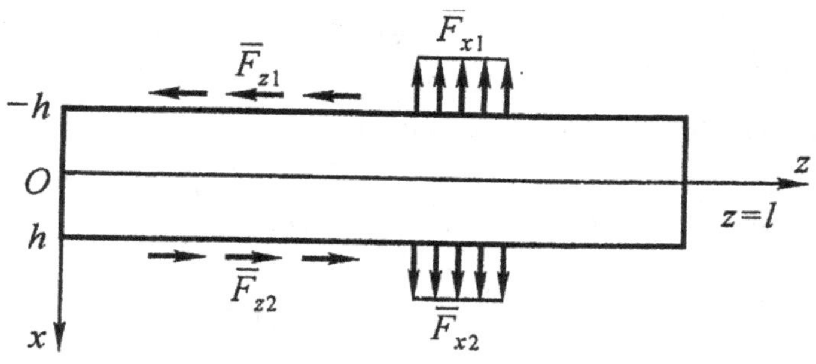

Fig. 5.1. Plane problem in rectangular domain.

variational principle

$$\delta\left\{\int_0^l\int_{-h}^h\left[\sigma_x\frac{\partial u}{\partial x}+\sigma_z\frac{\partial w}{\partial z}+\tau_{xz}\left(\frac{\partial u}{\partial z}+\frac{\partial w}{\partial x}\right)\right.\right.$$

$$\left.\left.-v_{\mathrm{c}}-F_xu-F_zw\right]\mathrm{d}x\mathrm{d}z+v_{\mathrm{e}}^1+v_{\mathrm{e}}^2\right\}=0 \tag{5.2.4}$$

where

$$v_{\mathrm{c}}=\frac{1}{2}\left(s_{11}\sigma_x^2+s_{33}\sigma_z^2+s_{55}\tau_{xz}^2\right)+s_{13}\sigma_x\sigma_z+s_{15}\sigma_x\tau_{xz}+s_{35}\sigma_z\tau_{xz} \tag{5.2.5}$$

$$v_{\mathrm{e}}^1=-\int_0^l\left\{\left[\overline{F}_{x2}u+\overline{F}_{z2}w\right]_{x=h}-\left[\overline{F}_{x1}u+\overline{F}_{z1}w\right]_{x=-h}\right\}\mathrm{d}z \tag{5.2.6}$$

$$v_{\mathrm{e}}^2=\begin{cases}-\displaystyle\int_{-h}^h\left[\sigma_z\left(w-\overline{w}\right)+\tau_{xz}\left(u-\overline{u}\right)\right]_{z=0}^l\mathrm{d}x & \text{on }\Gamma_u\\ -\displaystyle\int_{-h}^h\left[\overline{\sigma}_zw+\overline{\tau}_{xz}u\right]_{z=0}^l\mathrm{d}x & \text{on }\Gamma_\sigma\end{cases} \tag{5.2.7}$$

We are now ready to derive the Hamiltonian system. First we treat the longitudinal z-coordinate as the time coordinate and indicate differentiation with respect to z using a dot, i.e. $(\dot{}) = \partial/\partial z$. Next, the transverse stress σ_x should be eliminated. The variation of Eq. (5.2.4) with respect to σ_x is

$$\sigma_x=\frac{1}{s_{11}}\left(\frac{\partial u}{\partial x}-s_{13}\sigma-s_{15}\tau\right) \tag{5.2.8}$$

where σ_z, τ_{xz} are briefly denoted as σ, τ. Substituting Eq. (5.2.8) into Eq. (5.2.4) and eliminating σ_x, yield the mixed energy variational principle of Hamiltonian system

$$\delta\left\{\int_0^l\int_{-h}^h\left[\sigma\dot{w}+\tau\dot{u}-\mathscr{H}(w,u,\sigma,\tau)\right]\mathrm{d}x\mathrm{d}z+v_{\mathrm{e}}^1+v_{\mathrm{e}}^2\right\}=0 \tag{5.2.9}$$

where the Hamiltonian density function is

$$\mathscr{H}=\frac{1}{2s_{11}d}(b_{55}\sigma^2+b_{33}\tau^2-2b_{35}\sigma\tau)-\frac{1}{2s_{11}}\left(\frac{\partial u}{\partial x}\right)^2$$

$$-\tau\frac{\partial w}{\partial x}+\frac{1}{s_{11}}\frac{\partial u}{\partial x}(s_{13}\sigma+s_{15}\tau)+F_xu+F_zw \tag{5.2.10}$$

Obviously, the dual variables of displacements w, u are stresses σ, τ.

The variation of Eq. (5.2.9) yields the Hamiltonian dual equations as

$$\dot{\boldsymbol{v}} = \boldsymbol{H}\boldsymbol{v} + \boldsymbol{Q} \tag{5.2.11}$$

where $\boldsymbol{v}$ is the full state vector

$$\boldsymbol{v} = \{w, u, \sigma, \tau\}^{\mathrm{T}} \tag{5.2.12}$$

$\boldsymbol{H}$ is an operator matrix

$$\boldsymbol{H} = \begin{bmatrix} 0 & \dfrac{s_{13}}{s_{11}}\dfrac{\partial}{\partial x} & \dfrac{b_{55}}{s_{11}d} & -\dfrac{b_{35}}{s_{11}d} \\ -\dfrac{\partial}{\partial x} & \dfrac{s_{15}}{s_{11}}\dfrac{\partial}{\partial x} & -\dfrac{b_{35}}{s_{11}d} & \dfrac{b_{33}}{s_{11}d} \\ 0 & 0 & 0 & -\dfrac{\partial}{\partial x} \\ 0 & -\dfrac{1}{s_{11}}\dfrac{\partial^2}{\partial x^2} & \dfrac{s_{13}}{s_{11}}\dfrac{\partial}{\partial x} & \dfrac{s_{15}}{s_{11}}\dfrac{\partial}{\partial x} \end{bmatrix} \tag{5.2.13}$$

and $\boldsymbol{Q}$ is an inhomogeneous term related to body forces.

$$\boldsymbol{Q} = \{0, \quad 0, \quad -F_z, \quad -F_x\}^{\mathrm{T}} \tag{5.2.14}$$

The boundary conditions (5.2.2) on the two side boundary surfaces can be expressed as

$$\frac{1}{s_{11}}\left(\frac{\partial u}{\partial x} - s_{13}\sigma - s_{15}\tau\right) = \overline{F}_{x1}; \quad \tau = \overline{F}_{z1}, \quad \text{for } x = -h \tag{5.2.15a}$$

$$\frac{1}{s_{11}}\left(\frac{\partial u}{\partial x} - s_{13}\sigma - s_{15}\tau\right) = \overline{F}_{x2}; \quad \tau = \overline{F}_{z2}, \quad \text{for } x = h \tag{5.2.15b}$$

To discuss the property of operator matrix $\boldsymbol{H}$, we denote

$$\begin{aligned} \langle \boldsymbol{v}_1, \boldsymbol{v}_2 \rangle &\overset{\text{def}}{=} \int_{-h}^{h} \boldsymbol{v}_1^{\mathrm{T}} \boldsymbol{J} \boldsymbol{v}_2 \mathrm{d}x \\ &= \int_{-h}^{h} (w_1\sigma_2 + u_1\tau_2 - \sigma_1 w_2 - \tau_1 u_2)\mathrm{d}x \end{aligned} \tag{5.2.16}$$

where $\boldsymbol{J}$ is a unit symplectic matrix

$$\boldsymbol{J} = \begin{bmatrix} \boldsymbol{0} & \boldsymbol{I}_2 \\ -\boldsymbol{I}_2 & \boldsymbol{0} \end{bmatrix} \tag{5.2.17}$$

Obviously, Eq. (5.2.16) satisfies the four conditions (1.3.2) of symplectic inner product. Hence, the full state vector $\boldsymbol{v}$ forms a symplectic space in accordance with the definition (5.2.16) of symplectic inner product.

First, we discuss the corresponding homogeneous linear differential equations of Eq. (5.2.11)

$$\dot{\boldsymbol{v}} = \boldsymbol{H}\boldsymbol{v} \tag{5.2.18}$$

and the homogeneous boundary conditions of Eq. (5.2.15) on the side boundary surfaces ($x = \pm h$)

$$\frac{1}{s_{11}}\left(\frac{\partial u}{\partial x} - s_{13}\sigma\right) = 0; \quad \tau = 0 \quad \text{at } x = \pm h \tag{5.2.19}$$

It is obvious via integration by parts that if $\boldsymbol{v}_1$, $\boldsymbol{v}_2$ are continuously differentiable full state vectors satisfying the boundary conditions (5.2.19), we then have the identity

$$\langle \boldsymbol{v}_1, \boldsymbol{H}\boldsymbol{v}_2 \rangle = \langle \boldsymbol{v}_2, \boldsymbol{H}\boldsymbol{v}_1 \rangle \tag{5.2.20}$$

It states that the operator matrix $\boldsymbol{H}$ is the Hamiltonian operator matrix in the symplectic space.

Thus we have transformed the plane anisotropic elasticity problems into Hamiltonian system which can then be solved using the conventional procedure of Hamiltonian system.

Similar to the plane isotropic problems, the method of separation of variables can be applied to solve the system of Eqs. (5.2.18), i.e. assume that

$$\boldsymbol{v}(z, x) = \xi(z)\boldsymbol{\psi}(x) = \mathrm{e}^{\mu z}\boldsymbol{\psi}(x) \tag{5.2.21}$$

where μ is an eigenvalue and $\boldsymbol{\psi}(x)$ is the eigenvector function. The eigenvalue equation is

$$\boldsymbol{H}\boldsymbol{\psi}(x) = \mu\boldsymbol{\psi}(x) \tag{5.2.22}$$

which must satisfy the homogeneous boundary conditions (5.2.19) on the side boundary surfaces ($x = \pm h$).

As $\boldsymbol{H}$ is a Hamiltonian operator matrix, the eigenvector functions are adjoint symplectic orthogonal. Once the eigenvalue and eigenvector functions are determined, the expansion theorem can be applied to solve this problem.

5.3. Eigen-Solutions of Zero Eigenvalue

For problems with free homogeneous boundary conditions (5.2.19) on the side boundary surfaces ($x = \pm h$), there exist the basic eigen-solutions[1] and the Jordan form eigen-solutions of zero eigenvalue. Such solutions in elasticity have significant physical interpretation. To determine the eigen-solutions of zero eigenvalue, we solve the following differential equation

$$\boldsymbol{H}\boldsymbol{\psi}(x) = \mathbf{0} \tag{5.3.1}$$

which can be expanded into

$$\left.\begin{array}{lllll} 0 & +\dfrac{s_{13}}{s_{11}}\dfrac{\partial u}{\partial x} & +\dfrac{b_{55}}{s_{11}d}\sigma & -\dfrac{b_{35}}{s_{11}d}\tau = 0 \\ -\dfrac{\partial w}{\partial x} & +\dfrac{s_{15}}{s_{11}}\dfrac{\partial u}{\partial x} & -\dfrac{b_{35}}{s_{11}d}\sigma & +\dfrac{b_{33}}{s_{11}d}\tau = 0 \\ 0 & +0 & +0 & -\dfrac{\partial \tau}{\partial x} = 0 \\ 0 & -\dfrac{1}{s_{11}}\dfrac{\partial^2 u}{\partial x^2} & +\dfrac{s_{13}}{s_{11}}\dfrac{\partial \sigma}{\partial x} & +\dfrac{s_{15}}{s_{11}}\dfrac{\partial \tau}{\partial x} = 0 \end{array}\right\} \tag{5.3.2}$$

The eigen-solutions should also satisfy the homogeneous boundary conditions (5.2.19) on the side boundary surfaces.

Solving Eq. (5.3.2) and substituting into the boundary conditions (5.2.19) yield the basic eigen-solutions of zero eigenvalue as

$$\boldsymbol{\psi}_{0f}^{(0)} = \{1, \quad 0, \quad 0, \quad 0\}^{\mathrm{T}} \tag{5.3.3}$$

$$\boldsymbol{\psi}_{0s}^{(0)} = \{0, \quad 1, \quad 0, \quad 0\}^{\mathrm{T}} \tag{5.3.4}$$

It shows that there are two chains which are denoted by subscripts f and s, respectively. These two eigenvectors are the solutions of the original equation (5.2.18) and boundary conditions (5.2.19)

$$\boldsymbol{v}_{0f}^{(0)} = \boldsymbol{\psi}_{0f}^{(0)}, \quad \boldsymbol{v}_{0s}^{(0)} = \boldsymbol{\psi}_{0s}^{(0)} \tag{5.3.5}$$

which are physically interpreted as the rigid body translations along the z- and x-directions, respectively.

Besides the basic eigenvectors, the remaining eigenvectors are all Jordan form eigenvectors which form two chains. To obtain the first-order Jordan form eigen-solution on chain one, we solve the following equation with homogeneous boundary conditions (5.2.19)

$$\boldsymbol{H}\boldsymbol{\psi}_{0f}^{(1)} = \boldsymbol{\psi}_{0f}^{(0)} \tag{5.3.6}$$

The solution is

$$\begin{aligned}\boldsymbol{\psi}_{0f}^{(1)} &= \left\{w_{0f}^{(1)} \quad u_{0f}^{(1)} \quad \sigma_{0f}^{(1)} \quad \tau_{0f}^{(1)}\right\}^{\mathrm{T}} \\ &= \left\{\frac{s_{35}}{s_{33}}x, \quad \frac{s_{13}}{s_{33}}x, \quad \frac{1}{s_{33}}, \quad 0\right\}^{\mathrm{T}}\end{aligned} \tag{5.3.7}$$

The first-order Jordan form eigenvector $\boldsymbol{\psi}_{0f}^{(1)}$ is not the solution of the original equation which, however, can be used to construct solution to the original equation by

$$\boldsymbol{v}_{0f}^{(1)} = \boldsymbol{\psi}_{0f}^{(1)} + z\boldsymbol{\psi}_{0f}^{(0)} \tag{5.3.8}$$

The components of displacement and stress are

$$w = z + \frac{s_{35}}{s_{33}}x, \quad u = \frac{s_{13}}{s_{33}}x, \quad \sigma = \frac{1}{s_{33}}, \quad \tau = 0 \tag{5.3.9}$$

This solution is physically interpreted as the simple axial tension. As the material is anisotropic and in general $s_{35} \neq 0$, the deformed cross section remains a plane but is not parallel to the original cross section.

It is easy to verify that $\boldsymbol{\psi}_{0f}^{(1)}$ and $\boldsymbol{\psi}_{0s}^{(0)}$ are symplectic orthogonal while $\boldsymbol{\psi}_{0f}^{(1)}$ and $\boldsymbol{\psi}_{0f}^{(0)}$ are mutually symplectic adjoint

$$k_1 = \left\langle \boldsymbol{\psi}_{0f}^{(0)}, \ \boldsymbol{\psi}_{0f}^{(1)} \right\rangle = \frac{2h}{s_{33}} \neq 0 \tag{5.3.10}$$

Hence there is no second-order Jordan form eigen-solution on this Jordan chain and the chain is terminated. It should be noted here that k_1 has a special physical meaning, i.e. the extensional rigidity of cross section.

Similarly, the first-order Jordan form eigen-solution on chain two can be obtained by solving the following equation with homogeneous boundary conditions (5.2.19)

$$\boldsymbol{H}\boldsymbol{\psi}_{0s}^{(1)} = \boldsymbol{\psi}_{0s}^{(0)} \tag{5.3.11}$$

The solution is

$$\boldsymbol{\psi}_{0s}^{(1)} = \{-x, \quad 0, \quad 0, \quad 0\}^{\mathrm{T}} \tag{5.3.12}$$

Likewise, the first-order Jordan form eigenvector $\boldsymbol{\psi}_{0s}^{(1)}$ is not the solution of the original equation which, however, can be use to construct solution to the original equation by

$$\boldsymbol{v}_{0s}^{(1)} = \boldsymbol{\psi}_{0s}^{(1)} + z\boldsymbol{\psi}_{0s}^{(0)} \tag{5.3.13}$$

The components of displacement and stress are

$$w = -x, \quad u = z, \quad \sigma = 0, \quad \tau = 0 \tag{5.3.14}$$

The solution is physically interpreted as the in-plane rigid body rotation.

Through direct examination, we know that $\boldsymbol{\psi}_{0s}^{(1)}$ is symplectic orthogonal to both $\boldsymbol{\psi}_{0f}^{(0)}$ and $\boldsymbol{\psi}_{0s}^{(0)}$. Hence, there exists the second-order Jordan form eigen-solution which can be obtained by solving the following equation with homogeneous boundary conditions (5.2.19)

$$\boldsymbol{H}\boldsymbol{\psi}_{0s}^{(2)} = \boldsymbol{\psi}_{0s}^{(1)} \tag{5.3.15}$$

The solution is

$$\begin{aligned}\boldsymbol{\psi}_{0s}^{(2)} &= \{w_{0s}^{(2)}, \quad u_{0s}^{(2)}, \quad \sigma_{0s}^{(2)}, \quad \tau_{0s}^{(2)}\}^{\mathrm{T}} \\ &= \left\{-\frac{s_{35}}{2s_{33}}x^2 + g_1, \quad -\frac{s_{13}}{2s_{33}}x^2 + g_2, \quad -\frac{1}{s_{33}}x, \quad 0\right\}^{\mathrm{T}}\end{aligned} \tag{5.3.16}$$

Hence, the solution of the original equations can be constructed as

$$\boldsymbol{v}_{0s}^{(2)} = \boldsymbol{\psi}_{0s}^{(2)} + z\boldsymbol{\psi}_{0s}^{(1)} + \frac{1}{2}z^2\boldsymbol{\psi}_{0s}^{(0)} \tag{5.3.17}$$

The components of displacement and stress are

$$\left.\begin{aligned} w &= -\frac{s_{35}}{2s_{33}}x^2 + g_1 - xz \\ u &= -\frac{s_{13}}{2s_{33}}x^2 + g_2 + \frac{1}{2}z^2 \\ \sigma &= -\frac{1}{s_{33}}x \\ \tau &= 0 \end{aligned}\right\} \tag{5.3.18}$$

From the cross-sectional stress distribution, we know that there are only normal stresses but not shear stresses on the cross section. The normal stresses result in a constant force couple. Hence, Eq. (5.3.18) represents the solution of pure bending.

By examining, we know that $\boldsymbol{\psi}_{0s}^{(2)}$ is symplectic orthogonal to both $\boldsymbol{\psi}_{0f}^{(0)}$ and $\boldsymbol{\psi}_{0s}^{(0)}$ and, therefore, there exists the third-order Jordan form eigensolution. Solving the following equation with homogeneous boundary conditions (5.2.19)

$$\boldsymbol{H}\boldsymbol{\psi}_{0s}^{(3)} = \boldsymbol{\psi}_{0s}^{(2)} \tag{5.3.19}$$

yields the third-order Jordan form solution as

$$\boldsymbol{\psi}_{0s}^{(3)} = \left\{ w_{0s}^{(3)}, \quad u_{0s}^{(3)}, \quad \sigma_{0s}^{(3)}, \quad \tau_{0s}^{(3)} \right\}^{\mathrm{T}}$$
$$= \left\{ \begin{array}{l} \dfrac{(s_{33}s_{55} + s_{13}s_{33} - 2s_{35}^2)x^3 + (2s_{35}^2 - 3s_{33}s_{55})h^2 x}{6s_{33}^2} - g_2 x \\ \dfrac{(s_{15}s_{33} - 2s_{13}s_{35})x^3 + (2s_{13}s_{35} - 3s_{15}s_{33})h^2 x}{6s_{33}^2} \\ \dfrac{s_{35}}{3s_{33}^2}(h^2 - 3x^2) \\ \dfrac{1}{2s_{33}}(x^2 - h^2) \end{array} \right\} \tag{5.3.20}$$

Hence, the solution of the original equations can be constructed as

$$\boldsymbol{v}_{0s}^{(3)} = \boldsymbol{\psi}_{0s}^{(3)} + z\boldsymbol{\psi}_{0s}^{(2)} + \frac{1}{2}z^2\boldsymbol{\psi}_{0s}^{(1)} + \frac{1}{6}z^3\boldsymbol{\psi}_{0s}^{(0)} \tag{5.3.21}$$

The components of displacement and stress are

$$\left. \begin{aligned} w &= w_{0s}^{(3)} + z w_{0s}^{(2)} - \frac{1}{2}xz^2 \\ u &= u_{0s}^{(3)} + z u_{0s}^{(2)} + \frac{1}{6}z^3 \\ \sigma &= \sigma_{0s}^{(3)} + z\sigma_{0s}^{(2)} \\ \tau &= \tau_{0s}^{(3)} \end{aligned} \right\} \tag{5.3.22}$$

From the cross-sectional stress distribution, we know that the shear forces are constant. Hence, Eq. (5.3.22) represents the solution of constant shear force bending.

As $\psi_{0s}^{(3)}$ and $\psi_{0f}^{(0)}$ are symplectic orthogonal while $\psi_{0s}^{(3)}$ and $\psi_{0s}^{(0)}$ are mutually symplectic adjoint

$$k_2 = \langle \psi_{0s}^{(3)}, \psi_{0s}^{(0)} \rangle = -\langle \psi_{0s}^{(2)}, \psi_{0s}^{(1)} \rangle = \frac{2h^3}{3s_{33}} \neq 0 \tag{5.3.23}$$

Hence this Jordan chain is terminated. It should noted here that k_2 has a special physical meaning, i.e. the flexural rigidity of cross section.

We have obtained all six eigen-solutions of zero eigenvalue. By choosing appropriate constants g_1 and g_2

$$g_1 = -\frac{s_{35}h^2}{6s_{33}}; \quad g_2 = \frac{(3s_{13}s_{33} + 4s_{35}^2 - 6s_{33}s_{55})h^2}{30s_{33}^2} \tag{5.3.24}$$

we can ensure the symplectic orthogonality of $\psi_{0f}^{(1)}$ and $\psi_{0s}^{(2)}$, as well as $\psi_{0s}^{(2)}$ and $\psi_{0s}^{(3)}$. The other eigenvectors satisfy the adjoint symplectic orthonormal relation. Table 5.1 shows the adjoint symplectic orthonormal relation between the six eigen-solutions where 0 denotes the two quantities are naturally symplectic orthogonal; g_1 or g_2 denotes that symplectic orthogonality can be fulfilled by an appropriate choice of g_1 or g_2; and * denotes symplectic adjoint relation.

Here we again observe the adjoint solutions of axial translation, transverse translation and rigid body rotation are, respectively, the solutions of simple tension, constant shear force bending and pure bending. These six eigen-solutions of zero eigenvalue constitute a complete symplectic subspace and they are the basic solutions of the Saint–Venant problem.

Table 5.1. The adjoint symplectic orthonormal relation between eigen-solutions of zero eigenvalue for plane anisotropic elasticity problems.

	$\psi_{0f}^{(0)}$	$\psi_{0f}^{(1)}$	$\psi_{0s}^{(0)}$	$\psi_{0s}^{(1)}$	$\psi_{0s}^{(2)}$	$\psi_{0s}^{(3)}$
$\psi_{0f}^{(0)}$	0	*	0	0	0	0
$\psi_{0f}^{(1)}$		0	0	0	g_1	0
$\psi_{0s}^{(0)}$			0	0	0	*
$\psi_{0s}^{(1)}$				0	*	0
$\psi_{0s}^{(2)}$					0	g_2
$\psi_{0s}^{(3)}$						0

5.4. Analytical Solutions of Saint–Venant Problems

Consider a plane strip domain ($h \ll l$) with transverse and longitudinal loads. The Saint–Venant principle is always applied to such problems where the effect of self-equilibrium system of forces at two ends is localized in the vicinity of the two ends. It implies that the eigen-solutions of nonzero eigenvalues can be neglected. Only eigen-solutions of zero eigenvalue are adopted in the expansion theorem, as

$$\boldsymbol{v} = a_1\boldsymbol{\psi}_{0f}^{(0)} + a_2\boldsymbol{\psi}_{0f}^{(1)} + a_3\boldsymbol{\psi}_{0s}^{(0)} + a_4\boldsymbol{\psi}_{0s}^{(1)} + a_5\boldsymbol{\psi}_{0s}^{(2)} + a_6\boldsymbol{\psi}_{0s}^{(3)} \tag{5.4.1}$$

or

$$\left.\begin{aligned} w &= a_1 + a_2 w_{0f}^{(1)} - a_4 x + a_5 w_{0s}^{(2)} + a_6 w_{0s}^{(3)} \\ u &= a_2 u_{0f}^{(1)} + a_3 + a_5 u_{0s}^{(2)} + a_6 u_{0s}^{(3)} \\ \sigma &= a_2 \sigma_{0f}^{(1)} + a_5 \sigma_{0s}^{(2)} + a_6 \sigma_{0s}^{(3)} \\ \tau &= a_6 \tau_{0s}^{(3)} \end{aligned}\right\} \tag{5.4.1$'$}$$

where $a_i (i = 1, 2, \ldots, 6)$ are undetermined functions. Substituting Eq. (5.4.1) into the mixed energy variational principle of Hamiltonian system (5.2.9), after simplification and rearrangement, yields

$$\begin{aligned} \delta\Bigg\{ \int_0^l \Bigg(& k_1 a_2 \dot{a}_1 + k_2 a_5 \dot{a}_4 - k_2 a_6 \dot{a}_3 - \frac{1}{2} k_1 a_2^2 - \frac{1}{2} k_2 a_5^2 \\ & + k_2 a_4 a_6 - F_{\mathrm{N}} a_1 - W a_2 - F_{\mathrm{S}} a_3 - M a_4 - \theta a_5 - U a_6 \Bigg) \mathrm{d}z \\ & + (k_3 a_2 a_5 + k_4 a_5 a_6)\big|_{z=0}^{l} + \tilde{\boldsymbol{v}}_{\mathrm{e}}^2 \Bigg\} = 0 \end{aligned} \tag{5.4.2}$$

where

$$k_3 = -\frac{2 s_{35} h^3}{3 s_{33}^2}; \quad k_4 = \frac{2 s_{55} h^5}{15 s_{33}^2} \tag{5.4.3}$$

$$F_{\mathrm{N}} = \int_{-h}^{h} F_z \mathrm{d}x + \overline{F}_{z2} - \overline{F}_{z1} \quad \text{(axial force)} \tag{5.4.4a}$$

$$\begin{aligned} W = & \int_{-h}^{h} [w_{0f}^{(1)}(x) F_z + u_{0f}^{(1)}(x) F_x] \mathrm{d}x \\ & + w_{0f}^{(1)}(h)(\overline{F}_{z2} + \overline{F}_{z1}) + u_{0f}^{(1)}(h)(\overline{F}_{x2} + \overline{F}_{x1}) \end{aligned} \tag{5.4.4b}$$

$$F_S = \int_{-h}^{h} F_x \mathrm{d}x + \overline{F}_{x2} - \overline{F}_{x1} \quad \text{(shear force)} \tag{5.4.4c}$$

$$M = \int_{-h}^{h} (-x) F_z \mathrm{d}x - (\overline{F}_{z2} + \overline{F}_{z1})h \quad \text{(bending moment)} \tag{5.4.4d}$$

$$\begin{aligned}\theta = & \int_{-h}^{h} [w_{0s}^{(2)}(x)F_z + u_{0s}^{(2)}(x)F_x]\mathrm{d}x \\ & + w_{0s}^{(2)}(h)(\overline{F}_{z2} - \overline{F}_{z1}) + u_{0s}^{(2)}(h)(\overline{F}_{x2} - \overline{F}_{x1})\end{aligned} \tag{5.4.4e}$$

$$\begin{aligned}U = & \int_{-h}^{h} [w_{0s}^{(3)}(x)F_z + u_{0s}^{(3)}(x)F_x]\mathrm{d}x \\ & + w_{0s}^{(3)}(h)(\overline{F}_{z2} + \overline{F}_{z1}) + u_{0s}^{(3)}(h)(\overline{F}_{x2} + \overline{F}_{x1})\end{aligned} \tag{5.4.4f}$$

The boundary term $\boldsymbol{v}_{\mathrm{e}}^2$ is simplified to $\tilde{\boldsymbol{v}}_{\mathrm{e}}^2$. As the boundary conditions for specified forces, a_2, a_5, a_6 are not arbitrary variational quantities, the related boundary terms can be neglected and we have

$$\tilde{\boldsymbol{v}}_{\mathrm{e}}^2 = -[\overline{F}_{\mathrm{N}} a_1 + \overline{F}_{\mathrm{S}} a_3 + \overline{M} a_4]_{z=0}^{l} \tag{5.4.5}$$

where $\overline{F}_{\mathrm{N}}$, $\overline{F}_{\mathrm{S}}$ and $\overline{M}$ are respectively the known axial force, shear force and bending moment at both ends ($z = 0$ or l).

For specified displacements, the boundary conditions are

$$\begin{aligned}\tilde{\boldsymbol{v}}_{\mathrm{e}}^2 = & -[k_1 a_1 a_2 + k_2 a_4 a_5 + 2k_3 a_2 a_5 + 2k_4 a_5 a_6 \\ & - k_2 a_3 a_6 - \overline{W} a_2 - \overline{\theta} a_5 - \overline{U} a_6]_{z=0}^{l}\end{aligned} \tag{5.4.6}$$

where

$$\overline{W} = \int_{-h}^{h} \overline{w} \sigma_{0f}^{(1)} \mathrm{d}x \tag{5.4.7a}$$

$$\overline{\theta} = \int_{-h}^{h} \overline{w} \sigma_{0s}^{(2)} \mathrm{d}x \tag{5.4.7b}$$

$$\overline{U} = \int_{-h}^{h} (\overline{w} \sigma_{0s}^{(3)} + \overline{u} \tau_{0s}^{(3)}) \mathrm{d}x \tag{5.4.7c}$$

The variation of Eq. (5.4.2) yields the equations as

$$\left.\begin{aligned} k_1 \dot{a}_2 + F_{\mathrm{N}} = 0 \\ k_1 \dot{a}_1 - k_1 a_2 - W = 0 \end{aligned}\right\} \tag{5.4.8}$$

$$\left.\begin{aligned} k_2\dot{a}_6 - F_{\mathrm{S}} &= 0 \\ k_2\dot{a}_5 - k_2 a_6 + M &= 0 \\ k_2\dot{a}_4 - k_2 a_5 - \theta &= 0 \\ k_2\dot{a}_3 - k_2 a_4 + U &= 0 \end{aligned}\right\} \tag{5.4.9}$$

The six equations above are differential equations in the domain for Saint–Venant problems considering body forces and boundary conditions. The general solutions can be obtained by ordinary integration. In general, equilibrium equations in the domain and boundary conditions cannot be strictly satisfied. The satisfaction only implies the equilibrium of cross section. By using a particular solution to treat in advance the inhomogeneous terms in the domain and at both side bounding surfaces ($x = \pm h$), the problem can be transformed into the corresponding homogeneous equations. In this way, the solution strictly satisfies all differential equations in the domain and boundary conditions at both side bounding surfaces ($x = \pm h$). Hence, the exact elasticity solution can be established.

The integration constants resulted from integrating Eqs. (5.4.8) and (5.4.9) can be determined by the boundary conditions at both ends ($z = 0$ or l). The boundary conditions for cases with specified forces at the ends ($z = 0$ or l) are

$$\left.\begin{aligned} k_1 a_2 &= \overline{F}_{\mathrm{N}} \\ k_2 a_6 &= -\overline{F}_{\mathrm{S}} \\ k_2 a_5 &= \overline{M} \end{aligned}\right\} \quad \text{at } z = 0 \quad \text{or} \quad l \tag{5.4.10}$$

Obviously, these conditions can be physically interpreted as the equality of axial force, shear force and bending moment to the specified values, respectively.

The boundary conditions for cases with specified displacements at two ends ($z = 0$ or l) are

$$\left.\begin{aligned} k_1 a_1 + k_3 a_5 &= \overline{W} \\ k_2 a_4 + k_3 a_2 + k_4 a_6 &= \overline{\theta} \\ k_2 a_3 - k_4 a_5 &= -\overline{U} \end{aligned}\right\} \quad \text{at } z = 0 \quad \text{or} \quad l \tag{5.4.11}$$

Obviously, these conditions can be physically interpreted as the equality of equivalent displacement and angle of rotation to the specified values, respectively.

For clamped boundary conditions, $\overline{W} = \overline{\theta} = \overline{U} = 0$ in the expression above. The usual fixed boundary conditions of Saint–Venant problems are

$$w = u = \frac{\partial u}{\partial z} = 0 \quad \text{at } x = 0, \quad z = 0 \quad \text{or} \quad l \tag{5.4.12}$$

or

$$w = u = \frac{\partial w}{\partial x} = 0 \quad \text{at } x = 0, \quad z = 0 \quad \text{or} \quad l \tag{5.4.13}$$

Substitute the general solution of Eqs. (5.4.8) and (5.4.9) into the corresponding boundary conditions yields the analytical solution of Saint–Venant problems. The solutions of two classical problems are presented as follows.

1. Solution of eccentric tension

Let a cantilever be fixed at $z = 0$ with an eccentric axial force F acting at the other end $z = l$ with eccentricity e. Solving Eqs. (5.4.8) and (5.4.9) and substituting into the corresponding boundary conditions (5.4.10) yield

$$\left.\begin{array}{l} a_1 = \dfrac{Fz}{k_1} + f_1; \quad a_2 = \dfrac{F}{k_1}; \quad a_3 = -\dfrac{Fez^2}{2k_2} + f_2 z + f_3 \\ a_4 = -\dfrac{Fez}{k_2} + f_2; \quad a_5 = -\dfrac{Fe}{k_2}; \quad a_6 = 0 \end{array}\right\} \tag{5.4.14}$$

where constants f_1, f_2, f_3 represent translation of a rigid body. The constants can be determined by the clamped boundary conditions (5.4.11), (5.4.12) or (5.4.13). The stress field corresponding to the solution (5.4.14) is

$$\sigma_z = \frac{F}{2h} + \frac{3Fex}{2h^3}; \quad \sigma_x = \tau_{xz} = 0 \tag{5.4.15}$$

The stress distribution is the same as that of an isotropic rod. The deflection of the axis of beam is

$$u(0, z) = -\frac{3Fes_{33}}{4h^3} z^2 + \begin{cases} \dfrac{F[10zs_{33}s_{35} - e(4s_{35}^2 + 3s_{13}s_{33})]}{20hs_{33}} & \text{for boundary condition (5.4.11)} \\ 0 & \text{for boundary condition (5.4.12)} \\ \dfrac{Fzs_{35}}{2h} & \text{for boundary condition (5.4.13)} \end{cases} \tag{5.4.16}$$

The first part of the expression above is the solution of classical mechanics of material and it is the same as that of isotropic beam. The second part (rigid body translation) is resulted from the different assumptions of fixed end, which are high-order small quantities as compared with the first part. It shows that the different assumptions of fixed end have no effect on the stress distribution and deformation within the domain. They only cause small differences in rigid body translation.

2. Solution of cantilever with uniformly distributed load

Let an anisotropic cantilever with uniformly distributed load q acting on one side boundary surface $x = -h$. Assume that $z = 0$ is clamped while $z = l$ is free. First of all we solve the Jordan form $\boldsymbol{H}\boldsymbol{\psi} = \boldsymbol{\psi}_{0s}^{(3)}$ with inhomogeneous boundary conditions to give a particular solution of the original problem, as

$$\left.\begin{aligned}\tilde{\sigma}_z &= \frac{qx^3}{h^3}\left(\frac{s_{35}^2}{s_{33}^2} - \frac{2s_{13}+s_{55}}{4s_{33}}\right) + \frac{qx}{h}\left[\frac{9\,(2s_{13}+s_{55})}{20s_{33}} - \frac{4s_{35}^2}{5s_{33}^2}\right] \\ &\quad - \frac{qs_{13}}{2s_{33}} - \frac{qs_{35}}{2h^3s_{33}}z\left(h^2-3x^2\right) + \frac{3q}{4h^3}xz^2 \\ \tilde{\sigma}_x &= \frac{q}{4}\left(2 - 3\frac{x}{h} + \frac{x^3}{h^3}\right) \\ \tilde{\tau}_{xz} &= \frac{qs_{35}}{2h^3s_{33}}x\left(h^2-x^2\right) + \frac{3q}{4h^3}z\left(h^2-x^2\right)\end{aligned}\right\} \quad (5.4.17)$$

We further transform the original equation into a homogeneous one, and then solve the corresponding homogeneous equation. Finally the stress field in the domain is

$$\left.\begin{aligned}\sigma_z &= \frac{3q}{4h^3}x(l-z)^2 + \frac{qs_{35}}{2h^3s_{33}}(l-z)(h^2-3x^2) \\ &\quad - \left[q\left(\frac{s_{35}^2}{s_{33}^2} - \frac{2s_{13}+s_{55}}{4s_{33}}\right)\left(\frac{3x}{5h} - \frac{x^3}{h^3}\right)\right] \\ \sigma_x &= \left[\frac{q}{4}\left(2 - 3\frac{x}{h} + \frac{x^3}{h^3}\right)\right] \\ \tau_{xz} &= \frac{3q}{4h^3}(z-l)(h^2-x^2) + \left[\frac{qs_{35}}{2h^3s_{33}}x(h^2-x^2)\right]\end{aligned}\right\} \quad (5.4.18)$$

If the distribution of the external forces at the free end (form a equilibrium system of forces) is the same as Eq. (5.4.18), then Eq. (5.4.18) is the exact elasticity solution of the problem.

If we do not treat the inhomogeneous terms by applying Eq. (5.4.17) in advance but rather solve Eqs. (5.4.8) and (5.4.9) directly, we obtain an approximate solution, i.e. omit all terms in square brackets in Eq. (5.4.18). Obviously, for problems in plane strip domain ($h \ll l$), the terms in square brackets are higher-order small quantities. It clearly shows that the rational method by expanding eigenvectors of zero eigenvalue for solving the Saint–Venant problems is very effective and practical.

As only the eigen-solutions of zero eigenvalue are applied, the boundary conditions at both ends ($z = 0$ or l) cannot be satisfied strictly in general. We have to introduce the relaxed boundary conditions (5.4.10) or (5.4.11) where the effect is localized in the vicinity in accordance with the Saint–Venant principle. To strictly satisfy the boundary conditions we have to include the eigen-solutions of nonzero eigenvalues. Moreover, for complex problems such as a general rectangular domain or a short beam, the Saint–Venant principle is no longer applicable because the transverse dimension h is not a higher-order small quantity comparing to the longitudinal dimension l. Hence, we need to apply the eigen-solutions of nonzero eigenvalues in the expansion theorem in order to solve the problems.

5.5. Eigen-Solutions of Nonzero Eigenvalues

The eigenvalue equation for eigen-solutions of nonzero eigenvalues is Eq. (5.2.22). First, we should solve the eigenvalues $\tilde{\lambda}$, which satisfies the following equation

$$\det \begin{bmatrix} -\mu & \dfrac{s_{13}}{s_{11}}\tilde{\lambda} & \dfrac{b_{55}}{s_{11}d} & -\dfrac{b_{35}}{s_{11}d} \\ -\tilde{\lambda} & \dfrac{s_{15}}{s_{11}}\tilde{\lambda} - \mu & -\dfrac{b_{35}}{s_{11}d} & \dfrac{b_{33}}{s_{11}d} \\ 0 & 0 & -\mu & -\tilde{\lambda} \\ 0 & -\dfrac{1}{s_{11}}\tilde{\lambda}^2 & \dfrac{s_{13}}{s_{11}}\tilde{\lambda} & \dfrac{s_{15}}{s_{11}}\tilde{\lambda} - \mu \end{bmatrix} = 0 \tag{5.5.1}$$

Expanding the determinant yields the eigenvalue equation

$$s_{33}\tilde{\lambda}^4 - 2s_{35}\tilde{\lambda}^3\mu + (s_{55} + 2s_{13})\tilde{\lambda}^2\mu^2 - 2s_{15}\tilde{\lambda}\mu^3 + s_{11}\mu^4 = 0 \tag{5.5.2}$$

which has four roots

$$\tilde{\lambda}_i = \lambda_i\mu \quad (i = 1, 2, 3, 4) \tag{5.5.3}$$

For an ideal elastic body, we can prove that for λ_i there are only complex roots or pure imaginary roots but no real roots[2]. There are no repeated roots in general, i.e. there are two different pairs of complex conjugate roots. We only discuss the general case in this section while the other cases can be discussed in a similar way. If there are four different roots λ_i, the general solution of Eq. (5.2.22) can be expressed as

$$\left.\begin{aligned} w &= \sum_{i=1}^{4} A_i \exp(\lambda_i \mu x) \\ u &= \sum_{i=1}^{4} B_i \exp(\lambda_i \mu x) \\ \sigma &= \sum_{i=1}^{4} C_i \exp(\lambda_i \mu x) \\ \tau &= \sum_{i=1}^{4} D_i \exp(\lambda_i \mu x) \end{aligned}\right\} \tag{5.5.4}$$

where constants A_i, B_i, C_i, D_i are not independent. Substituting Eq. (5.5.4) into Eq. (5.2.22) and choosing D_i $(i = 1, 2, 3, 4)$ as independent constants yield

$$\left.\begin{aligned} A_i &= \frac{s_{35}\lambda_i - s_{33}\lambda_i^2 - s_{13}}{\lambda_i \mu} D_i \\ B_i &= \frac{s_{15}\lambda_i - s_{13}\lambda_i^2 - s_{11}}{\lambda_i^2 \mu} D_i \\ C_i &= -\lambda_i D_i \end{aligned}\right\} \quad (i = 1, 2, 3, 4) \tag{5.5.5}$$

Further substituting Eq. (5.5.5) into Eq. (5.5.4) yields

$$\left.\begin{aligned} w &= \sum_{i=1}^{4} \left[\frac{s_{35}\lambda_i - s_{33}\lambda_i^2 - s_{13}}{\lambda_i \mu} D_i \exp(\lambda_i \mu x) \right] \\ u &= \sum_{i=1}^{4} \left[\frac{s_{15}\lambda_i - s_{13}\lambda_i^2 - s_{11}}{\lambda_i^2 \mu} D_i \exp(\lambda_i \mu x) \right] \\ \sigma &= \sum_{i=1}^{4} [-\lambda_i D_i \exp(\lambda_i \mu x)] \\ \tau &= \sum_{i=1}^{4} [D_i \exp(\lambda_i \mu x)] \end{aligned}\right\} \tag{5.5.6}$$

From Eq. (5.2.8), we obtain

$$\sigma_x = \sum_{i=1}^{4} \left[-\frac{1}{\lambda_i} D_i \exp(\lambda_i \mu x) \right] \tag{5.5.7}$$

Substituting Eqs. (5.5.6) and (5.5.7) into the homogeneous boundary conditions (5.2.19) of side bounding surfaces yields

$$\left.\begin{aligned} \sum_{i=1}^{4} [D_i \exp(-\lambda_i \mu h)] &= 0 \\ \sum_{i=1}^{4} [-D_i \exp(-\lambda_i \mu h)]/\lambda_i &= 0 \\ \sum_{i=1}^{4} [D_i \exp(\lambda_i \mu h)] &= 0 \\ \sum_{i=1}^{4} [-D_i \exp(\lambda_i \mu h)]/\lambda_i &= 0 \end{aligned}\right\} \tag{5.5.8}$$

For nontrivial solution, the determinant of this coefficient matrix vanishes. Denote $\beta = 2\mu h$, by rearranging and simplifying, we obtain

$$\begin{vmatrix} 1 & 1 & 1 & 1 \\ \lambda_1 & \lambda_2 & \lambda_3 & \lambda_4 \\ \exp(\lambda_1 \beta) & \exp(\lambda_2 \beta) & \exp(\lambda_3 \beta) & \exp(\lambda_4 \beta) \\ \lambda_1 \exp(\lambda_1 \beta) & \lambda_2 \exp(\lambda_2 \beta) & \lambda_3 \exp(\lambda_3 \beta) & \lambda_4 \exp(\lambda_4 \beta) \end{vmatrix} = 0 \tag{5.5.9}$$

The equation above is the transcendental equation for solving nonzero eigenvalues in which numerical methods are required. Substituting the eigenvalues into Eq. (5.5.8) yields the trivial solution of D_i and, hence, the corresponding eigenvector functions are obtained.

All eigen-solutions of nonzero eigenvalues are covered in the Saint–Venant principle. Together with the eigen-solutions of zero eigenvalue, they constitute a complete adjoint symplectic orthonormal basis and the expansion theorem is then applicable. This is very important for the solution method. The solution method for solving isotropic elasticity problem can be applied to solve general anisotropic elasticity problems. Here we observe that there is no essential difference between the solution method of anisotropic problems and that of isotropic problems except derivation of the former is somewhat more complicated. This approach differs very much from the classical semi-inverse method because it is more rational. Therefore, the Hamiltonian system and symplectic mathematical methods have tremendous potential applications.

5.6. Introduction to Hamiltonian System for Generalized Plane Problems

Consider a homogeneous anisotropic infinite cylindrical solid with rectangular cross section. Let y-axis be the axial direction of this column and the external load be independent of y. This is a plane strain problem if y is the elastic principal axis. For general anisotropic materials, however, the displacement v along the y-direction will warp and, therefore, the problem is a **generalized plane strain problem**[3].

As the geometric properties, material properties and external loads are all independent of y-coordinate, all components of stress, strain and displacement are only functions of x, z. Displacement v has a term ε_{y0} which is linearly dependent on y but independent of x, z. The value of ε_{y0} can be determined from zero axial force on the cross section. For brevity, this term is not taken into consideration here.

For generalized plane strain problems, the strain-displacement relations (2.2.2) can be simplified as

$$\left.\begin{aligned} &\varepsilon_x = \frac{\partial u}{\partial x}; \quad \varepsilon_y = 0; \quad \varepsilon_z = \frac{\partial w}{\partial z} \\ &\gamma_{xy} = \frac{\partial v}{\partial x}; \quad \gamma_{xz} = \frac{\partial u}{\partial z} + \frac{\partial w}{\partial x}; \quad \gamma_{yz} = \frac{\partial v}{\partial z} \end{aligned}\right\} \tag{5.6.1}$$

These strain components are functions of x, z. The stress-strain relation is Eq. (2.3.1). Hence the deformation energy along the y-direction is

$$\begin{aligned} V_\varepsilon = \int_0^l \int_0^h \Big[&\frac{1}{2}c_{11}\varepsilon_x^2 + c_{13}\varepsilon_x\varepsilon_z + c_{14}\varepsilon_x\gamma_{xy} + c_{15}\varepsilon_x\gamma_{xz} + c_{16}\varepsilon_x\gamma_{yz} + \frac{1}{2}c_{33}\varepsilon_z^2 \\ &+ c_{34}\varepsilon_z\gamma_{xy} + c_{35}\varepsilon_z\gamma_{xz} + c_{36}\varepsilon_z\gamma_{yz} + \frac{1}{2}c_{44}\gamma_{xy}^2 + c_{45}\gamma_{xy}\gamma_{xz} + c_{46}\gamma_{xy}\gamma_{yz} \\ &+ \frac{1}{2}c_{55}\gamma_{xz}^2 + c_{56}\gamma_{xz}\gamma_{yz} + \frac{1}{2}c_{66}\gamma_{yz}^2\Big]\mathrm{d}x\mathrm{d}z = \int_0^l \int_0^h \mathscr{L}\,\mathrm{d}x\mathrm{d}z \end{aligned} \tag{5.6.2}$$

where strain components (5.6.1) have been substituted. Now the z-coordinate is treated as the time coordinate and an overdot denotes differentiation with respect to z, i.e. $(\dot{})=\mathrm{d}/\mathrm{d}x$. First we introduce the dual variables

$$\frac{\partial \mathscr{L}}{\partial \dot{u}}=c_{15}\frac{\partial u}{\partial x}+c_{35}\dot{w}+c_{45}\frac{\partial v}{\partial x}+c_{55}\dot{u}+c_{55}\frac{\partial w}{\partial x}+c_{56}\dot{v}=\tau_{xz} \tag{5.6.3a}$$

$$\frac{\partial \mathscr{L}}{\partial \dot{v}}=c_{16}\frac{\partial u}{\partial x}+c_{36}\dot{w}+c_{46}\frac{\partial v}{\partial x}+c_{56}\dot{u}+c_{56}\frac{\partial w}{\partial x}+c_{66}\dot{v}=\tau_{yz} \tag{5.6.3b}$$

$$\frac{\partial \mathscr{L}}{\partial \dot{w}}=c_{13}\frac{\partial u}{\partial x}+c_{33}\dot{w}+c_{34}\frac{\partial v}{\partial x}+c_{35}\dot{u}+c_{35}\frac{\partial w}{\partial x}+c_{36}\dot{v}=\sigma_{z} \tag{5.6.3c}$$

The primal variable $\boldsymbol{q}$ and the dual variable $\boldsymbol{p}$ are

$$\boldsymbol{q}=\{u,\quad v,\quad w\}^{\mathrm{T}},\quad \boldsymbol{p}=\{\tau_{xz},\quad \tau_{yz},\quad \sigma_z\}^{\mathrm{T}} \tag{5.6.4}$$

respectively. Denote

$$\boldsymbol{C}_d=\begin{bmatrix} c_{55} & c_{56} & c_{35}\\ c_{56} & c_{66} & c_{36}\\ c_{35} & c_{36} & c_{33}\end{bmatrix},\quad \boldsymbol{C}_e=\begin{bmatrix} c_{11} & c_{14} & c_{15}\\ c_{14} & c_{44} & c_{45}\\ c_{15} & c_{45} & c_{55}\end{bmatrix} \tag{5.6.5}$$

and

$$\boldsymbol{C}_t=\begin{bmatrix} c_{15} & c_{45} & c_{55}\\ c_{16} & c_{46} & c_{56}\\ c_{13} & c_{34} & c_{35}\end{bmatrix} \tag{5.6.6}$$

then the Lagrange function $\mathscr{L}$ can be expressed as

$$\mathscr{L}(\boldsymbol{q},\dot{\boldsymbol{q}})=\frac{1}{2}\dot{\boldsymbol{q}}^{\mathrm{T}}\boldsymbol{C}_d\dot{\boldsymbol{q}}+\dot{\boldsymbol{q}}^{\mathrm{T}}\boldsymbol{C}_t\left(\frac{\mathrm{d}\boldsymbol{q}}{\mathrm{d}x}\right)+\left(\frac{\mathrm{d}\boldsymbol{q}}{\mathrm{d}x}\right)^{\mathrm{T}}\boldsymbol{C}_e\left(\frac{\mathrm{d}\boldsymbol{q}}{\mathrm{d}x}\right) \tag{5.6.7}$$

Next, from Eq. (5.6.3), we obtain

$$\dot{\boldsymbol{q}}=-\boldsymbol{C}_d^{-1}\boldsymbol{C}_t\frac{\mathrm{d}\boldsymbol{q}}{\mathrm{d}x}+\boldsymbol{C}_d^{-1}\boldsymbol{p} \tag{5.6.8}$$

According to Legendre's transformation, the Hamiltonian function is

$$\begin{aligned}\mathscr{H}(\boldsymbol{q},\boldsymbol{p})&=\boldsymbol{p}^{\mathrm{T}}\dot{\boldsymbol{q}}-\mathscr{L}(\boldsymbol{q},\dot{\boldsymbol{q}})\\ &=\frac{1}{2}\boldsymbol{p}^{\mathrm{T}}\boldsymbol{C}_d^{-1}\boldsymbol{p}-\boldsymbol{p}^{\mathrm{T}}\boldsymbol{C}_d^{-1}\boldsymbol{C}_t\frac{\mathrm{d}\boldsymbol{q}}{\mathrm{d}x}-\left(\frac{\mathrm{d}\boldsymbol{q}}{\mathrm{d}x}\right)^{\mathrm{T}}\boldsymbol{B}_c\frac{\mathrm{d}\boldsymbol{q}}{\mathrm{d}x}\end{aligned} \tag{5.6.9}$$

where

$$\boldsymbol{B}_c = \boldsymbol{C}_e - \boldsymbol{C}_t^{\mathrm{T}} \boldsymbol{C}_d^{-1} \boldsymbol{C}_t \tag{5.6.10}$$

It is noted that matrices $\boldsymbol{C}_e$ and $\boldsymbol{C}_d$ are diagonal principal submatrices of the matrix of three-dimensional elastic constants. Both of them are symmetric and positive definite matrices and their inverse matrices exist. Furthermore, $\boldsymbol{B}_c$ is a symmetric matrix.

Having derived the Hamiltonian density function, we can express the Hamiltonian variational principle (or mixed energy variational principle) as

$$\delta\left\{\int_0^l \int_0^h [\boldsymbol{p}^{\mathrm{T}}\dot{\boldsymbol{q}} - \mathscr{H}(\boldsymbol{q},\boldsymbol{p}) - \boldsymbol{X}^{\mathrm{T}}\boldsymbol{q}]\mathrm{d}x\mathrm{d}z \right.$$
$$\left. - \int_0^l [(\overline{\boldsymbol{X}}_2^{\mathrm{T}}\boldsymbol{q})_{x=h} - (\overline{\boldsymbol{X}}_1^{\mathrm{T}}\boldsymbol{q})_{x=-h}]\mathrm{d}z \right\} = 0 \tag{5.6.11}$$

where $\boldsymbol{q}, \boldsymbol{p}$ are independent variable vector functions which has mutually independently variationals. The body force is $\boldsymbol{X}$ and the surface tractions are $\overline{\boldsymbol{X}}_1, \overline{\boldsymbol{X}}_2$. These external forces are in self-equilibrium, independent of y-coordinate and without components along the y-axis.

Performing variation on Eq. (5.6.11) and integrating by parts yield the Hamiltonian dual system of equations as

$$\begin{Bmatrix} \dot{\boldsymbol{q}} \\ \dot{\boldsymbol{p}} \end{Bmatrix} = \begin{bmatrix} \boldsymbol{A} & \boldsymbol{D} \\ \boldsymbol{B} & -\boldsymbol{A}^{\mathrm{T}} \end{bmatrix} \begin{Bmatrix} \boldsymbol{q} \\ \boldsymbol{p} \end{Bmatrix} - \begin{Bmatrix} \boldsymbol{0} \\ \boldsymbol{X} \end{Bmatrix} \tag{5.6.12}$$

where

$$\left.\begin{aligned} \boldsymbol{A} &= -\boldsymbol{C}_d^{-1}\boldsymbol{C}_t\frac{\partial}{\partial x}, \quad & \boldsymbol{A}^{\mathrm{T}} &= \boldsymbol{C}_t^{\mathrm{T}}\boldsymbol{C}_d^{-1}\frac{\partial}{\partial x} \\ \boldsymbol{B} &= -\boldsymbol{B}_c\frac{\partial^2}{\partial x^2}, \quad & \boldsymbol{D} &= \boldsymbol{C}_d^{-1} \end{aligned}\right\} \tag{5.6.13}$$

The boundary conditions are

$$\boldsymbol{C}_t^{\mathrm{T}}\boldsymbol{C}_d^{-1}\boldsymbol{p} + \boldsymbol{B}_c\frac{\partial \boldsymbol{q}}{\partial x} = \overline{\boldsymbol{X}}_2 \quad \text{at } x = h \tag{5.6.14a}$$

$$\boldsymbol{C}_t^{\mathrm{T}}\boldsymbol{C}_d^{-1}\boldsymbol{p} + \boldsymbol{B}_c\frac{\partial \boldsymbol{q}}{\partial x} = \overline{\boldsymbol{X}}_1 \quad \text{at } x = -h \tag{5.6.14b}$$

Introducing the full state vector $\boldsymbol{v}$ and operator matrix $\boldsymbol{H}$, we have

$$\boldsymbol{v} = \begin{Bmatrix} \boldsymbol{q} \\ \boldsymbol{p} \end{Bmatrix}, \quad \boldsymbol{H} = \begin{bmatrix} \boldsymbol{A} & \boldsymbol{D} \\ \boldsymbol{B} & -\boldsymbol{A}^{\mathrm{T}} \end{bmatrix} \tag{5.6.15}$$

The Hamiltonian dual equation (5.6.12) can be abbreviated to

$$\dot{\boldsymbol{v}} = \boldsymbol{H}\boldsymbol{v} + \boldsymbol{h}; \quad \boldsymbol{h}^{\mathrm{T}} = \{\boldsymbol{0}^{\mathrm{T}}, -\boldsymbol{X}^{\mathrm{T}}\} \tag{5.6.16}$$

The operator matrix $\boldsymbol{H}$ is independent of external load. Hence we investigate the homogeneous linear differential equation

$$\dot{\boldsymbol{v}} = \boldsymbol{H}\boldsymbol{v} \tag{5.6.17}$$

and the homogeneous boundary conditions on the side boundary surfaces

$$\boldsymbol{C}_t^{\mathrm{T}}\boldsymbol{C}_d^{-1}\boldsymbol{p} + \boldsymbol{B}_c\frac{\partial \boldsymbol{q}}{\partial x} = \boldsymbol{0} \quad \text{at } x = \pm h \tag{5.6.18}$$

Using an approach similar to the preceding few chapters, we can prove that the operator matrix $\boldsymbol{H}$ is a Hamiltonian operator matrix of symplectic geometric space.

At this point, we have derived the Hamiltonian system from the generalized plane problems. The system can be solved in a general way as discussed although the derivation is more complicated.

For example, the method of separation of variables can be applied to solve the homogeneous equation (5.6.17). Let

$$\boldsymbol{v}(z, x) = \xi(z)\boldsymbol{\psi}(x) \tag{5.6.19}$$

and substituting into Eq. (5.6.17) yield

$$\xi(z) = \mathrm{e}^{\mu z} \tag{5.6.20}$$

and the eigenvalue equation

$$\boldsymbol{H}\boldsymbol{\psi}(x) = \mu\boldsymbol{\psi}(x) \tag{5.6.21}$$

where μ is the undetermined eigenvalue and $\boldsymbol{\psi}(x)$ is the eigenvector which has to satisfy the homogeneous boundary conditions (5.6.18) on the side boundary surfaces ($x = \pm h$).

It has been mentioned repeatedly that the eigenvalue problem of Hamiltonian operator matrix has certain characteristics. The eigenvectors are adjoint symplectic orthonormal. The solution can be obtained via eigenvector expansion theorem.

A zero eigenvalue with Jordan form eigenvectors exists for a problem with free boundary conditions (5.6.18) on the side boundary surfaces ($x = \pm h$). For the present problem, there are eight eigenvectors of zero eigenvalue. These eigenvectors form three chains and they are adjoint symplectic orthonormal. Hence, they form a symplectic subspace. The details are omitted. Interested readers are referred to the related chapters in monograph[4].

References

1. Weian Yao, Hamiltonian system for plane anisotropic elasticity and analytical solutions of Saint–Venant problem, *Journal of Dalian University of Technology*, 1999, 39(5): 612–615 (in Chinese).
2. S.G. Lekhnitskii, *Theory of Elasticity of an Anisotropic Body*, [s.l.] Mir Publisher, 1981.
3. Zudao Luo and Sijian Li, *Mechanics of Anisotropic Materials*, Shanghai: Shanghai Jiaotong University Press, 1994 (in Chinese).
4. Wanxie Zhong, *A New Systematic Methodology for Theory of Elasticity*, Dalian University of Technology Press, 1995 (in Chinese).

Chapter 6

Saint–Venant Problems for Laminated Composite Plates

In this chapter, the theory of Hamiltonian system is introduced for the problems of laminated composite plates. A method for solving these problems is developed using expansion of eigenvector of Hamiltonian operator matrix in the transverse direction. All the six eigen-solutions of zero eigenvalue are obtained, and hence an analytical method for solving the Saint–Venant problems is established.

6.1. The Fundamental Equations

We will discuss in this chapter the plane stress elasticity problems of a n-layered laminated composite plate as shown in Fig. 6.1 where $x_0, x_1, \ldots, x_n$ etc. are known and $l \gg x_n$.

It is assumed here that each layer of the laminated plate is made of an orthotropic material with stress-strain relation as

$$\begin{Bmatrix} \varepsilon_{xi} \\ \varepsilon_{zi} \\ \gamma_{xzi} \end{Bmatrix} = \begin{bmatrix} s_{1i} & s_{2i} & 0 \\ s_{2i} & s_{4i} & 0 \\ 0 & 0 & s_{6i} \end{bmatrix} \begin{Bmatrix} \sigma_{xi} \\ \sigma_{zi} \\ \tau_{xzi} \end{Bmatrix} \quad (i = 1, 2, \ldots, n) \tag{6.1.1a}$$

or

$$\begin{Bmatrix} \sigma_{xi} \\ \sigma_{zi} \\ \tau_{xzi} \end{Bmatrix} = \begin{bmatrix} b_{1i} & b_{2i} & 0 \\ b_{2i} & b_{4i} & 0 \\ 0 & 0 & b_{6i} \end{bmatrix} \begin{Bmatrix} \varepsilon_{xi} \\ \varepsilon_{zi} \\ \gamma_{xzi} \end{Bmatrix} \quad (i = 1, 2, \ldots, n) \tag{6.1.1b}$$

where subscript i denotes the ith layer. For brevity, the subscript i will be often omitted subsequently unless stated otherwise due to possible confusion.

Then the complementary strain energy reads

$$v_{\mathrm{c}} = \frac{1}{2}(s_1\sigma_x^2 + s_4\sigma_z^2 + 2s_2\sigma_x\sigma_z + s_6\tau_{xz}^2) \tag{6.1.2}$$

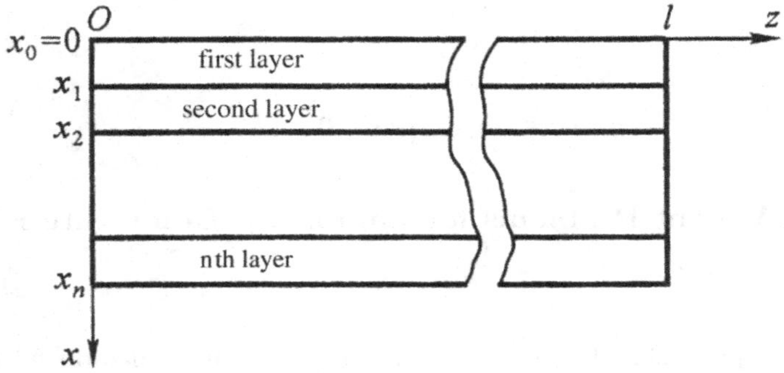

Fig. 6.1. A n-layered composite plate.

The geometric relations of strain-displacement is independent of material and they remain as

$$\varepsilon_z = \frac{\partial w}{\partial z}; \quad \varepsilon_x = \frac{\partial u}{\partial x}; \quad \gamma_{xz} = \frac{\partial u}{\partial z} + \frac{\partial w}{\partial x} \tag{6.1.3}$$

Similarly, the equilibrium equations remain as

$$\frac{\partial \sigma_x}{\partial x} + \frac{\partial \tau_{xz}}{\partial z} + F_x = 0 \quad \frac{\partial \tau_{xz}}{\partial x} + \frac{\partial \sigma_z}{\partial z} + F_z = 0 \tag{6.1.4}$$

where F_x, F_z are body forces along the x- and z-directions, respectively, in the domain.

In addition, there should be boundary conditions. It is assumed here that on both surfaces of the plate the forces are prescribed as

$$\sigma_x = \overline{F}_{x1}(z), \quad \tau_{xz} = \overline{F}_{z1}(z) \qquad \text{for } x = 0 \tag{6.1.5a}$$

$$\sigma_x = \overline{F}_{x2}(z), \quad \tau_{xz} = \overline{F}_{z2}(z) \qquad \text{for } x = x_n \tag{6.1.5b}$$

At the two ends $z = 0$ and l there are also the corresponding boundary conditions. For instance, the boundary conditions for specified forces are

$$\sigma_z = \overline{\sigma}(x), \quad \tau_{xz} = \overline{\tau}(x) \qquad \text{at } z = 0 \quad \text{or} \quad l \tag{6.1.6}$$

while the boundary conditions for specified displacements are

$$u = \overline{u}(x), \quad w = \overline{w}(x) \qquad \text{at } z = 0 \quad \text{or} \quad l \tag{6.1.7}$$

Laminated composite plates are different from homogeneous plates because there exists discontinuity of elastic properties in the body. The displacements and stresses at the interface should fulfill continuity condition of displacement and equilibrium condition of stresses, as

$$\left.\begin{aligned} u_i(x_i) &= u_{i+1}(x_i), & w_i(x_i) &= w_{i+1}(x_i) \\ \sigma_{xi}(x_i) &= \sigma_{x,i+1}(x_i), & \tau_{xzi}(x_i) &= \tau_{xz,i+1}(x_i) \end{aligned}\right\} \tag{6.1.8}$$

The governing differential equations and boundary conditions for laminated composite plates are presented above. We can, of course, establish the corresponding Hellinger–Reissner variational principle as

$$\delta\left\{\int_0^l \int_{x_0}^{x_n} \left[\sigma_x \frac{\partial u}{\partial x} + \sigma_z \frac{\partial w}{\partial z} + \tau_{xz}\left(\frac{\partial u}{\partial z} + \frac{\partial w}{\partial x}\right) - v_c - F_x u - F_z w\right] \mathrm{d}x\mathrm{d}z \right.$$
$$\left. - \int_0^l \left[(w\overline{F}_{z2} + u\overline{F}_{x2})_{x=x_n} - (w\overline{F}_{z1} + u\overline{F}_{x1})_{x=x_0}\right] \mathrm{d}z \right\} = 0 \tag{6.1.9}$$

The boundary conditions at the two ends ($z = 0$ or l), to be discussed later, have not been incorporated in Eq. (6.1.9). The variation of Eq. (6.1.9) yields all the governing differential equations and boundary conditions.

The problem for laminated composite plates is a classical subject which can be solved numerically by methods such as the finite element method, etc. However, the theory and analysis are not very complete. The classical solution approach always refers to the Lagrangian system with one kind of variable. This approach is difficult in expressing the continuous conditions of displacement and stress equilibrium (6.1.8) at the interface and it causes complication in the analytical solutions.

Because the symplectic system involves two kinds of variables, the continuity condition at the interface can be easily satisfied and thus analytical solution is possible. Although all layers of the laminated plates discussed here are assumed orthotropic, the method here is also applicable to anisotropic laminated plates where more complicated derivation will be involved.

6.2. Derivation of Hamiltonian System

In the following derivation of the Hamiltonian system[1,2], let z-coordinate be treated as the time coordinate and an overdot be used to indicate differentiation with respect to z, i.e. $(\dot{\ }) = \partial/\partial z$. Here, z-axis is longitudinal while x-axis is transverse, and the transverse stress σ_x should be eliminated. The variation of Eq. (6.1.9) with respect to σ_x is

$$\sigma_x = \frac{1}{s_1}\left(\frac{\partial u}{\partial x} - s_2\sigma\right) \tag{6.2.1}$$

where σ and τ are respectively the abbreviations of σ_z and τ_{xz}. Then the boundary conditions (6.1.5) can be expressed as

$$\frac{1}{s_1}\left(\frac{\partial u}{\partial x}-s_2\sigma\right)=\overline{F}_{x1}(z),\quad \tau=\overline{F}_{z1}(z)\qquad \text{at } x=0 \tag{6.2.2a}$$

$$\frac{1}{s_1}\left(\frac{\partial u}{\partial x}-s_2\sigma\right)=\overline{F}_{x2}(z),\quad \tau=\overline{F}_{z2}(z)\qquad \text{at } x=x_n \tag{6.2.2b}$$

and the continuous conditions of displacement and stress equilibrium at the interface are

$$\left.\begin{aligned} u_i(x_i)&=u_{i+1}(x_i)\\ w_i(x_i)&=w_{i+1}(x_i)\\ \frac{1}{s_{1i}}\left(\frac{\partial u_i}{\partial x}-s_{2i}\sigma_i\right)\Bigg|_{x=x_i}&=\frac{1}{s_{1,i+1}}\left(\frac{\partial u_{i+1}}{\partial x}-s_{2,i+1}\sigma_{i+1}\right)\Bigg|_{x=x_i}\\ \tau_i(x_i)&=\tau_{i+1}(x_i)\end{aligned}\right\} \tag{6.2.3}$$

Substituting Eq. (6.2.1) into Eq. (6.1.9) yields the Hamiltonian mixed energy variational principle as

$$\delta\left\{\int_0^l\sum_{i=1}^{n}\int_{x_{i-1}}^{x_i}[\sigma\dot{w}+\tau\dot{u}-\mathscr{H}(w,u,\sigma,\tau)]\mathrm{d}x\mathrm{d}z\right.$$
$$\left.-\int_0^l[(w\overline{F}_{z2}+u\overline{F}_{x2})_{x=x_n}-(w\overline{F}_{z1}+u\overline{F}_{x1})_{x=x_0}]\mathrm{d}z\right\}=0 \tag{6.2.4}$$

where the Hamiltonian density function is

$$\mathscr{H}=\frac{s_2}{s_1}\sigma\frac{\partial u}{\partial x}-\tau\frac{\partial w}{\partial x}+\frac{1}{2b_4}\sigma^2+\frac{1}{2}s_6\tau^2-\frac{1}{2s_1}\left(\frac{\partial u}{\partial x}\right)^2+F_xu+F_zw \tag{6.2.5}$$

It is clear the dual variables of displacements w and u are σ and τ respectively, and then the full state vector is

$$\boldsymbol{v}=\{w,\quad u;\quad \sigma,\quad \tau\}^{\mathrm{T}} \tag{6.2.6}$$

The variation of Eq. (6.2.4) yields the dual system of equations as

$$\dot{\boldsymbol{v}}=\boldsymbol{H}\boldsymbol{v}+\boldsymbol{Q} \tag{6.2.7}$$

where

$$\boldsymbol{H} = \begin{bmatrix} 0 & \dfrac{s_2}{s_1}\dfrac{\partial}{\partial x} & \dfrac{1}{b_4} & 0 \\ -\dfrac{\partial}{\partial x} & 0 & 0 & s_6 \\ 0 & 0 & 0 & -\dfrac{\partial}{\partial x} \\ 0 & -\dfrac{1}{s_1}\dfrac{\partial^2}{\partial x^2} & \dfrac{s_2}{s_1}\dfrac{\partial}{\partial x} & 0 \end{bmatrix} \qquad \boldsymbol{Q} = \begin{Bmatrix} 0 \\ 0 \\ -F_z \\ -F_x \end{Bmatrix} \tag{6.2.8}$$

To discuss the characteristics of operator matrix $\boldsymbol{H}$, we denote

$$\langle \boldsymbol{v}_1, \boldsymbol{v}_2 \rangle \overset{\text{def}}{=} \int_{x_0}^{x_n} \boldsymbol{v}_1^{\mathrm{T}} \boldsymbol{J} \boldsymbol{v}_2 \mathrm{d}x = \int_{x_0}^{x_n} (w_1\sigma_2 + u_1\tau_2 - \sigma_1 w_2 - \tau_1 u_2)\mathrm{d}x \tag{6.2.9}$$

where $\boldsymbol{J}$ is a unit symplectic matrix

$$\boldsymbol{J} = \begin{bmatrix} \boldsymbol{0} & \boldsymbol{I}_2 \\ -\boldsymbol{I}_2 & \boldsymbol{0} \end{bmatrix} \tag{6.2.10}$$

Obviously, Eq. (6.2.9) satisfies the four conditions of symplectic inner product (1.3.2). Hence, according to definition of symplectic inner product Eq. (6.2.9), the full state vector $\boldsymbol{v}$ forms a symplectic space.

We first discuss the corresponding homogeneous of Eq. (6.2.7)

$$\dot{\boldsymbol{v}} = \boldsymbol{H}\boldsymbol{v} \tag{6.2.11}$$

and the homogeneous boundary conditions on both side boundary surfaces

$$\frac{1}{s_1}\left(\frac{\partial u}{\partial x} - s_2\sigma\right) = 0, \quad \tau = 0 \qquad \text{for } x = x_0 \text{ and } x_n \tag{6.2.12}$$

By integrating by parts, if $\boldsymbol{v}_1, \boldsymbol{v}_2$ are continuously differentiable full state vectors satisfying the boundary conditions (6.2.12) on the surfaces and the continuity conditions (6.2.3) at the interface, we have an identity

$$\langle \boldsymbol{v}_1, \boldsymbol{H}\boldsymbol{v}_2 \rangle = \langle \boldsymbol{v}_2, \boldsymbol{H}\boldsymbol{v}_1 \rangle \tag{6.2.13}$$

It states that the operator matrix $\boldsymbol{H}$ is the Hamiltonian operator matrix in the symplectic space.

Thus, the Hamiltonian system for laminated composite plates has been derived. Subsequently, we apply the usual solution method of Hamiltonian system for solving the problem.

The method of separation of variables can be applied to solve the system of Eqs. (6.2.11). Assume that

$$\boldsymbol{v}(z,x) = e^{\mu z}\boldsymbol{\psi}(x) \tag{6.2.14}$$

where μ is the eigenvalue and $\boldsymbol{\psi}(x)$ the eigenvector. The eigenvalue equation is

$$\boldsymbol{H}\boldsymbol{\psi}(x) = \mu\boldsymbol{\psi}(x) \tag{6.2.15}$$

The solution has to satisfy the homogeneous boundary conditions (6.2.12) on the surfaces and the continuity conditions (6.2.3) at the interface.

As $\boldsymbol{H}$ is the Hamiltonian operator matrix, the eigenvectors are adjoint symplectic orthogonal. Once the eigenvalue and eigenvectors are established, the expansion theorem can be applied to solve this problem.

6.3. Eigen-Solutions of Zero Eigenvalue

For the free homogeneous boundary conditions (6.2.12) on the surfaces, there are the basic eigen-solutions and Jordan form eigen-solutions of zero eigenvalue[1,2]. These eigen-solutions in elasticity have important physical interpretations. To determine the basic eigen-solutions of zero eigenvalue, we need to solve the following differential equation

$$\boldsymbol{H}\boldsymbol{\psi}^{(0)} = \boldsymbol{0} \tag{6.3.1a}$$

while for the Jordan form eigen-solutions, we need to solve

$$\boldsymbol{H}\boldsymbol{\psi}^{(k)} = \boldsymbol{\psi}^{(k-1)} \tag{6.3.1b}$$

where $k = 1, 2, \ldots$ denotes the kth order Jordan form eigen-solution. Expanding Eq. (6.3.1) yields

$$\begin{array}{ccccl}
0 & +\dfrac{s_2}{s_1}\dfrac{\partial u^{(k)}}{\partial x} & +\dfrac{1}{b_4}\sigma^{(k)} & +0 & = 0[w^{(k-1)}] \\
-\dfrac{\partial w^{(k)}}{\partial x} & +0 & +0 & +s_6\tau^{(k)} & = 0[u^{(k-1)}] \\
0 & +0 & +0 & -\dfrac{\partial \tau^{(k)}}{\partial x} & = 0[\sigma^{(k-1)}] \\
0 & -\dfrac{1}{s_1}\dfrac{\partial^2 u^{(k)}}{\partial x^2} & +\dfrac{s_2}{s_1}\dfrac{\partial \sigma^{(k)}}{\partial x} & +0 & = 0[\tau^{(k-1)}]
\end{array} \tag{6.3.2}$$

The eigen-solutions have to satisfy the homogeneous boundary conditions (6.2.12) on the two surfaces and the conditions of continuous displacement and force equilibrium at the interface (6.2.3).

First, solving Eq. (6.3.1a) yields two basic eigen-solutions of zero eigenvalue

$$\boldsymbol{\psi}_f^{(0)} = \{1,\quad 0,\quad 0,\quad 0\}^{\mathrm{T}};\quad \boldsymbol{v}_f^{(0)} = \boldsymbol{\psi}_f^{(0)} \tag{6.3.3}$$

$$\boldsymbol{\psi}_s^{(0)} = \{0,\quad 1,\quad 0,\quad 0\}^{\mathrm{T}};\quad \boldsymbol{v}_s^{(0)} = \boldsymbol{\psi}_s^{(0)} \tag{6.3.4}$$

These are the solutions of the original problem (6.2.11) and they can be physically interpreted as the rigid displacements along the z- and x-directions, respectively. They are also the starting points of two Jordan chains.

Next, we seek to solve the first-order Jordan form eigen-solutions. Substituting Eq. (6.3.3) into Eq. (6.3.2) yields the first-order Jordan form eigen-solution on chain one as

$$\boldsymbol{\psi}_{fi}^{(1)} = \left\{0,\quad \frac{s_{2i}}{s_{4i}}x + r_i + d_1,\quad \frac{1}{s_{4i}},\quad 0\right\}^{\mathrm{T}} \quad i = 1, 2, \ldots, n \tag{6.3.5}$$

where

$$r_1 = 0,\quad r_{i+1} = r_i - x_i\left(\frac{s_{2(i+1)}}{s_{4(i+1)}} - \frac{s_{2i}}{s_{4i}}\right) \quad (i = 1, 2, \ldots, n-1) \tag{6.3.6}$$

Here d_1 is an undetermined constant. The solution of the original problem formed by Eq. (6.3.5) is

$$\boldsymbol{v}_f^{(1)} = \boldsymbol{\psi}_f^{(1)} + z\boldsymbol{\psi}_f^{(0)} \tag{6.3.7}$$

Obviously, it can be physically interpreted as the solution of simple tension.

As the vectors $\boldsymbol{\psi}_f^{(1)}$ and $\boldsymbol{\psi}_f^{(0)}$ are mutually symplectic adjoint, we have

$$k_1 = \langle \boldsymbol{\psi}_f^{(0)}, \boldsymbol{\psi}_f^{(1)} \rangle = \sum_{i=1}^{n} \frac{1}{s_{4i}}(x_i - x_{i-1}) \neq 0 \tag{6.3.8}$$

Therefore, the next order Jordon form of $\boldsymbol{\psi}_f^{(1)}$ does not exist. The constant k_1 has a particular physical interpretation, i.e. the extensional rigidity of cross section.

For chain two, substituting Eq. (6.3.4) into Eq. (6.3.2) yields the first-order Jordan form eigen-solution as

$$\boldsymbol{\psi}_s^{(1)} = \{d_2 - x,\quad 0,\quad 0,\quad 0\}^{\mathrm{T}} \tag{6.3.9}$$

where constant d_2 is determined by the symplectic orthogonality condition of $\boldsymbol{\psi}_s^{(1)}$ and $\boldsymbol{\psi}_f^{(1)}$, as

$$d_2 = \frac{c_1}{k_1} \tag{6.3.10}$$

in which

$$c_1 = \frac{1}{2}\sum_{i=1}^{n} \frac{1}{s_{4i}}(x_i^2 - x_{i-1}^2) \tag{6.3.11}$$

The constant d_2 also has a particular physical interpretation, i.e. the position of neutral axis in pure bending.

Solution of the original problem formed by Eq. (6.3.9) is

$$\boldsymbol{v}_s^{(1)} = \boldsymbol{\psi}_s^{(1)} + z\boldsymbol{\psi}_s^{(0)} \tag{6.3.12}$$

which, obviously, is the solution for rigid rotation.

As vector $\boldsymbol{\psi}_s^{(1)}$ is symplectic orthogonal to both $\boldsymbol{\psi}_f^{(0)}$ and $\boldsymbol{\psi}_s^{(0)}$, the next order Jordon form solution exists. Substituting Eq. (6.3.9) into Eq. (6.3.2) yields

$$\boldsymbol{\psi}_{si}^{(2)} = \left\{0,\ \ u_{si}^{(2)},\ \ \frac{d_2 - x}{s_{4i}},\ \ 0\right\}^{\mathrm{T}} \quad (i = 1, 2, \ldots, n) \tag{6.3.13}$$

where

$$u_{si}^{(2)} = -\frac{s_{2i}}{2s_{4i}}(x - d_2)^2 + q_i + d_3 \tag{6.3.14}$$

in which

$$q_1 = 0, \quad q_{i+1} = q_i + \frac{1}{2}(x_i - d_2)^2\left(\frac{s_{2(i+1)}}{s_{4(i+1)}} - \frac{s_{2i}}{s_{4i}}\right) \quad (i = 1, 2, \ldots, n-1) \tag{6.3.15}$$

The solution of the original problem is expressed as

$$\boldsymbol{v}_s^{(2)} = \boldsymbol{\psi}_s^{(2)} + z\boldsymbol{\psi}_s^{(1)} + \frac{1}{2}z^2\boldsymbol{\psi}_s^{(0)} \tag{6.3.16}$$

and the corresponding stress field is

$$\sigma_i = \frac{1}{s_{4i}}(d_2 - x) \quad (i = 1, 2, \ldots, n), \quad \tau = 0 \tag{6.3.17}$$

It implies that the shear stress vanishes and the bending moment is constant on the cross section. Hence, Eq. (6.3.16) is the solution for pure bending. From Eq. (6.3.17), we note that the normal stress on the cross section is no longer linear due to the differing material elastic constants of each layer, but rather it is piecewise linear. The cross section has axial displacement $w = z(d_2 - x)$ and it remains as a plane.

Because $\boldsymbol{\psi}_s^{(2)}$ and $\boldsymbol{\psi}_f^{(0)}$ are symplectic orthogonal by a proper choice of d_2, and also $\boldsymbol{\psi}_s^{(2)}$ is symplectic orthogonal to the basic eigen-solution $\boldsymbol{\psi}_s^{(0)}$, the next order Jordon form solution exists. Substituting Eq. (6.3.13) into Eq. (6.3.2) yields

$$\boldsymbol{\psi}_{si}^{(3)} = \left\{ w_{si}^{(3)}, \quad 0, \quad 0, \quad \frac{1}{2s_{4i}}(x - d_2)^2 + p_i \right\} \quad (i = 1, 2, \ldots, n) \tag{6.3.18}$$

where

$$w_{si}^{(3)} = \frac{s_{6i} + s_{2i}}{6s_{4i}}(x - d_2)^3 + (x - d_2)(s_{6i}p_i - q_i - d_3) + t_i + d_4 \tag{6.3.19}$$

in which

$$p_1 = -\frac{1}{2s_{41}}d_2^2,$$

$$p_{i+1} = p_i - \frac{1}{2}(x_i - d_2)^2\left(\frac{1}{s_{4(i+1)}} - \frac{1}{s_{4i}}\right), \quad (i = 1, 2, \ldots, n-1) \tag{6.3.20}$$

$$t_1 = 0, \quad t_{i+1} = t_i - \frac{1}{6}(x_i - d_2)^3\left(\frac{s_{6(i+1)} + s_{2(i+1)}}{s_{4(i+1)}} - \frac{s_{6i} + s_{2i}}{s_{4i}}\right)$$
$$- (x_i - d_2)(s_{6(i+1)}p_{i+1} - q_{i+1} - s_{6i}p_i + q_i), \quad (i = 1, 2, \ldots, n-1) \tag{6.3.21}$$

The solution of the original problem is expressed as

$$\boldsymbol{v}_s^{(3)} = \boldsymbol{\psi}_s^{(3)} + z\boldsymbol{\psi}_s^{(2)} + \frac{1}{2}z^2\boldsymbol{\psi}_s^{(1)} + \frac{1}{6}z^3\boldsymbol{\psi}_s^{(0)} \tag{6.3.22}$$

and the corresponding stress field is

$$\sigma_i = \frac{z}{s_{4i}}(d_2 - x), \quad \tau_i = \frac{1}{2s_{4i}}(x - d_2)^2 + p_i \quad (i = 1, 2, \ldots, n) \tag{6.3.23}$$

It implies that the shear stress on the cross section is constant. Hence Eq. (6.3.22) is the solution for constant shear bending. In addition, it can be shown that the shear stress at $x = 0$ and $x = x_n$ vanishes while it reaches the maximum at the neutral axis $x = d_2$ and furthermore, the cross section does not remain a plane. Warping of cross section is not negligible especially when the elastic properties between the inner and outer layers differ markedly.

As vectors $\boldsymbol{\psi}_s^{(3)}$ and $\boldsymbol{\psi}_s^{(0)}$ are mutually symplectic adjoint, the next order Jordon form eigen-solution does not exist. The symplectic orthogonality of $\boldsymbol{\psi}_s^{(3)}$ to both $\boldsymbol{\psi}_s^{(2)}$ and $\boldsymbol{\psi}_f^{(1)}$ and the decoupling between the two chains can be fulfilled by proper choice of constants d_3, d_1 and d_4, as

$$d_1 = \frac{c_2}{k_2}, \quad d_3 = \frac{c_4 + c_5}{2k_2}, \quad d_4 = -\frac{c_3}{k_1} \tag{6.3.24}$$

where

$$\begin{aligned} c_2 = \sum_{i=1}^{n} \Bigg\{ & \frac{s_{2i}}{8s_{4i}^2}[(x_i - d_2)^4 - (x_{i-1} - d_2)^4] \\ & + \frac{1}{6s_{4i}}\left(r_i + \frac{s_{2i}}{s_{4i}}d_2\right)[(x_i - d_2)^3 - (x_{i-1} - d_2)^3] \\ & + \frac{s_{2i}}{2s_{4i}}p_i[(x_i - d_2)^2 - (x_{i-1} - d_2)^2] \\ & + p_i\left(r_i + \frac{s_{2i}}{s_{4i}}d_2\right)(x_i - x_{i-1})\Bigg\} \end{aligned} \tag{6.3.25}$$

$$\begin{aligned} c_3 = \sum_{i=1}^{n} \frac{1}{s_{4i}} \Bigg\{ & \frac{s_{6i} + s_{2i}}{24s_{4i}}[(x_i - d_2)^4 - (x_{i-1} - d_2)^4] + \frac{1}{2}(s_{6i}p_i - q_i) \\ & \times [(x_i - d_2)^2 - (x_{i-1} - d_2)^2] + t_i(x_i - x_{i-1})\Bigg\} \end{aligned} \tag{6.3.26}$$

$$\begin{aligned} c_4 = \sum_{i=1}^{n} \frac{1}{s_{4i}} \Bigg\{ & \frac{s_{6i} + s_{2i}}{30s_{4i}}[(x_i - d_2)^5 - (x_{i-1} - d_2)^5] \\ & + \frac{1}{3}(s_{6i}p_i - q_i)[(x_i - d_2)^3 - (x_{i-1} - d_2)^3] \\ & + \frac{1}{2}t_i[(x_i - d_2)^2 - (x_{i-1} - d_2)^2]\Bigg\} \end{aligned} \tag{6.3.27}$$

$$\begin{aligned} c_5 = \sum_{i=1}^{n} \Bigg\{ & -\frac{1}{20}\frac{s_{2i}}{s_{4i}^2}[(x_i - d_2)^5 - (x_{i-1} - d_2)^5] \\ & + \frac{1}{6s_{4i}}(q_i - s_{2i}p_i)[(x_i - d_2)^3 - (x_{i-1} - d_2)^3] \\ & + p_i q_i(x_i - x_{i-1})\Bigg\} \end{aligned} \tag{6.3.28}$$

Table 6.1. The adjoint symplectic orthogonality between the eigen-solutions of zero eigenvalue.

	$\psi_f^{(0)}$	$\psi_f^{(1)}$	$\psi_s^{(0)}$	$\psi_s^{(1)}$	$\psi_s^{(2)}$	$\psi_s^{(3)}$
$\psi_f^{(0)}$	0	*	0	0	d_2	0
$\psi_f^{(1)}$		0	0	d_2	0	d_1, d_4
$\psi_s^{(0)}$			0	0	0	*
$\psi_s^{(1)}$				0	*	0
$\psi_s^{(2)}$					0	d_3
$\psi_s^{(3)}$						0

$$k_2 = \sum_{i=1}^{n} \frac{1}{3s_{4i}}[(x_i - d_2)^3 - (x_{i-1} - d_2)^3] \tag{6.3.29}$$

in which k_2 is physically interpreted as the flexural rigidity of the cross section.

Table 6.1 presents the adjoint symplectic orthogonality between the six eigen-solutions of zero eigenvalue where * denotes symplectic adjoint relation; 0 denotes symplectic orthogonality; and $d_i (i = 1, 2, 3, 4)$ denote that symplectic orthogonality can be satisfied through proper choice of d_i. Hence the six solutions $\psi_f^{(0)}, \psi_f^{(1)}, \psi_s^{(0)}, \psi_s^{(1)}, \psi_s^{(2)}$ and $\psi_s^{(3)}$ constitute all the eigenvectors of zero eigenvalue for laminated composite plates, and they form a set of adjoint symplectic orthogonal basis for symplectic subspace corresponding to zero eigenvalue.

The distribution of axial displacements $w_s^{(3)}$ and shear stresses $\tau_s^{(3)}$ corresponding to the Jordan form eigen-solution $\psi_s^{(3)}$ for laminated composite plate are presented in the following two examples.

Example 6.1 A symmetrically laminated composite plate is composed of three orthotropic layers with geometric parameters

$$n = 3, \quad x_0 = 0, \quad x_1 = 1.0, \quad x_2 = 3.0, \quad x_3 = 4.0.$$

The material properties of the outer layers ($i = 1, 3$) are

$$b_{1i} = b_{4i} = 1.0989, \quad b_{2i} = 0.32967, \quad b_{6i} = 0.384615.$$

and the material properties of the inner layer ($i = 2$) are

$$b_{12} = b_{42} = 0.10989, \quad b_{22} = 0.032967, \quad b_{62} = 0.0384615.$$

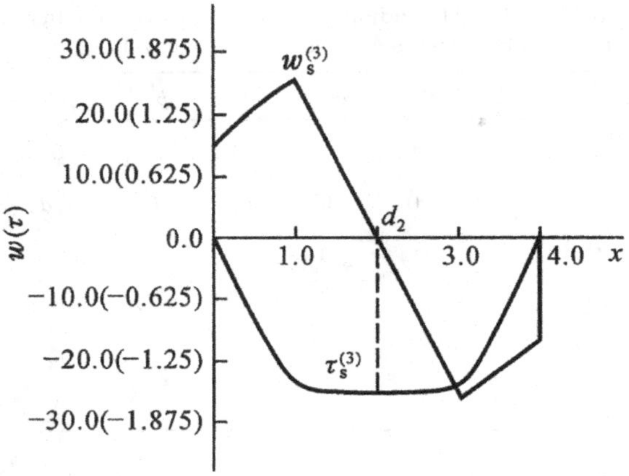

Fig. 6.2. The distribution of $w_s^{(3)}$ and $\tau_s^{(3)}$ for Example 6.1.

The distribution of axial displacements and shear stress corresponding to the Jordan form eigen-solution $\psi_s^{(3)}$ is illustrated in Fig. 6.2.

Example 6.2 An unsymmetrically laminated composite plate is composed of three orthotropic layers with geometric parameters

$$n = 3, \quad x_0 = 0, \quad x_1 = 1.0, \quad x_2 = 3.0, \quad x_3 = 4.5.$$

The material properties of the first layer are

$$b_{11} = 1.0, \quad b_{21} = 0.25, \quad b_{41} = 0.7, \quad b_{61} = 0.2.$$

the material properties of the second layer are

$$b_{12} = 0.1, \quad b_{22} = 0.02, \quad b_{42} = 0.09, \quad b_{62} = 0.034.$$

and the material properties of the third layer are

$$b_{13} = 0.5, \quad b_{23} = 0.15, \quad b_{43} = 0.3, \quad b_{63} = 0.15.$$

The distribution of axial displacement and shear stress corresponding to the Jordan form eigen-solution $\psi_s^{(3)}$ is illustrated in Fig. 6.3.

From Eq. (6.3.19) and the distribution of $w_s^{(3)}$ in the illustrations, it is noted there is a sudden change for $w_s^{(3)}$ at the interface on the cross section of the laminated plates especially when the elastic properties differ considerably. Hence the assumption of plane conditions before and after deformation is not applicable.

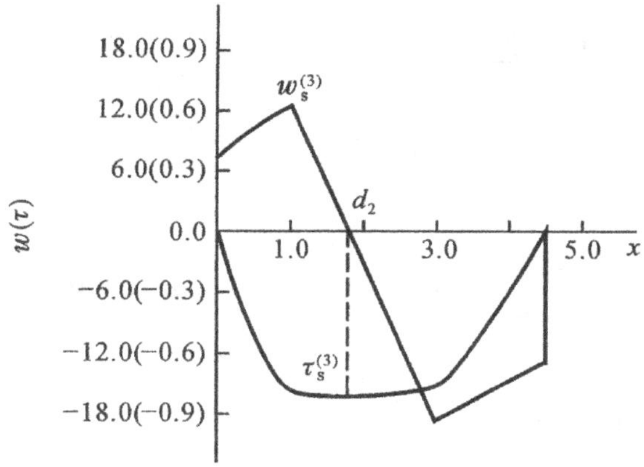

Fig. 6.3. The distribution of $w_s^{(3)}$ and $\tau_s^{(3)}$ for Example 6.2.

6.4. Analytical Solutions of Saint–Venant Problem

The eigenvectors of zero eigenvalue for laminated composite plates do not have exponential functions. They are not sensitive to the self-equilibrium forces on the cross section. The effect of non-self-equilibrium external loading on the cross section propagates to regions far away through these solutions. Hence, it is possible to neglect those eigen-solutions of nonzero eigenvalues according to Saint–Venant principle, i.e. only the eigen-solutions of zero eigenvalue are included in the expansion theorem, or

$$\boldsymbol{v} = a_1\boldsymbol{\psi}_f^{(0)} + a_2\boldsymbol{\psi}_f^{(1)} + a_3\boldsymbol{\psi}_s^{(0)} + a_4\boldsymbol{\psi}_s^{(1)} + a_5\boldsymbol{\psi}_s^{(2)} + a_6\boldsymbol{\psi}_s^{(3)} \tag{6.4.1}$$

Substituting into the variational formula (6.2.4) and rearranging yield

$$\delta\left\{\int_0^l (k_1a_2\dot{a}_1 + k_2a_5\dot{a}_4 - k_2a_6\dot{a}_3 - \frac{1}{2}k_1a_2^2 - \frac{1}{2}k_2a_5^2 + k_2a_4a_6 - F_{\mathrm{N}}a_1 \right.$$
$$\left. -Wa_2 - F_{\mathrm{S}}a_3 - Ma_4 - \theta a_5 - Ua_6)\mathrm{d}z + [k_3a_5a_6]_{z=0}^l + \tilde{v}_e^2\right\} = 0 \tag{6.4.2}$$

where

$$F_{\mathrm{N}} = \int_{x_0}^{x_n} F_z\mathrm{d}x + \overline{F}_{z2} - \overline{F}_{z1} \quad \text{(axial force)} \tag{6.4.3}$$

$$W = \sum_{i=1}^{n} \int_{x_{i-1}}^{x_i} \left(\frac{s_{2i}}{s_{4i}} x + r_i + d_1 \right) F_x \mathrm{d}x$$
$$+ \left[\frac{s_{2n}}{s_{4n}} x_n + r_n + d_1 \right] \overline{F}_{x2} - d_1 \overline{F}_{x1} \tag{6.4.4}$$

$$F_{\mathrm{S}} = \int_{x_0}^{x_n} F_x \mathrm{d}x + \overline{F}_{x2} - \overline{F}_{x1} \quad \text{(shear force)} \tag{6.4.5}$$

$$M = \int_{x_0}^{x_n} (d_2 - x) F_z \mathrm{d}x + (d_2 - x_n) \overline{F}_{z2} - d_2 \overline{F}_{z1} \quad \text{(bending moment)} \tag{6.4.6}$$

$$\theta = \sum_{i=1}^{n} \int_{x_{i-1}}^{x_i} u_{si}^{(2)}(x) F_x \mathrm{d}x + u_{sn}^{(2)}(x_n) \overline{F}_{x2} - u_{s1}^{(2)}(0) \overline{F}_{x1} \tag{6.4.7}$$

$$U = \sum_{i=1}^{n} \int_{x_{i-1}}^{x_i} w_{si}^{(3)}(x) F_z \mathrm{d}x + w_{sn}^{(3)}(x_n) \overline{F}_{z2} - w_{s1}^{(3)}(0) \overline{F}_{z1} \tag{6.4.8}$$

$$k_3 = \frac{1}{2}(c_5 - c_4) \tag{6.4.9}$$

The boundary conditions at the two ends ($z = 0$ or l) have been considered in the variational principle (6.4.2). Similar to the preceding chapter, for boundary conditions (6.1.6) corresponding to specified surface tractions at the ends, we have

$$\tilde{v}_e^2 = -[\overline{F}_{\mathrm{N}} a_1 + \overline{F}_{\mathrm{S}} a_3 + \overline{M} a_4]_{z=0}^{l} \tag{6.4.10}$$

where

$$\overline{F}_{\mathrm{N}} = \int_0^{x_n} \overline{\sigma} \mathrm{d}x; \quad \overline{F}_{\mathrm{S}} = \int_0^{x_n} \overline{\tau} \mathrm{d}x; \quad \overline{M} = \int_0^{x_n} (d_2 - x) \overline{\sigma} \mathrm{d}x \tag{6.4.11}$$

while for boundary conditions (6.1.7) corresponding to specified displacements, we have

$$\tilde{v}_e^2 = -[k_1 a_1 a_2 + k_2 a_4 a_5 - k_2 a_3 a_6 + 2k_3 a_5 a_6 - \overline{W} a_2 - \overline{\theta} a_5 - \overline{U} a_6]_{z=0}^{l} \tag{6.4.12}$$

where

$$\overline{W} = \sum_{i=1}^{n} \int_{x_{i-1}}^{x_i} \left[\frac{\overline{w}}{s_{4i}} \right] \mathrm{d}x \tag{6.4.13}$$

$$\overline{\theta} = \sum_{i=1}^{n} \int_{x_{i-1}}^{x_i} \left[\frac{\overline{w}}{s_{4i}}(d_2 - x) \right] \mathrm{d}x \tag{6.4.14}$$

$$\overline{U} = \sum_{i=1}^{n} \int_{x_{i-1}}^{x_i} \overline{u} \left[\frac{1}{2s_{4i}}(d_2 - x)^2 + p_i \right] \mathrm{d}x \tag{6.4.15}$$

The variation of Eq. (6.4.2) yields

$$\delta a_1 : \quad k_1 \dot{a}_2 + F_{\mathrm{N}} = 0 \tag{6.4.16}$$

$$\delta a_2 : \quad k_1 \dot{a}_1 - k_1 a_2 - W = 0 \tag{6.4.17}$$

$$\delta a_3 : \quad k_2 \dot{a}_6 - F_{\mathrm{S}} = 0 \tag{6.4.18}$$

$$\delta a_4 : \quad k_2 \dot{a}_5 - k_2 a_6 + M = 0 \tag{6.4.19}$$

$$\delta a_5 : \quad k_2 \dot{a}_4 - k_2 a_5 - \theta = 0 \tag{6.4.20}$$

$$\delta a_6 : \quad k_2 \dot{a}_3 - k_2 a_4 + U = 0 \tag{6.4.21}$$

The boundary conditions at ends $z = 0, l$ are required for solving the equations above. The boundary conditions at both ends ($z = 0$ or l) can be provided by the variation of Eq. (6.4.2). For example, the boundary conditions for specified force are

$$\left.\begin{aligned} k_1 a_2 - \overline{F}_{\mathrm{N}} &= 0 \\ k_2 a_6 + \overline{F}_{\mathrm{S}} &= 0 \\ k_2 a_5 - \overline{M} &= 0 \end{aligned}\right\} \quad \text{at } z = 0 \text{ or } l \tag{6.4.22}$$

while the boundary conditions for specified displacements are

$$\left.\begin{aligned} k_1 a_1 &= \overline{W} \\ k_2 a_4 + k_3 a_6 &= \overline{\theta} \\ k_2 a_3 - k_3 a_5 &= -\overline{U} \end{aligned}\right\} \quad \text{at } z = 0 \text{ or } l \tag{6.4.23}$$

Equations (6.4.16) to (6.4.21) can be solved readily with the boundary conditions above. The analytical solutions for some classical Saint–Venant problems are presented as follows. Here we assume a cantilevered plate with left end fixed and right end free. The plate domain and both boundary surfaces are free of external load.

1. Solution for simple tension

Let an axial force F be applied at $x = d_2$ on the right end, the solution is

$$a_1 = \frac{F}{k_1} z, \quad a_2 = \frac{F}{k_1}, \quad a_3 = a_4 = a_5 = a_6 = 0 \tag{6.4.24}$$

The cross section remains plane and is parallel to the original plane before deformation. The distribution of normal stresses can be represented as a step function. From Eqs. (6.4.16) to (6.4.21), it is obvious that the conclusion above remains valid if the plate is only subjected to distributed longitudinal forces and the corresponding resultant force on the cross section acts along $x = d_2$.

2. Solution for pure bending

Let a concentrated couple m be applied to the right end of the plate, the solution is

$$a_3 = m\left(\frac{z^2}{2k_2} + \frac{k_3}{k_2^3}\right), \quad a_4 = \frac{m}{k_2} z, \quad a_5 = \frac{m}{k_2}, \quad a_1 = a_2 = a_6 = 0 \tag{6.4.25}$$

The cross section remains plane and rotates at an angle about axis $x = d_2$. The normal stress is a piecewise linear function. Similarly, the conclusion above remains valid if all components of forces on the cross section results in a resultant couple.

3. Solution of constant shear bending

Let a shear force F be applied to the right end of the plate, the solution is

$$\left.\begin{aligned}
a_1 &= a_2 = 0 \\
a_3 &= \frac{F}{k_2}\left[-\frac{1}{6}z^3 + \frac{1}{2}lz^2 + \frac{k_3}{k_2}(l + z)\right] \\
a_4 &= \frac{F}{k_2}\left(lz - \frac{1}{2}z^2 + \frac{k_3}{k_2}\right) \\
a_5 &= \frac{F}{k_2}(l - z) \\
a_6 &= -\frac{F}{k_2}
\end{aligned}\right\} \tag{6.4.26}$$

Here, the cross section does not remain plane after deformation due to the existence of $w_s^{(3)}$. The shear stress on the cross section vanishes at both boundary surfaces ($x = 0$ or $x = x_n$), it is convexly distributed and reaches the maximum at the neutral axis $x = d_2$.

In this section, an analytical method by expansion of symplectic subspace of zero eigenvalue is presented for solving Saint–Venant problems of multi-layered laminated composite plates. However, it should be emphasized that considering only the eigen-solutions of zero eigenvalue is not suitable if the thickness of plate x_n is not a higher-order small quantity as compared with the length of plate l, or if the stress singularity at the end, such as the stress at the layer interface, is analyzed[3]. For such problems, it is necessary to add the eigen-solutions of nonzero eigenvalues into the variational principle and then obtain the solutions by expansion. In other words, the nonzero eigenvalues and the corresponding eigen-solutions should first be solved, and then the expanding theorem is applied. The details are omitted here.

References

1. Wanxie Zhong and Weian Yao, Analytical solutions on Saint–Venant problem of layered plates, *Acta Mechanica Sinica*, 1997, 29(5): 617–626 (in Chinese).
2. Wanxie Zhong and Weian Yao, The Saint–Venant solutions of multi-layered composite plates, *Advances in Structural Engineering*, 1997, 1(2): 127–133.
3. Zudao Luo and Sijian Li, *Mechanics of Anisotropic Materials*, Shanghai: Shanghai Jiaotong University Press, 1994 (in Chinese).

Chapter 7

Solutions for Plane Elasticity in Polar Coordinates

This chapter discusses the Hamiltonian system of plane elasticity problems in polar coordinates. By treating the radial coordinate and circumferential coordinate, respectively, as the time coordinate, two different forms of the Hamiltonian systems are established, and thus an analytical method for solving plane elasticity problems in circular and annular domains is presented. Special attention is focused in the Hamiltonian system with radical coordinate treated as time coordinate.

7.1. Plane Elasticity Equations in Polar Coordinates

The solution of plane elasticity problems in rectangular coordinates has been discussed in Chapter 4. For problems in circular, annular or wedge domain, it is more convenient to deal in polar coordinates. An arbitrary point in a polar coordinate system can be represented by the distance between this point and the origin ρ (radius) and the angle φ between the ρ-direction and a certain axis, for example, the x-axis, as shown in Fig. 7.1.

Consider an element *abcd* formed by two radial planes separated by an angle $\mathrm{d}\varphi$ and two cylindrical surfaces with radii ρ and $\rho + \mathrm{d}\rho$, respectively, as shown in Fig. 7.1. The stresses acting on the element are also shown in the figure.

Neglecting any body forces and for an infinitesimal $\mathrm{d}\varphi$, we have

$$\sin\frac{\mathrm{d}\varphi}{2} \approx \frac{\mathrm{d}\varphi}{2}, \quad \cos\frac{\mathrm{d}\varphi}{2} \approx 1 \tag{7.1.1}$$

Projecting all forces acting on the element on the central radial axis ρ yields

$$\left(\sigma_\rho + \frac{\partial \sigma_\rho}{\partial \rho}\mathrm{d}\rho\right)(\rho + \mathrm{d}\rho)\,\mathrm{d}\varphi - \sigma_\rho \rho\,\mathrm{d}\varphi$$

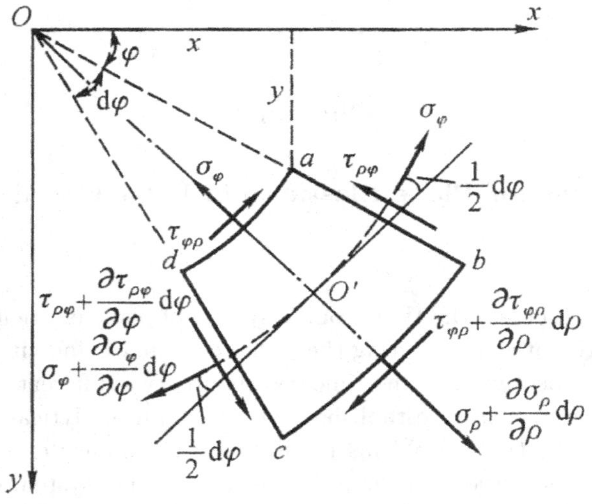

Fig. 7.1. Plane problem in polar coordinates.

$$
\begin{aligned}
&-\left(\sigma_\varphi + \frac{\partial \sigma_\varphi}{\partial \varphi}\mathrm{d}\varphi\right)\mathrm{d}\rho\frac{\mathrm{d}\varphi}{2} - \sigma_\varphi \mathrm{d}\rho\frac{\mathrm{d}\varphi}{2} \\
&+\left(\tau_{\rho\varphi} + \frac{\partial \tau_{\rho\varphi}}{\partial \varphi}\mathrm{d}\varphi\right)\mathrm{d}\rho \cdot 1 - \tau_{\rho\varphi}\mathrm{d}\rho \cdot 1 = 0 \qquad (7.1.2a)
\end{aligned}
$$

while projecting on the tangent of circumferential curve yields

$$
\begin{aligned}
&\left(\sigma_\varphi + \frac{\partial \sigma_\varphi}{\partial \varphi}\mathrm{d}\varphi\right)\mathrm{d}\rho \cdot 1 - \sigma_\varphi \mathrm{d}\rho \cdot 1 \\
&\quad + \left(\tau_{\rho\varphi} + \frac{\partial \tau_{\rho\varphi}}{\partial \varphi}\mathrm{d}\varphi\right)\mathrm{d}\rho\frac{\mathrm{d}\varphi}{2} + \tau_{\rho\varphi}\mathrm{d}\rho\frac{\mathrm{d}\varphi}{2} \\
&\quad + \left(\tau_{\rho\varphi} + \frac{\partial \tau_{\rho\varphi}}{\partial \rho}\mathrm{d}\rho\right)(\rho + \mathrm{d}\rho)\mathrm{d}\varphi - \tau_{\rho\varphi}\rho\mathrm{d}\varphi = 0 \qquad (7.1.2b)
\end{aligned}
$$

Simplifying the above expressions and neglecting all third-order infinitesimal quantities yield the equilibrium equation in polar coordinates as

$$
\frac{\partial \sigma_\rho}{\partial \rho} + \frac{1}{\rho}\frac{\partial \tau_{\rho\varphi}}{\partial \varphi} + \frac{\sigma_\rho - \sigma_\varphi}{\rho} = 0 \qquad (7.1.3a)
$$

$$
\frac{\partial \tau_{\rho\varphi}}{\partial \rho} + \frac{1}{\rho}\frac{\partial \sigma_\varphi}{\partial \varphi} + \frac{2\tau_{\rho\varphi}}{\rho} = 0 \qquad (7.1.3b)
$$

Then we derive the geometric equations in polar coordinates. The displacement components along the radial axis ρ and circumferential curve φ are denoted as u_ρ and u_φ, respectively. As the radial displacement of edge *ad* of the element *abcd* is u_ρ while that of edge *bc* is $u_\rho + (\partial u_\rho/\partial \rho)\mathrm{d}\rho$, the radial strain is

$$\varepsilon_\rho = \frac{\left(u_\rho + \frac{\partial u_\rho}{\partial \rho}\mathrm{d}\rho\right) - u_\rho}{\mathrm{d}\rho} = \frac{\partial u_\rho}{\partial \rho} \tag{7.1.4}$$

In general, there are two parts for the circumferential strain ε_φ as follows:

(1) A component due to radial displacement u_ρ

$$\frac{(\rho + u_\rho)\mathrm{d}\varphi - \rho \mathrm{d}\varphi}{\rho \mathrm{d}\varphi} = \frac{u_\rho}{\rho} \tag{7.1.5}$$

(2) A component due to circumferential displacement u_φ

$$\frac{\left(u_\varphi + \frac{\partial u_\varphi}{\partial \varphi}\mathrm{d}\varphi\right) - u_\varphi}{\rho \mathrm{d}\varphi} = \frac{1}{\rho}\frac{\partial u_\varphi}{\partial \varphi} \tag{7.1.6}$$

Hence, the resultant circumferential strain is

$$\varepsilon_\varphi = \frac{1}{\rho}\left(u_\rho + \frac{\partial u_\varphi}{\partial \varphi}\right) \tag{7.1.7}$$

Now we consider shear strain. Let the element *abcd* be deformed into *a′b′c′d′*.

It is obvious from Fig. 7.2 that the shear strain $\gamma_{\rho\varphi}$ is

$$\gamma_{\rho\varphi} = \gamma + (\beta - \alpha) \tag{7.1.8}$$

where γ denotes the rotation angle of edge *ad* due to the radial displacement u_ρ, as

$$\gamma = \frac{\partial u_\rho}{\rho \partial \varphi} \tag{7.1.9}$$

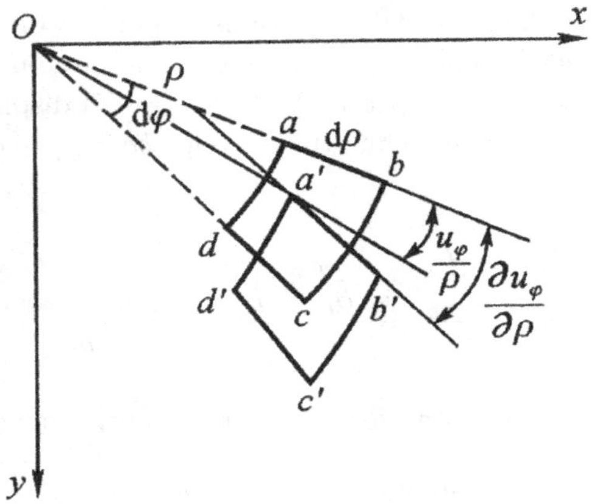

Fig. 7.2. Shear strain in polar coordinate.

β denotes the rotation angle of edge ab due to the circumferential displacement u_φ, as

$$\beta = \frac{\partial u_\varphi}{\partial \rho} \tag{7.1.10}$$

and α is the change of angle from a to a', as

$$\alpha = \frac{u_\varphi}{\rho} \tag{7.1.11}$$

Hence, the shear strain is

$$\gamma_{\rho\varphi} = \frac{\partial u_\rho}{\rho \partial \varphi} + \frac{\partial u_\varphi}{\partial \rho} - \frac{u_\varphi}{\rho} \tag{7.1.12}$$

From Eqs. (7.1.4), (7.1.7) and (7.1.12), the strain-displacement relations in polar coordinates are

$$\left.\begin{aligned} \varepsilon_\rho &= \frac{\partial u_\rho}{\partial \rho} \\ \varepsilon_\varphi &= \frac{1}{\rho}\left(u_\rho + \frac{\partial u_\varphi}{\partial \varphi}\right) \\ \gamma_{\rho\varphi} &= \frac{\partial u_\rho}{\rho \partial \varphi} + \frac{\partial u_\varphi}{\partial \rho} - \frac{u_\varphi}{\rho} \end{aligned}\right\} \tag{7.1.13}$$

Subsequently, we consider the strain-stress relations in polar coordinates for plane stress problems. They are identical with the relations in rectangular coordinates, i.e.

$$\left.\begin{aligned} \varepsilon_\rho &= \frac{1}{E}(\sigma_\rho - \nu\sigma_\varphi) \\ \varepsilon_\varphi &= \frac{1}{E}(\sigma_\varphi - \nu\sigma_\rho) \\ \gamma_{\rho\varphi} &= \frac{2(1+\nu)}{E}\tau_{\rho\varphi} \end{aligned}\right\} \tag{7.1.14}$$

For plane strain problems, the strain-stress relations remain the same form as the above expressions, except E, ν must be interpreted differently, see Eq. (4.1.13) for details.

7.2. Variational Principle for a Circular Sector

Many useful solutions of elasticity can be described in polar coordinates. The typical domain is a circular sector

$$R_1 \le \rho \le R_2, \quad -\alpha \le \varphi \le \alpha \tag{7.2.1}$$

as illustrated in Fig. 7.3.

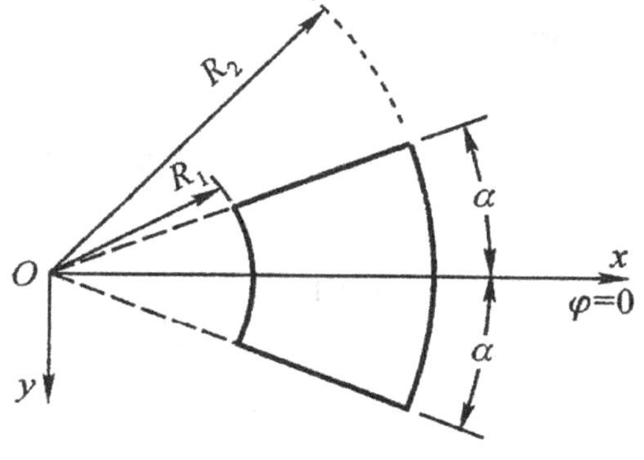

Fig. 7.3. A circular sector in polar coordinates.

For problems in circular sector domain, the corresponding Hellinger–Reissner variational principle is

$$\delta \int_{-\alpha}^{\alpha} \int_{R_1}^{R_2} \left[\sigma_\rho \frac{\partial u_\rho}{\partial \rho} + \frac{\sigma_\varphi}{\rho} \left(u_\rho + \frac{\partial u_\varphi}{\partial \varphi} \right) + \tau_{\rho\varphi} \left(\frac{\partial u_\varphi}{\partial \rho} - \frac{u_\varphi}{\rho} + \frac{1}{\rho} \frac{\partial u_\rho}{\partial \varphi} \right) \right.$$

$$\left. - \frac{1}{2E} (\sigma_\rho^2 + \sigma_\varphi^2 - 2\nu \sigma_\rho \sigma_\varphi + 2(1+\nu)\tau_{\rho\varphi}^2) \right] \rho \mathrm{d}\rho \mathrm{d}\varphi = 0 \qquad (7.2.2)$$

where u_ρ, u_φ, σ_ρ, σ_φ, $\tau_{\rho\varphi}$ are treated as mutually independent variational quantities. The variation of Eq. (7.2.2) yields the equilibrium equations (7.1.3) and the stress-strain relations (7.1.14) in terms of displacement components. In Eq. (7.2.2), free boundary conditions are treated as the natural boundary conditions of variation. If there are displacement boundary conditions, the corresponding boundary terms should be added to the expression above.

The Hamiltonian system and the method of separation of variables are valid for problems in polar coordinates. However, it is necessary to perform substitution of variables first. Here we introduce the transformation

$$\xi = \ln \rho, \quad \text{i.e. } \rho = \mathrm{e}^\xi \qquad (7.2.3)$$

and denote

$$\xi_1 = \ln R_1, \quad \xi_2 = \ln R_2 \qquad (7.2.4)$$

Thus the variational principle (7.2.2) can be written as

$$\delta \int_{-\alpha}^{\alpha} \int_{\xi_1}^{\xi_2} \left[\frac{\sigma_\rho}{\rho} \frac{\partial u_\rho}{\partial \xi} + \frac{\sigma_\varphi}{\rho} \left(u_\rho + \frac{\partial u_\varphi}{\partial \varphi} \right) + \frac{\tau_{\rho\varphi}}{\rho} \left(\frac{\partial u_\varphi}{\partial \xi} - u_\varphi + \frac{\partial u_\rho}{\partial \varphi} \right) \right.$$

$$\left. - \frac{1}{2E} (\sigma_\rho^2 + \sigma_\varphi^2 - 2\nu \sigma_\rho \sigma_\varphi + 2(1+\nu)\tau_{\rho\varphi}^2) \right] \rho^2 \mathrm{d}\xi \mathrm{d}\varphi = 0 \qquad (7.2.5)$$

With the new variables introduced as follows

$$S_\rho = \rho \sigma_\rho, \quad S_\varphi = \rho \sigma_\varphi, \quad S_{\rho\varphi} = \rho \tau_{\rho\varphi} \qquad (7.2.6)$$

the variational equation (7.2.5) can be expressed as

$$\delta \int_{-\alpha}^{\alpha} \int_{\xi_1}^{\xi_2} \left[S_\rho \frac{\partial u_\rho}{\partial \xi} + S_\varphi \left(u_\rho + \frac{\partial u_\varphi}{\partial \varphi} \right) + S_{\rho\varphi} \left(\frac{\partial u_\varphi}{\partial \xi} - u_\varphi + \frac{\partial u_\rho}{\partial \varphi} \right) \right.$$

$$\left. - \frac{1}{2E}(S_\rho^2 + S_\varphi^2 - 2\nu S_\rho S_\varphi + 2(1+\nu)S_{\rho\varphi}^2) \right] \mathrm{d}\xi\, \mathrm{d}\varphi = 0 \qquad (7.2.7)$$

Hence there are only multipliers with constant coefficients in the equation of variation (7.2.7). Although it is much easier, there are still five independent variables $u_\rho, u_\varphi, S_\rho, S_\varphi, S_{\rho\varphi}$. Furthermore, the new domain becomes

$$\xi_1 \le \xi \le \xi_2, \quad -\alpha \le \varphi \le \alpha \qquad (7.2.8)$$

It is equivalent to the rectangular domain in rectangular coordinates. Although the two coordinate directions are symmetrical in a rectangular domain, the simulated radial ξ-direction and the simulated circumferential φ-direction have different characteristics. Hence, it is necessary to consider the Hamiltonian systems corresponding to two different directions: the radial Hamiltonian system with ξ-coordinate treated as time coordinate and the circumferential Hamiltonian system with φ-coordinate treated as time coordinate.

7.3. Hamiltonian System with Radial Coordinate Treated as "Time"

If ξ is treated as time coordinate, φ becomes the transverse direction and, therefore, the transverse forces should be eliminated. The variation of Eq. (7.2.7) with respect to S_φ yields

$$S_\varphi = E\left(u_\rho + \frac{\partial u_\varphi}{\partial \varphi} \right) + \nu S_\rho \qquad (7.3.1)$$

Substituting into Eq. (7.2.7) and eliminating S_φ yield the Hamiltonian mixed energy variational principle as

$$\delta \int_{-\alpha}^{\alpha} \int_{\xi_1}^{\xi_2} \left[S_\rho \frac{\partial u_\rho}{\partial \xi} + S_{\rho\varphi} \frac{\partial u_\varphi}{\partial \xi} + S_\rho \nu \left(u_\rho + \frac{\partial u_\varphi}{\partial \varphi} \right) + S_{\rho\varphi} \left(\frac{\partial u_\rho}{\partial \varphi} - u_\varphi \right) \right.$$

$$\left. + \frac{1}{2} E \left(u_\rho + \frac{\partial u_\varphi}{\partial \varphi} \right)^2 - \frac{1}{2E}((1-\nu^2)S_\rho^2 + 2(1+\nu)S_{\rho\varphi}^2) \right] \mathrm{d}\xi \mathrm{d}\varphi = 0$$

$$(7.3.2)$$

The dual variables of displacements u_ρ, u_φ are $S_\rho, S_{\rho\varphi}$ respectively. Denoting

$$\boldsymbol{q} = \{\, u_\rho, \quad u_\varphi \,\}^{\mathrm{T}}, \quad \boldsymbol{p} = \{\, S_\rho, \quad S_{\rho\varphi} \,\}^{\mathrm{T}} \tag{7.3.3}$$

and representing differentiation with respect to ξ as a dot, Eq. (7.3.2) can be expressed as

$$\delta \int_{-\alpha}^{\alpha} \int_{\xi_1}^{\xi_2} [\boldsymbol{p}^{\mathrm{T}} \dot{\boldsymbol{q}} - \mathscr{H}(\boldsymbol{q}, \boldsymbol{p})] \mathrm{d}\xi \mathrm{d}\varphi = 0 \tag{7.3.4}$$

where the Hamiltonian density function is

$$\begin{aligned} \mathscr{H}(\boldsymbol{q}, \boldsymbol{p}) = & -S_\rho \nu \left(u_\rho + \frac{\partial u_\varphi}{\partial \varphi} \right) - S_{\rho\varphi} \left(\frac{\partial u_\rho}{\partial \varphi} - u_\varphi \right) \\ & - \frac{1}{2} E \left(u_\rho + \frac{\partial u_\varphi}{\partial \varphi} \right)^2 + \frac{1}{2E} \left[(1 - \nu^2) S_\rho^2 + 2(1 + \nu) S_{\rho\varphi}^2 \right] \end{aligned} \tag{7.3.5}$$

This is the variational principle expression of Hamiltonian system for field problems. Expanding the variational equation (7.3.4) yields the Hamiltonian dual system of equations.

$$\left.\begin{aligned} \dot{\boldsymbol{q}} &= \boldsymbol{A} \boldsymbol{q} + \boldsymbol{D} \boldsymbol{p} \\ \dot{\boldsymbol{p}} &= \boldsymbol{B} \boldsymbol{q} - \boldsymbol{A}^{\mathrm{T}} \boldsymbol{p} \end{aligned}\right\} \tag{7.3.6}$$

where the operator matrices are

$$\left.\begin{aligned} & \boldsymbol{A} = \begin{bmatrix} -\nu & -\nu \dfrac{\partial \cdot}{\partial \varphi} \\ -\dfrac{\partial \cdot}{\partial \varphi} & 1 \end{bmatrix}, \quad \boldsymbol{A}^{\mathrm{T}} = \begin{bmatrix} -\nu & \dfrac{\partial \cdot}{\partial \varphi} \\ \nu \dfrac{\partial \cdot}{\partial \varphi} & 1 \end{bmatrix} \\ & \boldsymbol{D} = \begin{bmatrix} \dfrac{1 - \nu^2}{E} & 0 \\ 0 & \dfrac{2(1 + \nu)}{E} \end{bmatrix} \\ & \boldsymbol{B} = \begin{bmatrix} E & E \dfrac{\partial \cdot}{\partial \varphi} \\ -E \dfrac{\partial \cdot}{\partial \varphi} & -E \dfrac{\partial^2 \cdot}{\partial \varphi^2} \end{bmatrix} \end{aligned}\right\} \tag{7.3.7}$$

Introducing the full state vector

$$\boldsymbol{v} = \{ \boldsymbol{q}^{\mathrm{T}} \quad \boldsymbol{p}^{\mathrm{T}} \}^{\mathrm{T}} = \{ u_\rho, \quad u_\varphi; \quad S_\rho, \quad S_{\rho\varphi} \}^{\mathrm{T}} \tag{7.3.8}$$

then Eq. (7.3.6) can be written as

$$\dot{\boldsymbol{v}} = \boldsymbol{H}\boldsymbol{v} \tag{7.3.9}$$

where the Hamiltonian operator is

$$\boldsymbol{H} = \begin{bmatrix} \boldsymbol{A} & \boldsymbol{D} \\ \boldsymbol{B} & -\boldsymbol{A}^{\mathrm{T}} \end{bmatrix} \tag{7.3.10}$$

No external load is included in the derivation above and the equations are homogeneous. Besides, the boundary conditions at two ends ($\xi = \xi_1$ or $\xi = \xi_2$) are not involved. In the Hamiltonian variational principle (7.3.4) the free boundary conditions are treated as natural boundary conditions. The free boundary conditions at $\varphi = \pm\alpha$ are

$$E\left(u_\rho + \frac{\partial u_\varphi}{\partial \varphi}\right) + \nu S_\rho = 0, \quad S_{\rho\varphi} = 0 \quad \text{at } \varphi = \pm\alpha \tag{7.3.11}$$

To discuss the property of operator $\boldsymbol{H}$, we introduce a unit symplectic matrix

$$\boldsymbol{J} = \begin{bmatrix} \boldsymbol{0} & \boldsymbol{I}_2 \\ -\boldsymbol{I}_2 & \boldsymbol{0} \end{bmatrix} \tag{7.3.12}$$

Denote

$$\begin{aligned} \langle \boldsymbol{v}_1, \boldsymbol{v}_2 \rangle &\overset{\text{def}}{=} \int_{-\alpha}^{\alpha} \boldsymbol{v}_1^{\mathrm{T}} \boldsymbol{J} \boldsymbol{v}_2 \mathrm{d}\varphi \\ &= \int_{-\alpha}^{\alpha} (u_{\rho 1} S_{\rho 2} + u_{\varphi 1} S_{\rho\varphi 2} - S_{\rho 1} u_{\rho 2} - S_{\rho\varphi 1} u_{\varphi 2}) \mathrm{d}\varphi \end{aligned} \tag{7.3.13}$$

Obviously Eq. (7.3.13) satisfies the four conditions of symplectic inner product (1.3.2). Hence, the full state vectors $\boldsymbol{v}$ form a symplectic space according to the definition of symplectic inner product (7.3.13).

Based on integration by parts, it is easy to verify that

$$\begin{aligned} \langle \boldsymbol{v}_1, \boldsymbol{H}\boldsymbol{v}_2 \rangle = \langle \boldsymbol{v}_2, \boldsymbol{H}\boldsymbol{v}_1 \rangle &+ \left[u_{\varphi 2} \left(E u_{\rho 1} + E \frac{\partial u_{\varphi 1}}{\partial \varphi} + \nu S_{\rho 1} \right) + u_{\rho 2} S_{\rho\varphi 1} \right]_{-\alpha}^{\alpha} \\ &- \left[u_{\varphi 1} \left(E u_{\rho 2} + E \frac{\partial u_{\varphi 2}}{\partial \varphi} + \nu S_{\rho 2} \right) + u_{\rho 1} S_{\rho\varphi 2} \right]_{-\alpha}^{\alpha} \end{aligned} \tag{7.3.14}$$

For continuously differentiable full state vectors $\boldsymbol{v}_1, \boldsymbol{v}_2$ satisfying the boundary conditions (7.3.11), we have the identity

$$\langle \boldsymbol{v}_1, \boldsymbol{H}\boldsymbol{v}_2 \rangle = \langle \boldsymbol{v}_2, \boldsymbol{H}\boldsymbol{v}_1 \rangle \tag{7.3.15}$$

Hence, the operator $\boldsymbol{H}$ is a Hamiltonian operator matrix in symplectic space.

Similar to the cases in Chapter 4, the dual equation (7.3.9) with boundary conditions (7.3.11) is a linear Hamiltonian system. Therefore, the superposition principle is applicable and the method of separation of variables is particularly effective. Let

$$\boldsymbol{v}(\xi, \varphi) = \mathrm{e}^{\mu\xi}\boldsymbol{\psi}(\varphi) \tag{7.3.16}$$

where μ is the unknown eigenvalue and $\boldsymbol{\psi}(x)$ is the eigenvector which is a function of φ. The eigenvalue equation is

$$\boldsymbol{H}\boldsymbol{\psi}(\varphi) = \mu\boldsymbol{\psi}(\varphi) \tag{7.3.17}$$

The eigenvector $\boldsymbol{\psi}(\varphi)$ is required to satisfy the boundary conditions (7.3.11).

At this point, $\boldsymbol{H}$ has been proved as a Hamiltonian operator matrix and the properties of Hamiltonian operator matrix have been repeatedly elaborated in various previous chapters, i.e.

(1) If μ is an eigenvalue of a Hamiltonian matrix, $-\mu$ is also an eigenvalue.

As a Hamiltonian eigen-problem with infinite dimensions is discussed here, there are infinite eigenvalues which can be divided into two sets:

$$(\alpha)\ \mu_i, \mathrm{Re}(\mu_i) < 0 \text{ or } \mathrm{Re}(\mu_i) = 0 \wedge \mathrm{Im}(\mu_i) < 0 \quad (i = 1, 2, \ldots) \tag{7.3.18a}$$

$$(\beta)\ \mu_{-i} = -\mu_i \tag{7.3.18b}$$

The eigenvalues in the (α)-set are further arranged in an ascending order according to $|\mu_i|$ and then the (β)-set can be arranged correspondingly.

(2) The eigenvectors of Hamiltonian operator matrix are adjoint symplectic orthogonal. Let $\boldsymbol{\psi}_i$ and $\boldsymbol{\psi}_j$ be the eigenvectors corresponding to the eigenvalues μ_i and μ_j, respectively. For $\mu_i + \mu_j \neq 0$, the eigenvectors are symplectic orthogonal

$$\langle \boldsymbol{\psi}_i, \boldsymbol{\psi}_j \rangle = \int_{-\alpha}^{\alpha} \boldsymbol{\psi}_i^{\mathrm{T}} \boldsymbol{J} \boldsymbol{\psi}_j \mathrm{d}\varphi = 0 \tag{7.3.19}$$

The eigenvector, which is symplectic adjoint with ψ_i, is the eigenvector (or the Jordan form eigenvector) corresponding to eigenvalue $-\mu_i$.

To obtain eigen-solutions of nonzero eigenvalues, the eigen equation (7.3.17) is expanded as

$$\left.\begin{array}{lllll} -(\mu+\nu)u_\rho & -\nu\dfrac{\mathrm{d}u_\varphi}{\mathrm{d}\varphi} & +\dfrac{1-\nu^2}{E}S_\rho & +0 & =0 \\ -\dfrac{\mathrm{d}u_\rho}{\mathrm{d}\varphi} & +(1-\mu)u_\varphi & +0 & +\dfrac{2(1+\nu)}{E}S_{\rho\varphi} & =0 \\ Eu_\rho & +E\dfrac{\mathrm{d}u_\varphi}{\mathrm{d}\varphi} & +(\nu-\mu)S_\rho & -\dfrac{\mathrm{d}S_{\rho\varphi}}{\mathrm{d}\varphi} & =0 \\ -E\dfrac{\mathrm{d}u_\rho}{\mathrm{d}\varphi} & -E\dfrac{\mathrm{d}^2u_\varphi}{\mathrm{d}\varphi^2} & -\nu\dfrac{\mathrm{d}S_\rho}{\mathrm{d}\varphi} & -(1+\mu)S_{\rho\varphi} & =0 \end{array}\right\} \tag{7.3.20}$$

This is a set of ordinary differential equations with respect to φ. The eigenvalue λ with respect to the φ-direction should first be solved by using the eigenvalue equation

$$\det\begin{bmatrix} -(\mu+\nu) & -\nu\lambda & (1-\nu^2)/E & 0 \\ -\lambda & (1-\mu) & 0 & 2(1+\nu)/E \\ E & E\lambda & \nu-\mu & -\lambda \\ -E\lambda & -E\lambda^2 & -\nu\lambda & -(1+\mu) \end{bmatrix}=0 \tag{7.3.21}$$

Expanding the determinant yields the characteristic polynomial

$$\lambda^4+2(1+\mu^2)\lambda^2+(1-\mu^2)^2=0 \tag{7.3.22}$$

and the solutions are

$$\lambda_{1,2}=\pm(1+\mu)\mathrm{i},\quad \lambda_{3,4}=\pm(1-\mu)\mathrm{i} \tag{7.3.23}$$

The general solutions for different μ are different. Here μ is yet to be determined.

(1) For $\mu \neq 0, \pm 1$, Eq. (7.3.23) give four distinct roots and the general solution is

$$\left.\begin{aligned} u_\rho &= A_1\cos[(1+\mu)\varphi] + B_1\sin[(1+\mu)\varphi] \\ &\quad + C_1\cos[(1-\mu)\varphi] + D_1\sin[(1-\mu)\varphi] \\ u_\varphi &= A_2\sin[(1+\mu)\varphi] + B_2\cos[(1+\mu)\varphi] \\ &\quad + C_2\sin[(1-\mu)\varphi] + D_2\cos[(1-\mu)\varphi] \\ S_\rho &= A_3\cos[(1+\mu)\varphi] + B_3\sin[(1+\mu)\varphi] \\ &\quad + C_3\cos[(1-\mu)\varphi] + D_3\sin[(1-\mu)\varphi] \\ S_{\rho\varphi} &= A_4\sin[(1+\mu)\varphi] + B_4\cos[(1+\mu)\varphi] \\ &\quad + C_4\sin[(1-\mu)\varphi] + D_4\cos[(1-\mu)\varphi] \end{aligned}\right\} \tag{7.3.24}$$

The constants are not independent and they are required to satisfy Eq. (7.3.20), i.e.

$$\left.\begin{array}{ccccc} -(\mu+\nu)A_1 & -\nu(1+\mu)A_2 & +\dfrac{1-\nu^2}{E}A_3 & +0 & =0 \\ (1+\mu)A_1 & +(1-\mu)A_2 & +0 & +\dfrac{2(1+\nu)}{E}A_4 & =0 \\ EA_1 & +E(1+\mu)A_2 & +(\nu-\mu)A_3 & -(1+\mu)A_4 & =0 \\ E(1+\mu)A_1 & +E(1+\mu)^2A_2 & +\nu(1+\mu)A_3 & -(1+\mu)A_4 & =0 \end{array}\right\} \tag{7.3.25}$$

$$\left.\begin{array}{ccccc} -(\mu+\nu)B_1 & +\nu(1+\mu)B_2 & +\dfrac{1-\nu^2}{E}B_3 & +0 & =0 \\ -(1+\mu)B_1 & +(1-\mu)B_2 & +0 & +\dfrac{2(1+\nu)}{E}B_4 & =0 \\ EB_1 & -E(1+\mu)B_2 & +(\nu-\mu)B_3 & +(1+\mu)B_4 & =0 \\ -E(1+\mu)B_1 & +E(1+\mu)^2B_2 & -\nu(1+\mu)B_3 & -(1+\mu)B_4 & =0 \end{array}\right\} \tag{7.3.26}$$

$$\left.\begin{array}{ccccc} -(\mu+\nu)C_1 & -\nu(1-\mu)C_2 & +\dfrac{1-\nu^2}{E}C_3 & +0 & =0 \\ (1-\mu)C_1 & +(1-\mu)C_2 & +0 & +\dfrac{2(1+\nu)}{E}C_4 & =0 \\ EC_1 & +E(1-\mu)C_2 & +(\nu-\mu)C_3 & -(1-\mu)C_4 & =0 \\ E(1-\mu)C_1 & +E(1-\mu)^2C_2 & +\nu(1-\mu)C_3 & -(1+\mu)C_4 & =0 \end{array}\right\} \tag{7.3.27}$$

and

$$\left.\begin{array}{lllll} -(\mu+\nu)D_1 & +\nu(1-\mu)D_2 & +\dfrac{1-\nu^2}{E}D_3 & +0 & =0 \\ -(1-\mu)D_1 & +(1-\mu)D_2 & +0 & +\dfrac{2(1+\nu)}{E}D_4 & =0 \\ ED_1 & -E(1-\mu)D_2 & +(\nu-\mu)D_3 & +(1-\mu)D_4 & =0 \\ -E(1-\mu)D_1 & +E(1-\mu)^2D_2 & -\nu(1-\mu)D_3 & -(1+\mu)D_4 & =0 \end{array}\right\} \tag{7.3.28}$$

There is a redundant equation in each of the four sets of equations and hence there is only one independent coefficient in each of the four sets of coefficients A_i, B_i, C_i, D_i. Let A_1, B_2, C_1, D_2 be chosen as the independent coefficients, solving the four sets of equations yields

$$\left.\begin{aligned} A_2 &= -A_1 \\ A_3 &= \frac{E\mu}{1+\nu}A_1 \\ A_4 &= -\frac{E\mu}{1+\nu}A_1 \end{aligned}\right\}; \qquad \left.\begin{aligned} C_2 &= \frac{-3+\nu-\mu-\nu\mu}{3-\nu-\mu-\nu\mu}C_1 \\ C_3 &= \frac{E\mu(3-\mu)}{3-\nu-\mu-\nu\mu}C_1 \\ C_4 &= \frac{E\mu(1-\mu)}{3-\nu-\mu-\nu\mu}C_1 \end{aligned}\right\} \tag{7.3.29}$$

and

$$\left.\begin{aligned} B_1 &= B_2 \\ B_3 &= \frac{E\mu}{1+\nu}B_2 \\ B_4 &= \frac{E\mu}{1+\nu}B_2 \end{aligned}\right\}; \qquad \left.\begin{aligned} D_1 &= \frac{3-\nu-\mu-\nu\mu}{3-\nu+\mu+\nu\mu}D_2 \\ D_3 &= \frac{E\mu(3-\mu)}{3-\nu+\mu+\nu\mu}D_2 \\ D_4 &= \frac{-E\mu(1-\mu)}{3-\nu+\mu+\nu\mu}D_2 \end{aligned}\right\} \tag{7.3.30}$$

(2) For $\mu = \pm 1$, Eq. (7.3.23) yield roots $\pm 2\mathrm{i}$ and zero where zero is a double root. The general solution is

$$\left.\begin{aligned} u_\rho &= A_1\cos(2\varphi) + B_1\sin(2\varphi) + C_1 + D_1\varphi \\ u_\varphi &= A_2\sin(2\varphi) + B_2\cos(2\varphi) + C_2\varphi + D_2 \\ S_\rho &= A_3\cos(2\varphi) + B_3\sin(2\varphi) + C_3 + D_3\varphi \\ S_{\rho\varphi} &= A_4\sin(2\varphi) + B_4\cos(2\varphi) + C_4\varphi + D_4 \end{aligned}\right\} \tag{7.3.31}$$

Similarly, these coefficients are not independent and they are required to satisfy Eq. (7.3.20). Substituting Eq. (7.3.31) into Eq. (7.3.20) yields $C_2 = C_4 = D_1 = D_3 = 0$. Thus, the general solution (7.3.31) for $\mu = \pm 1$

can still be presented in the form of Eq. (7.3.24). The relations between the coefficients remain as Eqs. (7.3.29) and (7.3.30). Hence, the general solution of eigen-solution for $\mu \neq 0$ is given by Eq. (7.3.24) in all subsequent discussion.

(3) The case for $\mu = 0$ is actually not included in the grouping in Eq. (7.3.18). Zero eigenvalue is a special case which has particular physical interpretation and should be discussed separately. Such cases have been repeatedly discussed in the previous few chapters.

The general solution of eigen-solution correspond to nonzero eigenvalues has been discussed above. It should be noted that the general solution (7.3.24) is also required to satisfy, in addition to the relations in Eqs. (7.3.29) and (7.3.30), the boundary conditions (7.3.11).

For elasticity problems of homogeneous materials, the solutions can be divided into two parts: the symmetric deformation and the anti-symmetric deformation with respect to $\varphi = 0$. The symmetric conditions are

$$u_\varphi = 0, \quad S_{\rho\varphi} = 0 \quad \text{at } \varphi = 0 \tag{7.3.32}$$

while the anti-symmetric conditions are

$$E\frac{\partial u_\varphi}{\partial \varphi} + \nu S_\rho = 0, \quad u_\rho = 0 \quad \text{at } \varphi = 0 \tag{7.3.33}$$

Obviously in the general solution (7.3.24), sets A and C correspond to the eigen-solutions of symmetric deformation while sets B and D correspond to the eigen-solutions of anti-symmetric deformation.

Before presenting the eigen-solutions, the expansion theorem of eigenvectors is introduced first.

Every full state vector $\boldsymbol{v}$ can be represented by eigen-solutions, that is

$$\boldsymbol{v} = \sum_{i=1}^{\infty} (a_i \boldsymbol{\psi}_i + b_i \boldsymbol{\psi}_{-i}) \tag{7.3.34}$$

where a_i and b_i are undetermined factors. It should be emphasized that Eq. (7.3.34) includes all eigenvectors corresponding to Eq. (7.3.18) and zero eigenvalue. In addition, normal adjoint symplectic orthonormalization of the eigenvectors have been completed, i.e.

$$\langle \boldsymbol{\psi}_i, \boldsymbol{\psi}_{-j} \rangle = \delta_{ij}, \quad \langle \boldsymbol{\psi}_i, \boldsymbol{\psi}_j \rangle = \langle \boldsymbol{\psi}_{-i}, \quad \boldsymbol{\psi}_{-j} \rangle = 0 \quad (i, j = 1, 2, \ldots) \tag{7.3.35}$$

where δ_{ij} is the Kronecker delta.

For a Hamiltonian operator matrix of infinite dimension, there is a completeness problem for basis which should be demonstrated strictly in terms of mathematics. However, it can be treated as a complete system in elasticity solution methodology and the proof of completeness is left for mathematicians[3].

7.4. Eigen-Solutions for Symmetric Deformation in Radial Hamiltonian System

The eigenvalue equation for symmetric deformation is Eq. (7.3.17) and the boundary conditions are the symmetric conditions (7.3.32) and the free boundary conditions at $\varphi = \alpha$

$$E\left(u_\rho + \frac{\partial u_\varphi}{\partial \varphi}\right) + \nu S_\rho = 0, \quad S_{\rho\varphi} = 0 \quad \text{at } \varphi = \alpha \tag{7.4.1}$$

7.4.1. *Eigen-Solutions of Zero Eigenvalue*

As repeatedly emphasized in the previous few chapters, zero eigenvalue is associated with special cases which should first be analyzed. The basic equation for eigen-solutions of zero eigenvalue is

$$\boldsymbol{H}\boldsymbol{\psi}_0^{(s0)} = \boldsymbol{0} \tag{7.4.2}$$

Expanding the above equation yields

$$\left.\begin{array}{lllll}
-\nu u_{\rho 0}^{(s0)} & -\nu \dfrac{\mathrm{d}u_{\varphi 0}^{(s0)}}{\mathrm{d}\varphi} & +\dfrac{1-\nu^2}{E} S_{\rho 0}^{(s0)} & +0 & = 0 \\[2ex]
-\dfrac{\mathrm{d}u_{\rho 0}^{(s0)}}{\mathrm{d}\varphi} & +u_{\varphi 0}^{(s0)} & +0 & +\dfrac{2(1+\nu)}{E} S_{\rho\varphi 0}^{(s0)} & = 0 \\[2ex]
E u_{\rho 0}^{(s0)} & +E\dfrac{\mathrm{d}u_{\varphi 0}^{(s0)}}{\mathrm{d}\varphi} & +\nu S_{\rho 0}^{(s0)} & -\dfrac{\mathrm{d}S_{\rho\varphi 0}^{(s0)}}{\mathrm{d}\varphi} & = 0 \\[2ex]
-E\dfrac{\mathrm{d}u_{\rho 0}^{(s0)}}{\mathrm{d}\varphi} & -E\dfrac{\mathrm{d}^2 u_{\varphi 0}^{(s0)}}{\mathrm{d}\varphi^2} & -\nu \dfrac{\mathrm{d}S_{\rho 0}^{(s0)}}{\mathrm{d}\varphi} & -S_{\rho\varphi 0}^{(s0)} & = 0
\end{array}\right\} \tag{7.4.2$'$}$$

From the last two equations we obtain

$$\frac{\mathrm{d}^2 S_{\rho\varphi 0}^{(s0)}}{\mathrm{d}\varphi^2} + S_{\rho\varphi 0}^{(s0)} = 0 \tag{7.4.3}$$

The general solution is

$$S_{\rho\varphi 0}^{(s0)} = c_1 \cos\varphi + c_2 \sin\varphi \tag{7.4.4}$$

Substituting Eq. (7.4.4) into the third of Eq. (7.4.2) yields

$$E\left(u_{\rho 0}^{(s0)}+\frac{\mathrm{d}u_{\varphi 0}^{(s0)}}{\mathrm{d}\varphi}\right)+\nu S_{\rho 0}^{(s0)}=-c_1\sin\varphi+c_2\cos\varphi \tag{7.4.5}$$

Further substituting Eqs. (7.4.4) and (7.4.5) into the symmetric conditions (7.3.32) and the boundary conditions (7.4.1) leads to

$$c_1=c_2=0 \tag{7.4.6}$$

and hence

$$S_{\rho\varphi 0}^{(s0)}=E\left(u_{\rho 0}^{(s0)}+\frac{\mathrm{d}u_{\varphi 0}^{(s0)}}{\mathrm{d}\varphi}\right)+\nu S_{\rho 0}^{(s0)}=0 \tag{7.4.7}$$

Solving simultaneously Eq. (7.4.7) and the first of Eq. (7.4.2), we obtain the solution

$$S_{\rho 0}^{(s0)}=0 \tag{7.4.8}$$

and

$$u_{\rho 0}^{(s0)}+\frac{\mathrm{d}u_{\varphi 0}^{(s0)}}{\mathrm{d}\varphi}=0 \tag{7.4.9}$$

Substituting $S_{\rho\varphi 0}^{(s0)}=0$ into the second of Eq. (7.4.2) yields

$$\frac{\mathrm{d}u_{\rho 0}^{(s0)}}{\mathrm{d}\varphi}-u_{\varphi 0}^{(s0)}=0 \tag{7.4.10}$$

Solving simultaneously Eqs. (7.4.9) and (7.4.10) yields

$$\left.\begin{aligned} u_{\rho 0}^{(s0)}&=c_3\cos\varphi+c_4\sin\varphi\\ u_{\varphi 0}^{(s0)}&=-c_3\sin\varphi+c_4\cos\varphi\end{aligned}\right\} \tag{7.4.11}$$

Substituting the solution into the symmetric conditions (7.3.32) yields $c_4=0$. Therefore the symmetric basic eigenvector of zero eigenvalue is

$$\boldsymbol{\psi}_0^{(s0)}=\left\{\cos\varphi,\quad -\sin\varphi,\quad 0,\quad 0\right\}^{\mathrm{T}} \tag{7.4.12}$$

The eigenvector (7.4.12) is the solution $\boldsymbol{v}_0^{(s0)}=\boldsymbol{\psi}_0^{(s0)}$ of the original problem and it is physically interpreted as the unit rigid body translation along the symmetry axis.

The first-order Jordan form eigen-solution exists because there is only one chain for the eigen-solution of zero eigenvalue. The corresponding equation is

$$\boldsymbol{H}\boldsymbol{\psi}_0^{(s1)} = \boldsymbol{\psi}_0^{(s0)} \tag{7.4.13}$$

Similar to the derivation earlier, we obtain $S_{\rho\varphi 0}^{(s1)} = 0$ from the last two equations in Eq. (7.4.13) and the corresponding boundary conditions.

Substitute $S_{\rho\varphi 0}^{(s1)} = 0$ into the third equation of the expanded Eq. (7.4.13) and solve simultaneously with the first equation, the solutions are

$$S_{\rho 0}^{(s1)} = E\cos\varphi \tag{7.4.14}$$

and

$$u_{\rho 0}^{(s1)} + \frac{\mathrm{d}u_{\varphi 0}^{(s1)}}{\mathrm{d}\varphi} = -\nu\cos\varphi \tag{7.4.15}$$

Substituting $S_{\rho\varphi 0}^{(s1)} = 0$ into the second equation of Eq. (7.4.13) yields

$$\frac{\mathrm{d}u_{\rho 0}^{(s1)}}{\mathrm{d}\varphi} - u_{\varphi 0}^{(s1)} = \sin\varphi \tag{7.4.16}$$

Solving the simultaneous equations of Eqs. (7.4.15) and (7.4.16) yields

$$\left.\begin{aligned} u_{\rho 0}^{(s1)} &= \frac{1-\nu}{2}\varphi\sin\varphi + c_3\cos\varphi + c_4\sin\varphi \\ u_{\varphi 0}^{(s1)} &= \frac{1-\nu}{2}\varphi\cos\varphi - \frac{1+\nu}{2}\sin\varphi - c_3\sin\varphi + c_4\cos\varphi \end{aligned}\right\} \tag{7.4.17}$$

Substituting the solution into the symmetric conditions (7.3.32) yields $c_4 = 0$. Besides, term c_3 is a basic eigen-solution which can be superposed arbitrarily. Therefore the symmetric first-order Jordan form eigenvector of zero eigenvalue is

$$\boldsymbol{\psi}_0^{(s1)} = \left\{\frac{1-\nu}{2}\varphi\sin\varphi,\quad \frac{1-\nu}{2}\varphi\cos\varphi - \frac{1+\nu}{2}\sin\varphi,\quad E\cos\varphi,\quad 0\right\}^{\mathrm{T}} \tag{7.4.18}$$

Although this vector is not the direct solution of the original problem (7.3.9), the solution can be constituted according to Eq. (7.4.18) as

$$\boldsymbol{v}_0^{(s1)} = \boldsymbol{\psi}_0^{(s1)} + \xi\boldsymbol{\psi}_0^{(s0)} \tag{7.4.19}$$

and the corresponding stress field is

$$\sigma_\rho = \frac{1}{\rho} E \cos\varphi, \quad \sigma_\varphi = 0, \quad \tau_{\rho\varphi} = 0 \tag{7.4.20}$$

Obviously a resultant force along the symmetry axis at two ends ($\xi = \xi_1$ or $\xi = \xi_2$) is resulted as

$$\left.\begin{aligned} F_{\mathrm{N}} &= \int_{-\alpha}^{\alpha} (-\sigma_\rho \cos\varphi + \tau_{\rho\varphi} \sin\varphi)\rho \mathrm{d}\varphi = -\frac{E}{2}(2\alpha + \sin 2\alpha) \neq 0 \\ F_{\mathrm{S}} &= \int_{-\alpha}^{\alpha} (-\sigma_\rho \sin\varphi - \tau_{\rho\varphi} \cos\varphi)\rho \mathrm{d}\varphi = 0 \\ M &= \int_{-\alpha}^{\alpha} (\tau_{\rho\varphi}\rho^2) \mathrm{d}\varphi = 0 \end{aligned}\right\} \tag{7.4.21}$$

For $\xi_1 \to -\infty (R_1 = 0)$, we have

$$\tilde{\boldsymbol{v}}_0^{(s1)} = \frac{1}{F_{\mathrm{N}}} \boldsymbol{v}_0^{(s1)} \tag{7.4.22}$$

This is the solution of an elastic wedge acted by a unit concentrated force at the apex $R_1 = 0$ and along the symmetry axis (see Fig. 7.4). It is consistent with the published solution in the open literature. For an angle $\alpha = \pi/2$, the solution (7.4.22) becomes the solution of a semi-infinite plane acted by a concentrated force.

As eigenvectors $\boldsymbol{\psi}_0^{(s1)}$ and $\boldsymbol{\psi}_0^{(s0)}$ are symplectic adjoint, i.e.

$$\langle \boldsymbol{\psi}_0^{(s0)}, \boldsymbol{\psi}_0^{(s1)} \rangle = \int_{-\alpha}^{\alpha} \boldsymbol{\psi}_0^{(s0)\mathrm{T}} \boldsymbol{J} \boldsymbol{\psi}_0^{(s1)} \mathrm{d}\varphi = \frac{1}{2} E(2\alpha + \sin 2\alpha) \neq 0 \tag{7.4.23}$$

Equation (7.4.18) is the dual solution of Eq. (7.4.12), and thus the Jordan form chain of symmetric eigen-solution of zero eigenvalue is terminated.

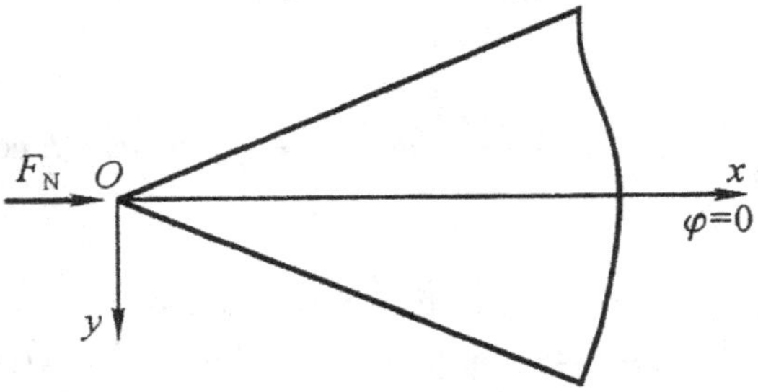

Fig. 7.4. An elastic wedge acted by a concentrated force at the apex and along the symmetry axis.

7.4.2. Eigen-Solutions of Nonzero Eigenvalues

From Eqs. (7.3.24) and (7.3.29), the general eigen-solution symmetric with respect to $\varphi = 0$ for nonzero eigenvalues is

$$\left.\begin{aligned}
u_\rho &= A_1 \cos[(1+\mu)\varphi] + C_1 \cos[(1-\mu)\varphi] \\
u_\varphi &= -A_1 \sin[(1+\mu)\varphi] + \frac{-3+\nu-\mu-\nu\mu}{3-\nu-\mu-\nu\mu} C_1 \sin[(1-\mu)\varphi] \\
S_\rho &= \frac{E\mu}{1+\nu} A_1 \cos[(1+\mu)\varphi] + \frac{E\mu(3-\mu)}{3-\nu-\mu-\nu\mu} C_1 \cos[(1-\mu)\varphi] \\
S_{\rho\varphi} &= -\frac{E\mu}{1+\nu} A_1 \sin[(1+\mu)\varphi] + \frac{E\mu(1-\mu)}{3-\nu-\mu-\nu\mu} C_1 \sin[(1-\mu)\varphi]
\end{aligned}\right\} \tag{7.4.24}$$

Although the general solution (7.4.24) satisfies the differential equation (7.3.17) within the domain and the symmetric conditions (7.3.32), the boundary conditions (7.4.1) have not been satisfied. Substituting Eq. (7.4.24) into Eq. (7.4.1) yields

$$\left.\begin{aligned}
-\frac{E\mu}{1+\nu} A_1 \cos[(1+\mu)\alpha] + \frac{E\mu(1+\mu)}{3-\nu-\mu-\nu\mu} C_1 \cos[(1-\mu)\alpha] = 0 \\
-\frac{E\mu}{1+\nu} A_1 \sin[(1+\mu)\alpha] + \frac{E\mu(1-\mu)}{3-\nu-\mu-\nu\mu} C_1 \sin[(1-\mu)\alpha] = 0
\end{aligned}\right\} \tag{7.4.25}$$

The determinant vanishes as A_1 and C_1 are not zero simultaneously. Thus the transcendental equation for nonzero eigenvalue μ with respect to symmetric deformation is derived as

$$\sin(2\mu\alpha) + \mu \sin(2\alpha) = 0 \tag{7.4.26}$$

Apparently $-\mu$ must be an eigenvalue if μ is an eigenvalue and it validates one of the characteristics in Eqs. (7.3.18) for symplectic eigenvalue problems. In addition, it can be proved that there is no pure imaginary root.

It is possible to apply the Newton method to solve Eq. (7.4.26). For nonzero eigenvalue, the transcendental equation (7.4.26) can be written as

$$\frac{\sin x}{x} = -\frac{\sin 2\alpha}{2\alpha} \quad \text{where } x = 2\mu\alpha \tag{7.4.27}$$

There solution can be divided into two cases: $\alpha > \pi/2$ and $\alpha < \pi/2$. The right-hand-side is positive for the former while it is negative for the latter.

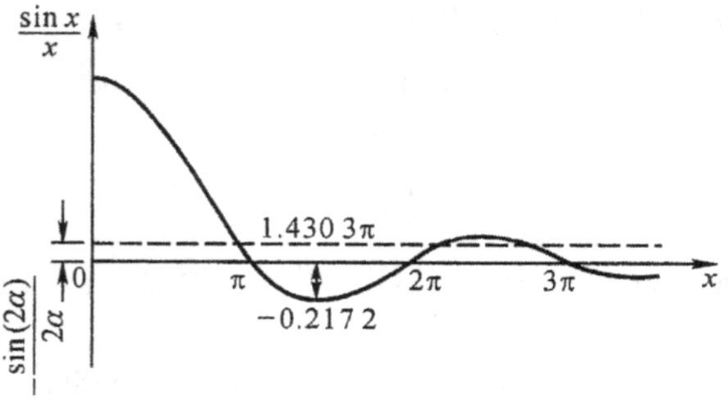

Fig. 7.5. The curve of function $\sin x/x$.

For $\alpha > \pi/2$, there exist stress singularity when $\rho \to 0$ and hence a real root in $0 < x < \pi$.

According to the characteristics of the transcendental equation above, discussion for roots x in the first quadrant of the complex plane is only needed. As there are complex roots for relatively large $|x|$, the existence of complex roots should be estimated. From the function curve of $\sin x/x$ as illustrated in Fig. 7.5, there is a root in $\pi/2 < \alpha \leq \pi$. For $\eta = -\sin 2\alpha/2\alpha < 0.1284$, there are two real roots in $2\pi < x < 3\pi$, otherwise there are complex roots in $2\pi < \mathrm{Re}(x) < 3\pi$. To solve for complex roots, the asymptotic method can be applied to obtain an initial value for iteration.

Substituting $x = 2n\pi + a + ib$ $(n = 1, 2, \ldots; a > 0, b > 0)$ into

$$\eta x = \sin x \approx \frac{\mathrm{i}}{2}\mathrm{e}^{-\mathrm{i}x} \tag{7.4.28}$$

and grouping the real part and imaginary part yield

$$2\eta b \approx \mathrm{e}^b \cos a; \quad 2\eta(2n\pi + a) \approx \mathrm{e}^b \sin a \tag{7.4.29}$$

As b is relatively larger, we obtain

$$a \approx \frac{\pi}{2} - \varepsilon, \quad b \approx \ln(4\eta n\pi + 2\eta a) \tag{7.4.30}$$

where ε is a higher-order small quantity. Based on the asymptotic solution (7.4.30), the Newton method can be applied to obtain the complex eigenvalues.

For $\alpha = \pi$, it is equivalent to the solution of a crack. Here the solutions of Eq. (7.4.27) are $x = \pi, 2\pi, 3\pi, \ldots$, which is equivalent to $\mu = 1/2, 1, 3/2, \ldots$. Noticing that $S_\rho = \rho\sigma_\rho, S_{\rho\varphi} = \rho\sigma_{\rho\varphi}$, there is singularity $\rho^{\mu-1}$ at the apex for $\rho \to 0$.

Table 7.1. The eigenvalue $2\mu\alpha$ for symmetric deformation of a sector domain.

$\alpha/(^\circ)$	1 2 3	4 5	6 7	8 9	10 11	12 13
180	π 2π 3π	4π 5π	6π 7π	8π 9π	10π 11π	12π 13π
170	-3.313595 $2\pi + 0.395112$ $+2.603832$	$4\pi + 0.887387$ $+2.131078$	$6\pi + 1.516289$ $\pm 0.584204i$	$8\pi + 1.521668$ $\pm 0.995763i$	$10\pi + 1.525936$ $\pm 1.257540i$	$12\pi + 1.529418$ $\pm 1.456962i$
160	-3.471163 $2\pi + 0.992428$ $+1.910136$	$4\pi + 1.474761$ $\pm 1.065548i$	$6\pi + 1.489566$ $\pm 1.498203i$	$8\pi + 1.499908$ $\pm 1.788172i$	$10\pi + 1.507612$ $\pm 2.009485i$	$12\pi + 1.513610$ $\pm 2.189284i$
150	-3.601216 $2\pi + 1.419095$ $\pm 0.742802i$	$4\pi + 1.453587$ $\pm 1.491782i$	$6\pi + 1.473863$ $\pm 1.887954i$	$8\pi + 1.487460$ $\pm 2.165877i$	$10\pi + 1.497315$ $\pm 2.381868i$	$12\pi + 1.504837$ $\pm 2.558770i$
120	-3.704010 $2\pi + 1.397618$ $\pm 1.056610i$	$4\pi + 1.439762$ $\pm 1.734300i$	$6\pi + 1.463809$ $\pm 2.119570i$	$8\pi + 1.479576$ $\pm 2.394128i$	$10\pi + 1.490837$ $\pm 2.608348i$	$12\pi + 1.499344$ $\pm 2.784264i$

The roots of several eigenvalues computed for different angles α are presented in Table 7.1. Only roots for $\mathrm{Re}\,(2\mu\alpha) > 0$ are listed in the table.

After solving the roots of eigenvalues, the ratio of A_1 to C_1 can be obtained from Eq. (7.4.25). They are then substituted into Eq. (7.4.24) to give the expressions of eigenvectors which are the basic components of expansion solution.

7.5. Eigen-Solutions for Anti-Symmetric Deformation in Radial Hamiltonian System

The differential equation for anti-symmetric deformation and the free boundary conditions on $\varphi = \alpha$ are the same as those of symmetric deformation, i.e. Eqs. (7.3.17) and (7.4.1) while the symmetric conditions (7.3.32) should be replaced by the anti-symmetric conditions (7.3.33).

7.5.1. *Eigen-Solutions of Zero Eigenvalue*

The basic equation for anti-symmetric eigen-solution of zero eigenvalue is

$$\boldsymbol{H}\boldsymbol{\psi}_0^{(a0)} = \boldsymbol{0} \tag{7.5.1}$$

From the last two equations of Eq. (7.5.1), we obtain

$$S_{\rho\varphi 0}^{(a0)} = c_1 \cos\varphi + c_2 \sin\varphi \tag{7.5.2}$$

Substituting the above equation into the third of Eq. (7.5.1) yields

$$E\left(u_{\rho 0}^{(a0)} + \frac{\mathrm{d}u_{\varphi 0}^{(a0)}}{\mathrm{d}\varphi}\right) + \nu S_{\rho 0}^{(a0)} = -c_1 \sin\varphi + c_2 \cos\varphi \tag{7.5.3}$$

Substituting Eqs. (7.5.2) and (7.5.3) into the boundary conditions (7.4.1) and the anti-symmetric conditions (7.3.33) yields $c_1 = c_2 = 0$, and subsequently $S_{\rho\varphi 0}^{(a0)} = 0$.

Then substituting Eqs. (7.5.2) and (7.5.3) into Eq. (7.5.1) and solving the simultaneous equations yield

$$S_{\rho 0}^{(a0)} = 0 \tag{7.5.4}$$

and

$$\left.\begin{aligned} u_{\rho 0}^{(a0)} &= c_3 \cos\varphi + c_4 \sin\varphi \\ u_{\varphi 0}^{(a0)} &= -c_3 \sin\varphi + c_4 \cos\varphi \end{aligned}\right\} \tag{7.5.5}$$

Substituting the above equation into the anti-symmetric conditions (7.3.33) yields $c_3 = 0$, hence the basic eigenvector of zero eigenvalue for anti-symmetric deformation is

$$\boldsymbol{\psi}_0^{(a0)} = \{\sin\varphi,\ \cos\varphi,\ 0,\ 0\}^{\mathrm{T}} \tag{7.5.6}$$

The eigenvector (7.5.6) is the solution of the original problem, $\boldsymbol{v}_0^{(a0)} = \boldsymbol{\psi}_0^{(a0)}$, and it can be physically interpreted as the unit rigid body translation along the direction perpendicular to the symmetry axis.

Similarly, the first-order Jordan form solution exists because there is only one chain for the eigen-solution of zero eigenvalue. The corresponding equation is

$$\boldsymbol{H}\boldsymbol{\psi}_0^{(a1)} = \boldsymbol{\psi}_0^{(a0)} \tag{7.5.7}$$

Similar to the derivation above, from the last two equations of Eq. (7.5.7) and the boundary conditions we obtain

$$S_{\rho\varphi 0}^{(a1)} = 0 \quad \text{and} \quad E\left(u_{\rho 0}^{(a1)} + \frac{\mathrm{d}u_{\varphi 0}^{(a1)}}{\mathrm{d}\varphi}\right) + \nu S_{\rho 0}^{(a1)} = 0 \tag{7.5.8}$$

Substitute Eq. (7.5.8) into the first and the second equations of the expanded expression of Eq. (7.5.7), the solutions are

$$S_{\rho 0}^{(a1)} = E\sin\varphi \tag{7.5.9}$$

and

$$u_{\rho 0}^{(a1)} + \frac{\mathrm{d}u_{\varphi 0}^{(a1)}}{\mathrm{d}\varphi} = -\nu\sin\varphi, \qquad \frac{\mathrm{d}u_{\rho 0}^{(a1)}}{\mathrm{d}\varphi} - u_{\varphi 0}^{(a1)} = -\cos\varphi \tag{7.5.10}$$

Solving the simultaneously Eq. (7.5.10) yields

$$\left.\begin{aligned} u_{\rho 0}^{(a1)} &= -\frac{1-\nu}{2}\varphi\cos\varphi + c_3\cos\varphi + c_4\sin\varphi \\ u_{\varphi 0}^{(a1)} &= \frac{1-\nu}{2}\varphi\sin\varphi + \frac{1+\nu}{2}\cos\varphi - c_3\sin\varphi + c_4\cos\varphi \end{aligned}\right\} \tag{7.5.11}$$

Substituting the solutions into the anti-symmetric conditions (7.3.33) yields $c_3 = 0$. Because term c_4 is a basic eigen-solution which can be superposed arbitrarily, therefore, the anti-symmetric first-order Jordan form eigenvector of zero eigenvalue is

$$\boldsymbol{\psi}_0^{(a1)} = \left\{-\frac{1-\nu}{2}\varphi\cos\varphi,\quad \frac{1-\nu}{2}\varphi\sin\varphi + \frac{1+\nu}{2}\cos\varphi,\quad E\sin\varphi,\quad 0\right\}^{\mathrm{T}} \tag{7.5.12}$$

The solution of the original problem in Eq. (7.3.9) thus constituted is

$$\boldsymbol{v}_0^{(a1)} = \boldsymbol{\psi}_0^{(a1)} + \xi \boldsymbol{\psi}_0^{(a0)} \tag{7.5.13}$$

and the corresponding stress field is

$$\sigma_\rho = \frac{1}{\rho} E \sin\varphi, \quad \sigma_\varphi = 0, \quad \tau_{\rho\varphi} = 0 \tag{7.5.14}$$

It results in a resultant force perpendicular to the symmetry axis at the origin as

$$\left.\begin{aligned} F_{\mathrm{N}} &= \int_{-\alpha}^{\alpha} [-\sigma_\rho \cos\varphi + \tau_{\rho\varphi} \sin\varphi]\rho \mathrm{d}\varphi = 0 \\ F_{\mathrm{S}} &= \int_{-\alpha}^{\alpha} [-\sigma_\rho \sin\varphi - \tau_{\rho\varphi} \cos\varphi]\rho \mathrm{d}\varphi = -\frac{1}{2} E[2\alpha - \sin(2\alpha)] \neq 0 \\ M &= \int_{-\alpha}^{\alpha} [\tau_{\rho\varphi}\rho^2] \mathrm{d}\varphi = 0 \end{aligned}\right\} \tag{7.5.15}$$

For $\xi_1 \to -\infty (R_1 = 0)$, we have

$$\tilde{\boldsymbol{v}}_0^{(a1)} = \frac{1}{F_{\mathrm{S}}} \boldsymbol{v}_0^{(a1)} \tag{7.5.16}$$

That is the solution of an elastic wedge acted by a unit concentrated force at the apex $R_1 = 0$ and along the direction perpendicular to the symmetry axis (see Fig. 7.6).

As eigenvectors $\boldsymbol{\psi}_0^{(a1)}$ and $\boldsymbol{\psi}_0^{(a0)}$ are symplectic adjoint, i.e.

$$\langle \boldsymbol{\psi}_0^{(a0)}, \quad \boldsymbol{\psi}_0^{(a1)} \rangle = \frac{1}{2} E[2\alpha - \sin(2\alpha)] \neq 0 \tag{7.5.17}$$

Equation (7.5.12) is the dual solution of Eq. (7.5.6). Thus the Jordan form chain for anti-symmetric eigen-solution of zero eigenvalue is terminated.

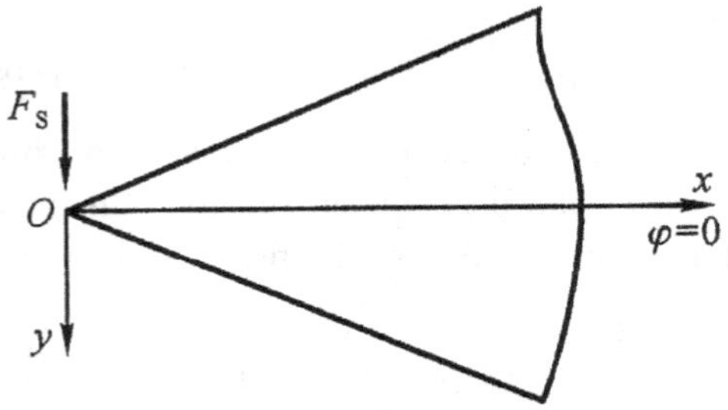

Fig. 7.6. An elastic wedge acted by a concentrated force at the apex and normal to the symmetry axis.

At this point, we should review the anti-symmetric solution in strip domain in Chapter 4. The shortness of Jordan form chain of eigenvectors of zero eigenvalue indicates the missing of some solutions. Here, it is noticed that $\boldsymbol{v}_0^{(a0)}$ and $\boldsymbol{v}_0^{(a1)}$ correspond to the solutions for lateral translation of rigid body and constant shear bending in a strip domain. The corresponding solutions for rotation of rigid body and pure bending are missing. The eigen-solutions of eigenvalues $\mu = \pm 1$ in a sector domain to be discussed in the following sections provide an answer to the query.

7.5.2. *Eigen-Solutions of $\mu = \pm 1$*

From Eqs. (7.3.24) and (7.3.30), the general solution for eigen-solutions of nonzero eigenvalues corresponding to anti-symmetric deformation with respect to $\varphi = 0$ is

$$\left.\begin{aligned} u_\rho &= B_2 \sin[(1+\mu)\varphi] + \frac{3-\nu-\mu-\nu\mu}{3-\nu+\mu+\nu\mu} D_2 \sin[(1-\mu)\varphi] \\ u_\varphi &= B_2 \cos[(1+\mu)\varphi] + D_2 \cos[(1-\mu)\varphi] \\ S_\rho &= \frac{E\mu}{1+\nu} B_2 \sin[(1+\mu)\varphi] + \frac{E\mu(3-\mu)}{3-\nu+\mu+\nu\mu} D_2 \sin[(1-\mu)\varphi] \\ S_{\rho\varphi} &= \frac{E\mu}{1+\nu} B_2 \cos[(1+\mu)\varphi] - \frac{E\mu(1-\mu)}{3-\nu+\mu+\nu\mu} D_2 \cos[(1-\mu)\varphi] \end{aligned}\right\} \tag{7.5.18}$$

Although the general solution (7.5.18) satisfies differential equations (7.3.17) in the domain and the anti-symmetric conditions (7.3.33), it has not satisfy the boundary conditions (7.4.1). Substituting Eq. (7.5.18) into Eq. (7.4.1) yields

$$\left.\begin{aligned} -\frac{E\mu}{1+\nu} B_2 \sin[(1+\mu)\alpha] + \frac{E\mu(1+\mu)}{3-\nu+\mu+\nu\mu} D_2 \sin[(1-\mu)\alpha] = 0 \\ \frac{E\mu}{1+\nu} B_2 \cos[(1+\mu)\alpha] - \frac{E\mu(1-\mu)}{3-\nu+\mu+\nu\mu} D_2 \cos[(1-\mu)\alpha] = 0 \end{aligned}\right\} \tag{7.5.19}$$

The determinant vanishes as B_2 and D_2 are not zero simultaneously. Hence the transcendental equation for nonzero eigenvalue μ with respect to anti-symmetric deformation is

$$\sin(2\mu\alpha) - \mu \sin(2\alpha) = 0 \tag{7.5.20}$$

Obviously, $-\mu$ must be an eigenvalue if μ is an eigenvalue. This property validates one of the characteristics of symplectic eigenvalue problems (7.3.18).

Apparently, $\mu = \pm 1$ is a root of Eq. (7.5.20) for an arbitrary α. Hence, the eigen-solutions of eigenvalues $\mu = \pm 1$ exist for anti-symmetric deformation.

For $\mu = 1$, we obtain $B_2 = 0$ from Eq. (7.5.19). As D_2 is an arbitrary constant, $D_2 = 1$ is assumed without loss of generality. Then the basic eigen-solution of $\mu = 1$ is

$$\boldsymbol{\psi}_1^{(a0)} = \{0, \quad 1, \quad 0, \quad 0\}^{\mathrm{T}} \tag{7.5.21}$$

The solution corresponding to the original problem (7.3.9) is

$$\boldsymbol{v}_1^{(a0)} = \mathrm{e}^{\xi}\boldsymbol{\psi}_1^{(a0)} = \{0, \quad \rho, \quad 0, \quad 0\}^{\mathrm{T}} \tag{7.5.22}$$

and it is physically interpreted as rotation of rigid body about the origin.

For $\mu = -1$, we obtain from Eq. (7.5.19)

$$B_2 = \frac{1+\nu}{1-\nu} D_2 \cos(2\alpha) \tag{7.5.23}$$

Without loss of generality, it is assumed $D_2 = 1 - \nu$ and the eigen-solution of $\mu = -1$ is

$$\boldsymbol{\psi}_{-1}^{(a0)} = \begin{Bmatrix} 2\sin(2\varphi) \\ (1+\nu)\cos(2\alpha) + (1-\nu)\cos(2\varphi) \\ -2E\sin(2\varphi) \\ -E\cos(2\alpha) + E\cos(2\varphi) \end{Bmatrix} \tag{7.5.24}$$

The solution corresponding to the original problem is

$$\boldsymbol{v}_{-1}^{(a0)} = \exp(-\xi)\boldsymbol{\psi}_{-1}^{(a0)} = \rho^{-1}\boldsymbol{\psi}_{-1}^{(a0)} \tag{7.5.25}$$

and the stress field corresponding to this eigen-solution is

$$\left.\begin{aligned} \sigma_\rho &= -\frac{2}{\rho^2} E\sin(2\varphi) \\ \sigma_\varphi &= 0 \\ \tau_{\rho\varphi} &= \frac{1}{\rho^2} E[\cos(2\varphi) - \cos(2\alpha)] \end{aligned}\right\} \tag{7.5.26}$$

This stress field results in a concentrated couple of forces at the apex of the elastic wedge.

$$\left.\begin{aligned} F_{\rm N} &= \int_{-\alpha}^{\alpha} (-\sigma_\rho \cos\varphi + \tau_{\rho\varphi} \sin\varphi)\rho {\rm d}\varphi = 0 \\ F_{\rm S} &= \int_{-\alpha}^{\alpha} (-\sigma_\rho \sin\varphi - \tau_{\rho\varphi} \cos\varphi)\rho {\rm d}\varphi = 0 \\ M &= \int_{-\alpha}^{\alpha} (\tau_{\rho\varphi}\rho^2){\rm d}\varphi = E(\sin(2\alpha) - 2\alpha\cos(2\alpha)) \end{aligned}\right\} \tag{7.5.27}$$

For general cases, i.e. $\alpha \neq \tilde{\alpha}$ $(\tan(2\tilde{\alpha}) = 2\tilde{\alpha}, \tilde{\alpha} \approx 0.715\pi), M \neq 0$, then the eigen-solution is equivalent to a concentrated couple acting at the apex of wedge. Thus $\boldsymbol{v}_1^{(a0)}$ and $\boldsymbol{v}_{-1}^{(a0)}$ are, respectively, the rotation of rigid body and pure bending solutions in a sector domain. They correspond to the rotation of rigid body and pure bending solutions in strip domain. Here, $\boldsymbol{\psi}_1^{(a0)}$ and $\boldsymbol{\psi}_{-1}^{(a0)}$ are symplectic adjoint, as

$$\langle \boldsymbol{\psi}_1^{(a0)},\ \boldsymbol{\psi}_{-1}^{(a1)} \rangle = E[\sin(2\alpha) - 2\alpha\cos(2\alpha)] = M \neq 0 \tag{7.5.28}$$

and the solution of the wedge with a unit concentrated couple acting at the apex (see Fig. 7.7) is

$$\boldsymbol{v} = \frac{1}{M}\boldsymbol{v}_{-1}^{(a0)} = \frac{1}{\rho M}\boldsymbol{\psi}_{-1}^{(a0)} \tag{7.5.29}$$

In particular, we notice from Eq. (7.5.27) that $M = 0$ when $\alpha = \tilde{\alpha}$ and $\boldsymbol{\psi}_1^{(a0)}$ and $\boldsymbol{\psi}_{-1}^{(a0)}$ are no longer symplectic adjoint but rather they are symplectic orthogonal. Hence, there exist eigenvectors symplectic adjoint with $\boldsymbol{\psi}_1^{(a0)}$ and $\boldsymbol{\psi}_{-1}^{(a0)}$, respectively. However, it is rather puzzling here because the stress components given by Eq. (7.5.29) become infinite. What is then

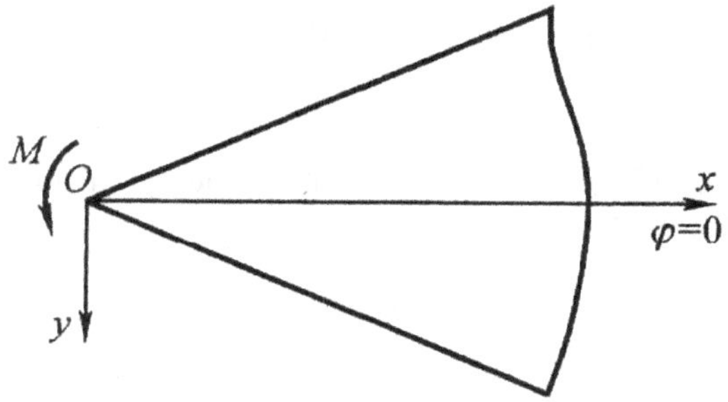

Fig. 7.7. The elastic wedge acted by a concentrated couple at the apex.

the solution of the elastic wedge with a concentrated couple at the apex in such a special instance? This is just the so-called **paradox** of an elastic wedge. Although it is possible to apply a semi-inverse method to obtain the solution of this paradox[4,5], this solution methodology is rather irrational and cannot be applied directly to other similar problems. In addition, the solution is unable to provide a reasonable explanation for the paradox as to what actually happens.

It should also be noted from Eq. (7.5.20) that for $\alpha \neq \tilde{\alpha}$, $\mu = \pm 1$ are nonrepeated single roots in general. Hence, there are no Jordan form eigen-solutions for $\mu = \pm 1$ and the eigenvectors $\boldsymbol{\psi}_1^{(a0)}$ and $\boldsymbol{\psi}_{-1}^{(a0)}$ are the eigenvectors of $\mu = \pm 1$. However, for $\alpha = \tilde{\alpha}$, $\mu = \pm 1$ are double roots of Eq. (7.5.20). In accordance with the derivation above there is only one eigen-solution chain for each of $\mu = \pm 1$. Hence, there exist other Jordan form eigenvectors of eigenvalues $\mu = \pm 1$ in addition to the eigenvectors $\boldsymbol{\psi}_1^{(a0)}$ and $\boldsymbol{\psi}_{-1}^{(a0)}$.

By virtue of the method of mathematical physics, the equation for the Jordan form eigenvectors of $\mu = 1$ can be obtained by solving the following equation

$$\boldsymbol{H}\boldsymbol{\psi}_1^{(a1)} = \boldsymbol{\psi}_1^{(a1)} + \boldsymbol{\psi}_1^{(a0)} \tag{7.5.30}$$

Neglecting the details of derivation, the general solution of Eq. (7.5.30) is

$$\left.\begin{aligned} u_\rho &= c_2\cos(2\varphi) + c_3\sin(2\varphi) - \frac{1-\nu}{2}\varphi - (1-\nu)c_0 \\ u_\varphi &= -c_2\sin(2\varphi) + c_3\cos(2\varphi) + c_1 \\ S_\rho &= \frac{E}{1+\nu}\left[c_2\cos(2\varphi) + c_3\sin(2\varphi) - \frac{1+\nu}{2}\varphi - (1+\nu)c_0\right] \\ S_{\rho\varphi} &= \frac{E}{1+\nu}\left[-c_2\sin(2\varphi) + c_3\cos(2\varphi) + \frac{1+\nu}{4}\right] \end{aligned}\right\} \tag{7.5.31}$$

Substituting Eq. (7.5.31) into anti-symmetric conditions (7.3.33) and boundary conditions (7.4.1) yields the coefficients as

$$c_0 = c_2 = 0, \quad c_3 = -\frac{1+\nu}{4\cos(2\tilde{\alpha})} \tag{7.5.32}$$

and c_1 denotes the basic eigenvector $\boldsymbol{\psi}_1^{(a0)}$ which can be superposed arbitrarily. Thus for $\alpha = \tilde{\alpha}$, the Jordan form eigen-solution corresponding to

$\mu = 1$ is

$$\psi_1^{(a1)} = \frac{1}{4\cos(2\tilde{\alpha})}\left\{\begin{array}{c} -(1+\nu)\sin(2\varphi) - 2(1-\nu)\varphi\cos(2\tilde{\alpha}) \\ -(1+\nu)\cos(2\varphi) \\ -E[\sin(2\varphi) + 2\varphi\cos(2\tilde{\alpha})] \\ E[\cos(2\tilde{\alpha}) - \cos(2\varphi)] \end{array}\right\} \tag{7.5.33}$$

For $\mu = -1$, the Jordan form eigen-solution can be obtained by soling the following equation

$$\boldsymbol{H}\boldsymbol{\psi}_{-1}^{(a1)} = -\boldsymbol{\psi}_{-1}^{(a1)} + \boldsymbol{\psi}_{-1}^{(a0)} \tag{7.5.34}$$

Neglecting the details of derivation, the eigen-solution satisfying the anti-symmetric conditions (7.3.33) and the boundary conditions (7.4.1) is

$$\boldsymbol{\psi}_{-1}^{(a1)} = \left\{\begin{array}{l} -2\varphi\cos(2\varphi) + \dfrac{5-\nu}{2}\sin(2\varphi) + (1+\nu)\varphi\cos(2\tilde{\alpha}) \\ (2-\nu)\cos(2\varphi) + (1-\nu)\varphi\sin(2\varphi) + (1+\nu)(1+2\tilde{\alpha}^2)\cos(2\tilde{\alpha}) \\ -\dfrac{1}{2}E\sin(2\varphi) + E\varphi[2\cos(2\varphi) - \cos(2\tilde{\alpha})] \\ E[\varphi\sin(2\varphi) - \tilde{\alpha}\sin(2\tilde{\alpha})] \end{array}\right\} \tag{7.5.35}$$

Apparently the solution (7.5.35) can be superposed arbitrarily with any basic eigenvectors. The solution of the original problem corresponding to Eq. (7.5.35) is

$$\boldsymbol{v}_{-1}^{(a1)} = \exp(-\xi)(\boldsymbol{\psi}_{-1}^{(a1)} + \xi\boldsymbol{\psi}_{-1}^{(a0)}) = \rho^{-1}(\boldsymbol{\psi}_{-1}^{(a1)} + \boldsymbol{\psi}_{-1}^{(a0)}\ln\rho) \tag{7.5.36}$$

and the corresponding stress field is

$$\left.\begin{aligned} \sigma_\rho &= \frac{E}{\rho^2}\left\{-\frac{1}{2}\sin(2\varphi) + \varphi[2\cos(2\varphi) - \cos(2\tilde{\alpha})] - 2\sin(2\varphi)\ln\rho\right\} \\ \sigma_\varphi &= \frac{E}{\rho^2}\left[\varphi\cos(2\tilde{\alpha}) - \frac{1}{2}\sin(2\varphi)\right] \\ \tau_{\rho\varphi} &= \frac{E}{\rho^2}\{\varphi\sin(2\varphi) - \tilde{\alpha}\sin(2\tilde{\alpha}) + [\cos(2\varphi) - \cos(2\tilde{\alpha})]\ln\rho\} \end{aligned}\right\} \tag{7.5.37}$$

The solution results in a concentrated couple of forces at the apex of the elastic wedge as

$$\tilde{M} = \int_{-\tilde{\alpha}}^{\tilde{\alpha}} \tau_{\rho\varphi}\rho^2 \mathrm{d}\varphi = -2E\tilde{\alpha}^2 \sin(2\tilde{\alpha}) \neq 0 \tag{7.5.38}$$

Hence, the solution of paradox of elastic wedge with unit concentrated couple of forces acting at the apex is

$$\boldsymbol{v} = \frac{1}{\tilde{M}}\boldsymbol{v}_{-1}^{(a1)} = \frac{1}{\rho\tilde{M}}[\boldsymbol{\psi}_{-1}^{(a1)} + \boldsymbol{\psi}_{-1}^{(a0)} \ln\rho] \tag{7.5.39}$$

which corresponds to the special Jordan form solution in Hamiltonian system. The solution for the paradox with forces acting on two sides can be obtained in the same way[6].

Obviously, $\boldsymbol{\psi}_1^{(a0)}$ and $\boldsymbol{\psi}_{-1}^{(a1)}$ as well as $\boldsymbol{\psi}_1^{(a1)}$ and $\boldsymbol{\psi}_{-1}^{(a0)}$ are mutually symplectic adjoint, as

$$\langle\boldsymbol{\psi}_1^{(a0)},\quad \boldsymbol{\psi}_{-1}^{(a1)}\rangle = \langle\boldsymbol{\psi}_{-1}^{(a0)},\quad \boldsymbol{\psi}_1^{(a1)}\rangle = \tilde{M} \neq 0 \tag{7.5.40}$$

whereas $\boldsymbol{\psi}_1^{(a1)}$ and $\boldsymbol{\psi}_{-1}^{(a1)}$ can be made symplectic orthogonal by a proper choice of constant c_1. Consequently, we have obtained all eigen-solutions for $\mu = \pm 1$ for angle of the wedge $\alpha = \tilde{\alpha}$.

It is interesting to note that $\boldsymbol{\psi}_1^{(a0)}$ and $\boldsymbol{\psi}_{-1}^{(a0)}$ are symplectic adjoint for $\alpha \neq \tilde{\alpha}$ while $\boldsymbol{\psi}_1^{(a0)}$ and $\boldsymbol{\psi}_{-1}^{(a1)}$ are symplectic adjoint for $\alpha = \tilde{\alpha}$. It implies that the eigen-solution for rotation of rigid body is always symplectic adjoint with the eigen-solution of pure bending acted by a concentrated couple. It further implies there are special dual relations between the eigen-solutions.

7.5.3. *Eigen-Solutions of General Nonzero Eigenvalues*

There are infinite eigenvalues for Eq. (7.5.20) in addition to the roots $\mu = \pm 1$. Besides, it can be proved that there is no pure imaginary root for Eq. (7.5.20). Based on the characteristics of eigenvalue equation, it is only necessary to determine the real roots for $\mu > 0$ and the complex roots in the first quadrant of the complex domain. For nonzero eigenvalues, the transcendental equation (7.5.20) can be written as

$$\frac{\sin x}{x} = \frac{\sin(2\alpha)}{2\alpha} \quad \text{where } x = 2\mu\alpha \tag{7.5.41}$$

From the viewpoint of singular solution, we are more concern with the root for $|\mu| < 1$. First, we consider the wedge domain $\alpha < \pi/2$ where the right-hand-side of Eq. (7.5.41) is positive. As the function $\sin x/x$ is monotonously decreasing in $0 \leq x < \pi$, there is no root for $|\mu| < 1$ and hence no root for $\text{Re}(\mu) < 1$.

Next, we consider the singularity in $\pi/2 < \alpha \leq \pi$ where the right-hand-side of Eq. (7.5.41) is negative. From Fig. 7.5, it is observed that there are only solutions of $\mu = 1$ for $2\alpha < 2\tilde{\alpha}$ ($\tan 2\tilde{\alpha} = 2\tilde{\alpha}, \tilde{\alpha} \approx 0.715\pi$); there exist real roots $\mu < 1$ for $2\tilde{\alpha} < 2\alpha \leq 2\pi$; and $\mu = 1/2, 1, 3/2, \ldots$ for $\alpha = \pi$. There are many complex roots of μ for other different values of α. The solution methodology is similar to that of symmetric deformation and, therefore, the details are omitted here. The numerical solutions are presented in Table 7.2.

The corresponding eigenvectors should be derived after obtaining the eigenvalues. The procedure is as follows: the ratio of B_2 to D_2 is obtained from Eq. (7.5.19), then $D_2 = 1$ is assume without the loss of generality, finally the constants are substituted into Eq. (7.5.18) to obtain the eigenvectors. These eigen-solutions are the basic components of the expansion method with external loads and boundary conditions at $\rho = R_1$ and $\rho = R_2$. The numerical computation can be accomplished by a computer program. As the procedure has been clarified, the synthesis of specific equations is thus neglected.

The case of a sector domain with free side boundaries has been discussed at length above. In fact the sides are not necessarily free. The solution methodology is also valid for cases with clamped sides or sides with mixed boundary conditions due to different materials.

For the case with external forces acting on the side edges, the solutions can be obtained by applying the eigenvector expansion method. Accordingly, a number of eigen-solutions have been covered by the Saint–Venant principle. The readers are referred to Chapter 4 for equality of the actual solution procedure for the expansion method and the problem in a strip domain. Through the solution procedure above, it is observed that some of the common solutions in elasticity courses correspond to the eigen-solutions of some special eigenvalues. The procedure for solution derivation in a Hamiltonian system is completely rational and, furthermore, the solutions for stress and displacement are obtained at the same time. Since the concept is clear and the methodology can be outlined explicitly, the symplectic approach has significant advantage. In addition, the expansion method can also be outlined explicitly.

The main advantage of an analytical method is its capability of deriving an exact solution. Because the method is only applicable to problems in regular domains, a finite-element method is thus required to do the computation for the whole structure. If the components (substructures)

Table 7.2. The eigenvalue $2\mu\alpha$ for anti-symmetric deformation of a sector domain.

$\alpha/(^\circ)$	1 2	3 4	5 6	7 8	9 10	11 12
180	π 2π	3π 4π	5π 6π	7π 8π	9π 10π	11π 12π
170	$2\pi - 2.948172$ -0.349066	$4\pi - 2.524356$ -0.749255	$6\pi - 1.736228$ -1.520924	$8\pi - 1.622449$ $\pm 0.821515i$	$10\pi - 1.617676$ $\pm 1.137183i$	$12\pi - 1.613832$ $\pm 1.362919i$
160	$2\pi - 2.718902$ -0.698132	$4\pi - 1.677034$ $\pm 0.706814i$	$6\pi - 1.658695$ $\pm 1.308939i$	$8\pi - 1.646442$ $\pm 1.654492i$	$10\pi - 1.637574$ $\pm 1.905229i$	$12\pi - 1.630808$ $\pm 2.103549i$
150	$2\pi - 2.456159$ -1.047247	$4\pi - 1.702609$ $\pm 1.201317i$	$6\pi - 1.676738$ $\pm 1.710587i$	$8\pi - 1.660332$ $\pm 2.036754i$	$10\pi - 1.648846$ $\pm 2.279830i$	$12\pi - 1.640282$ $\pm 2.474288i$
120	$2\pi - 2.094435$ -1.470587	$4\pi - 1.719585$ $\pm 1.459254i$	$6\pi - 1.688385$ $\pm 1.946271i$	$8\pi - 1.669172$ $\pm 2.266485i$	$10\pi - 1.655959$ $\pm 2.507054i$	$12\pi - 1.646228$ $\pm 2.700211i$

compose of regular domains, an analytic method can be applied to determine the relevant stiffness matrix which can then be linked to the parent structure through the relations of substructures. Although solutions to infinite domain problems in a finite-element method are rather inconvenient, the infinite domain can often be made up by a regular outer infinite domain and an inner local domain. The stiffness matrix of the regular outer infinite domain can be given by the expansion method of Hamiltonian system, and it is then linked to the inner finite domain. In this way, the stiffness matrix of the regular outer infinite domain is exact. This step is of great significance for solving this kind of problems numerically.

An important application of problems in sector domain is the computation of singularity in fracture mechanics. The properties of singularity are determined by the eigenvalue μ which depends on the material disposition adjacent to the tip of crack and is independent of field properties far away. The intensity of singularity is dependent to the surrounding structural and loading conditions. For such cases, a finite element method is required because the whole structure is complicated. Here the sectorial domain at the crack can be regarded as a super-element of the whole structure. The sectorial element can be solved by the analytical method. Here an element stiffness matrix for the circle with centre at the tip of crack should be given in order to link with the other parts of the structure. In this way, the stress intensity factor can be computed. The element stiffness matrix of sectorial domain can be computed by the eigenvector expansion method with the variational principle. In fact this eigenvector expansion method is also applicable even for the singularity problem of adhesive interface of different medium.

7.6. Hamiltonian System with Circumferential Coordinate Treated as "Time"

In the former three sections, a Hamiltonian system with ξ treated as the time coordinate and the transverse φ direction forces eliminated by Eq. (7.3.1) was derived based on the transformation (7.2.3). In fact, φ can also be treated as time and thus ξ becomes the transverse direction[7]. In a similar way, the transverse forces should be eliminated in accordance with the variational principle. The variation of Eq. (7.2.7) with respect to S_ρ yields

$$S_\rho = E\frac{\partial u_\rho}{\partial \xi} + \nu S_\varphi \tag{7.6.1}$$

Substituting into Eq. (7.2.7) and eliminating S_ρ yield (for case without external forces in the domain)

$$\delta \int_{-\alpha}^{\alpha} \int_{\xi_1}^{\xi_2} \left\{ S_{\rho\varphi} \frac{\partial u_\rho}{\partial \varphi} + S_\varphi \frac{\partial u_\varphi}{\partial \varphi} + S_\varphi \left(u_\rho + \nu \frac{\partial u_\rho}{\partial \xi} \right) - S_{\rho\varphi} \left(u_\varphi - \frac{\partial u_\varphi}{\partial \xi} \right) \right.$$
$$\left. + \frac{1}{2} E \left(\frac{\partial u_\rho}{\partial \xi} \right)^2 - \frac{1}{2E} \left[\left(1 - \nu^2\right) S_\varphi^2 + 2 \left(1 + \nu\right) S_{\rho\varphi}^2 \right] \right\} \mathrm{d}\xi \mathrm{d}\varphi = 0 \tag{7.6.2}$$

This is the mixed energy variational principle in a Hamiltonian system. The dual variables of the displacements u_ρ, u_φ are, respectively, $S_{\rho\varphi}, S_\varphi$. Let

$$\boldsymbol{q} = \left\{ u_\rho, \quad u_\varphi \right\}^{\mathrm{T}}, \quad \boldsymbol{p} = \left\{ S_{\rho\varphi}, \quad S_\varphi \right\}^{\mathrm{T}} \tag{7.6.3}$$

and denote differentiation with respect to φ by a dot, then Eq. (7.6.2) can be written as

$$\delta \int_{-\alpha}^{\alpha} \int_{\xi_1}^{\xi_2} \left[\boldsymbol{p}^{\mathrm{T}} \dot{\boldsymbol{q}} - \mathscr{H}(\boldsymbol{q}, \boldsymbol{p}) \right] \mathrm{d}\xi \mathrm{d}\varphi = 0 \tag{7.6.4}$$

where the Hamiltonian density function is

$$\mathscr{H}(\boldsymbol{q}, \boldsymbol{p}) = S_{\rho\varphi} \left(u_\varphi - \frac{\partial u_\varphi}{\partial \xi} \right) - S_\varphi \left(u_\rho + \nu \frac{\partial u_\rho}{\partial \xi} \right)$$
$$- \frac{1}{2} E \left(\frac{\partial u_\rho}{\partial \xi} \right)^2 + \frac{1}{2E} \left[(1 - \nu^2) S_\varphi^2 + 2(1 + \nu) S_{\rho\varphi}^2 \right] \tag{7.6.5}$$

This is a Hamiltonian system expression for field problems and it is in the form of variational principle. Expanding the variation expression yields the Hamiltonian dual equations

$$\left. \begin{aligned} \dot{\boldsymbol{q}} &= \boldsymbol{A}\boldsymbol{q} + \boldsymbol{D}\boldsymbol{p} \\ \dot{\boldsymbol{p}} &= \boldsymbol{B}\boldsymbol{q} - \boldsymbol{A}^{\mathrm{T}}\boldsymbol{p} \end{aligned} \right\} \tag{7.6.6}$$

where

$$\left. \begin{aligned} \boldsymbol{A} &= \begin{bmatrix} 0 & 1 - \dfrac{\partial \cdot}{\partial \xi} \\ -1 - \nu \dfrac{\partial \cdot}{\partial \xi} & 0 \end{bmatrix}, \quad \boldsymbol{A}^{\mathrm{T}} = \begin{bmatrix} 0 & -1 + \nu \dfrac{\partial \cdot}{\partial \xi} \\ 1 + \dfrac{\partial \cdot}{\partial \xi} & 0 \end{bmatrix} \\ \boldsymbol{D} &= \begin{bmatrix} \dfrac{2(1+\nu)}{E} & 0 \\ 0 & \dfrac{1-\nu^2}{E} \end{bmatrix}, \quad \boldsymbol{B} = \begin{bmatrix} -E \dfrac{\partial^2 \cdot}{\partial \xi^2} & 0 \\ 0 & 0 \end{bmatrix} \end{aligned} \right\} \tag{7.6.7}$$

In the Hamiltonian variational principle (7.6.4), the free boundary conditions are treated as variational natural boundary conditions. The free boundary conditions on two sides $\xi = \xi_1$ and $\xi = \xi_2$ are

$$E\frac{\partial u_\rho}{\partial \xi} + \nu S_\varphi = 0, \quad S_{\rho\varphi} = 0 \tag{7.6.8}$$

No external loads have been applied and no boundary conditions at $\varphi = \pm\alpha$ have been imposed in the derivation above. Hence, the resultant equations are a set of homogeneous equations.

Introduce a full state vector

$$\boldsymbol{v} = \begin{Bmatrix} \boldsymbol{q} \\ \boldsymbol{p} \end{Bmatrix} = \{u_\rho, \quad u_\varphi, \quad S_{\rho\varphi}, \quad S_\varphi\}^{\mathrm{T}} \tag{7.6.9}$$

then Eq. (7.6.6) can be rewritten as

$$\dot{\boldsymbol{v}} = \boldsymbol{H}\boldsymbol{v} \tag{7.6.10}$$

where the Hamiltonian operator matrix is

$$\boldsymbol{H} = \begin{bmatrix} \boldsymbol{A} & \boldsymbol{D} \\ \boldsymbol{B} & -\boldsymbol{A}^{\mathrm{T}} \end{bmatrix} \tag{7.6.11}$$

The dual equation (7.6.10) with the boundary conditions (7.6.8) is a linear system. Thus the superposition principle is applicable, and the method of separation of variables is particularly effective. Let

$$\boldsymbol{v}(\xi, \varphi) = \mathrm{e}^{\mu\varphi}\boldsymbol{\psi}(\xi) \tag{7.6.12}$$

where μ is the eigenvalue and $\boldsymbol{\psi}(\xi)$ the eigenvector which is a function of ξ. The eigenvalue equation is

$$\boldsymbol{H}\boldsymbol{\psi}(\xi) = \mu\boldsymbol{\psi}(\xi) \tag{7.6.13}$$

In addition, the eigenvector $\boldsymbol{\psi}(\xi)$ is required to satisfy the boundary conditions (7.6.8).

To discuss the properties of the operator matrix $\boldsymbol{H}$, denote

$$\langle \boldsymbol{v}_1, \boldsymbol{v}_2 \rangle = \int_{\xi_1}^{\xi_2} \boldsymbol{v}_1^{\mathrm{T}} \boldsymbol{J} \boldsymbol{v}_2 \mathrm{d}\xi \tag{7.6.14}$$

where $\boldsymbol{J}$ is a unit symplectic matrix defined in Eq. (7.3.12). Obviously Eq. (7.6.14) satisfies the four conditions of symplectic inner product (1.3.2). According to the definition of symplectic inner product (7.6.14), the full state vectors $\boldsymbol{v}$ form a symplectic space.

By integration by parts, if $\boldsymbol{v}_1, \boldsymbol{v}_2$ are continuously differentiable full state vectors which satisfy the boundary conditions (7.6.8), we have an identity

$$\langle \boldsymbol{v}_1, \boldsymbol{H}\boldsymbol{v}_2 \rangle \equiv \langle \boldsymbol{v}_2, \boldsymbol{H}\boldsymbol{v}_1 \rangle \tag{7.6.15}$$

Hence, the operator matrix $\boldsymbol{H}$ is a Hamiltonian operator matrix in the symplectic space. The eigenvalue problem of Hamiltonian operator matrix $\boldsymbol{H}$ has some special characteristics, i.e. $-\mu$ is also an eigenvalue if μ is an eigenvalue of $\boldsymbol{H}$. These eigenvalues can be divided into two sets:

$$(\alpha)\ \mu_i, \mathrm{Re}(\mu_i) < 0 \quad \text{or} \quad \mathrm{Re}(\mu_i) = 0 \wedge \mathrm{Im}(\mu_i) < 0 \quad (i = 1, 2, \ldots) \tag{7.6.16a}$$

$$(\beta)\ \mu_{-i} = -\mu_i \tag{7.6.16b}$$

Furthermore, the eigenvectors are adjoint symplectic orthonormal.

Having determined the eigenvectors, the eigenvector expansion method can be implemented. The relevant discussion is entirely the same as what presented in the foregoing sections and, therefore, the details are omitted here.

The following discussion is restricted to the solution of eigenvectors.

7.6.1. *Eigen-Solutions of Zero Eigenvalue*

The eigen-solutions of zero eigenvalue are not included in the classification in Eq. (7.6.16), but they are the most important eigen-solutions. The equations for eigen-solution of zero eigenvalue are listed as follows:

$$\left.\begin{array}{lllll} 0 & +u_\varphi - \dfrac{\mathrm{d}u_\varphi}{\mathrm{d}\xi} & +\dfrac{2(1+\nu)}{E}S_{\rho\varphi} & +0 & = 0 \\ -u_\rho - \nu\dfrac{\mathrm{d}u_\rho}{\mathrm{d}\xi} & +0 & +0 & +\dfrac{1-\nu^2}{E}S_\varphi & = 0 \\ -E\dfrac{\mathrm{d}^2u_\rho}{\mathrm{d}\xi^2} & +0 & +0 & +S_\varphi - \nu\dfrac{\mathrm{d}S_\varphi}{\mathrm{d}\xi} & = 0 \\ 0 & +0 & -S_{\rho\varphi} - \dfrac{\mathrm{d}S_{\rho\varphi}}{\mathrm{d}\xi} & +0 & = 0 \end{array}\right\} \tag{7.6.17}$$

From the fourth equation and the boundary conditions (7.6.8), we have

$$S_{\rho\varphi} = 0 \tag{7.6.18}$$

then substituting into the first of Eq. (7.6.17) yields

$$u_\varphi = \mathrm{e}^\xi = \rho \tag{7.6.19}$$

Solving simultaneously the second and third expression of Eq. (7.6.17) yields

$$u_\rho = c_1\mathrm{e}^\xi + c_2\mathrm{e}^{-\xi}, \quad S_\varphi = \frac{E}{1-\nu}c_1\mathrm{e}^\xi + \frac{E}{1+\nu}c_2\mathrm{e}^{-\xi} \tag{7.6.20}$$

then substituting the solution into boundary conditions (7.6.8) yields the integral constants

$$c_1 = c_2 = 0 \tag{7.6.21}$$

Hence, the basic eigenvector of zero eigenvalue is

$$\boldsymbol{\psi}_0^{(0)} = \left\{0, \quad \mathrm{e}^\xi, \quad 0, \quad 0\right\}^{\mathrm{T}} \tag{7.6.22}$$

It is the solution of the original problem (7.6.10), as

$$\boldsymbol{v}_0^{(0)} = \boldsymbol{\psi}_0^{(0)} \tag{7.6.23}$$

and it is physically interpreted as rigid body rotation.

The Jordan form eigenvector exists as there is only one chain for this eigenvector of zero eigenvalue. The corresponding equation is

$$\left.\begin{array}{llllr}
0 & +u_\varphi - \dfrac{\mathrm{d}u_\varphi}{\mathrm{d}\xi} & +\dfrac{2(1+\nu)}{E}S_{\rho\varphi} & +0 & =0 \\
-u_\rho - \nu\dfrac{\mathrm{d}u_\rho}{\mathrm{d}\xi} & +0 & +0 & +\dfrac{1-\nu^2}{E}S_\varphi & =\mathrm{e}^\xi \\
-E\dfrac{\mathrm{d}^2u_\rho}{\mathrm{d}\xi^2} & +0 & +0 & +S_\varphi - \nu\dfrac{\mathrm{d}S_\varphi}{\mathrm{d}\xi} & =0 \\
0 & +0 & -S_{\rho\varphi} - \dfrac{\mathrm{d}S_{\rho\varphi}}{\mathrm{d}\xi} & +0 & =0
\end{array}\right\} \tag{7.6.24}$$

Similar to the solution procedure of the basic eigen-solution (7.6.17), from the first and the fourth equations and the boundary conditions (7.6.8), we obtain

$$u_\varphi = c\mathrm{e}^\xi, \quad S_{\rho\varphi} = 0 \tag{7.6.25}$$

As $u_\varphi = c\mathrm{e}^\xi$ implies a basic eigen-solution which can be superposed arbitrarily, it is possible to assume $c = 0$ without loss of generality.

Similarly, solving simultaneously the second and the third expression of Eq. (7.6.24) yields

$$\left.\begin{aligned} u_\rho &= c_3 \mathrm{e}^{\xi} + c_4 \mathrm{e}^{-\xi} + \frac{1-\nu}{2}\xi \mathrm{e}^{\xi} \\ S_\varphi &= \frac{E}{1-\nu} c_3 \mathrm{e}^{\xi} + \frac{E}{1+\nu} c_4 \mathrm{e}^{-\xi} + \frac{E}{2}\mathrm{e}^{\xi}\left(\xi + \frac{2-\nu}{1-\nu}\right) \end{aligned}\right\} \tag{7.6.26}$$

where c_3, c_4 are undetermined constants. Substituting the solution into the boundary conditions (7.6.8), the constants are

$$\left.\begin{aligned} c_3 &= -\frac{1}{2}\left[1 + (1-\nu)\frac{R_2^2 \ln R_2 - R_1^2 \ln R_1}{R_2^2 - R_1^2}\right] \\ c_4 &= -\frac{(1+\nu)R_2^2 R_1^2}{2(R_2^2 - R_1^2)} \ln\left(\frac{R_2}{R_1}\right) \end{aligned}\right\} \tag{7.6.27}$$

Thus the first-order Jordan form eigenvector is

$$\boldsymbol{\psi}_0^{(1)} = \left\{\begin{matrix} c_3\rho + c_4\dfrac{1}{\rho} + \dfrac{1-\nu}{2}\rho\ln\rho \\ 0 \\ 0 \\ \dfrac{E}{1-\nu}c_3\rho + \dfrac{E}{1+\nu}c_4\dfrac{1}{\rho} + \dfrac{E}{2}\rho\left(\ln\rho + \dfrac{2-\nu}{1-\nu}\right) \end{matrix}\right\} \tag{7.6.28}$$

which is not the solution of the original problem (7.6.10). The actual solution is

$$\boldsymbol{v}_0^{(1)} = \boldsymbol{\psi}_0^{(1)} + \varphi\boldsymbol{\psi}_0^{(0)} \tag{7.6.29}$$

which is physically interpreted as the pure bending of a curved beam.

As the eigenvectors $\boldsymbol{\psi}_0^{(1)}$ and $\boldsymbol{\psi}_0^{(0)}$ are mutually symplectic adjoint,

$$\langle\boldsymbol{\psi}_0^{(0)}, \boldsymbol{\psi}_0^{(1)}\rangle = \frac{E[(R_2^2 - R_1^2)^2 - 4R_2^2R_1^2\ln^2(R_2/R_1)]}{8(R_2^2 - R_1^2)} > 0 \tag{7.6.30}$$

the second-order Jordan form eigenvector does not exist. This fact can be proved by directly solving the solution.

It is well known that there are six eigen-solutions of zero eigenvalue for plane problems in a rectangular domain. These solutions are not covered by the Saint–Venant principle. For problems in a sector domain, there are only two eigen-solutions of zero eigenvalue and this fact is similar to the

Hamiltonian system with radial coordinate treated as "time". Apparently, four eigen-solutions are missing. The eigen-solutions of zero eigenvalue do not decay with increasing or decreasing coordinate φ. Because the eigen-solutions of pure imaginary eigenvalues also do not decay with increasing or decreasing coordinate φ, they are basic solutions requiring further analysis.

7.6.2. *Eigen-Solutions of* $\mu = \pm\mathrm{i}$

To obtain the eigen-solutions of nonzero eigenvalues, the eigenvalue equation (7.6.13) should be expanded as

$$\left.\begin{array}{lllll} -\mu u_\rho & +u_\varphi - \dfrac{\mathrm{d}u_\varphi}{\mathrm{d}\xi} & +\dfrac{2(1+\nu)}{E}S_{\rho\varphi} & +0 & =0 \\ -u_\rho - \nu\dfrac{\mathrm{d}u_\rho}{\mathrm{d}\xi} & -\mu u_\varphi & +0 & +\dfrac{1-\nu^2}{E}S_\varphi & =0 \\ -E\dfrac{\mathrm{d}^2 u_\rho}{\mathrm{d}\xi^2} & +0 & -\mu S_{\rho\varphi} & +S_\varphi - \nu\dfrac{\mathrm{d}S_\varphi}{\mathrm{d}\xi} & =0 \\ 0 & +0 & -S_{\rho\varphi} - \dfrac{\mathrm{d}S_{\rho\varphi}}{\mathrm{d}\xi} & -\mu S_\varphi & =0 \end{array}\right\} \tag{7.6.31}$$

This is a system of simultaneous ordinary differential equations with respect to ξ which can be solved by firstly determining the eigenvalues λ in the ξ direction. The corresponding characteristic polynomial is

$$\det\begin{bmatrix} -\mu & 1-\lambda & 2(1+\nu)/E & 0 \\ -1-\nu\lambda & -\mu & 0 & (1-\nu^2)/E \\ -E\lambda^2 & 0 & -\mu & 1-\nu\lambda \\ 0 & 0 & -1-\lambda & -\mu \end{bmatrix} = 0 \tag{7.6.32}$$

Expanding the determinant yields

$$\lambda^4 - 2(1-\mu^2)\lambda^2 + (1+\mu^2)^2 = 0 \tag{7.6.33}$$

and the solutions are

$$\lambda_{1,2} = \pm(1+\mathrm{i}\mu), \quad \lambda_{3,4} = \pm(1-\mathrm{i}\mu) \tag{7.6.34}$$

The expressions of general solutions are different for different eigenvalues μ which are still undetermined. From the expressions in Eqs. (7.6.34), there is a repeated root 0 in addition to the roots 2 and -2 for $\mu = \pm\mathrm{i}$. Hence, the general solution is

$$\left.\begin{aligned} u_\rho &= A_1 + A_2\xi + A_3\mathrm{e}^{2\xi} + A_4\mathrm{e}^{-2\xi} \\ u_\varphi &= B_1 + B_2\xi + B_3\mathrm{e}^{2\xi} + B_4\mathrm{e}^{-2\xi} \\ S_{\rho\varphi} &= C_1 + C_2\xi + C_3\mathrm{e}^{2\xi} + C_4\mathrm{e}^{-2\xi} \\ S_\varphi &= D_1 + D_2\xi + D_3\mathrm{e}^{2\xi} + D_4\mathrm{e}^{-2\xi} \end{aligned}\right\} \tag{7.6.35}$$

The constants are not independent. Substituting Eq. (7.6.35) into Eq. (7.6.31) yields

$$\left.\begin{aligned} &A_2 = -\frac{(1+\nu)(3-\nu)}{E(1-\nu)}\mu C_1; \quad D_1 = \mu C_1; \quad C_2 = D_2 = 0 \\ &B_1 = \mu A_1 + \frac{(1+\nu)^2}{E(1-\nu)}C_1, \quad B_2 = \frac{(1+\nu)(3-\nu)}{E(1-\nu)}C_1 \end{aligned}\right\} \tag{7.6.36}$$

and

$$\left.\begin{aligned} &A_3 = \frac{1-3\nu}{2E}\mu C_3, \quad B_3 = \frac{5+\nu}{2E}C_3, \quad D_3 = 3\mu C_3 \\ &A_4 = -\frac{1+\nu}{2E}\mu C_4, \quad B_4 = -\frac{1+\nu}{2E}C_4, \quad D_4 = -\mu C_4 \end{aligned}\right\} \tag{7.6.37}$$

Substituting Eqs. (7.6.35)–(7.6.37) into the boundary conditions (7.6.8) yields

$$C_1 = C_3 = C_4 = 0 \tag{7.6.38}$$

The eigen-solution corresponding to eigenvalue $\mu = \mathrm{i}$ is

$$\boldsymbol{\psi}_{\mathrm{i}}^{(0)} = \{1, \quad \mathrm{i}, \quad 0, \quad 0\}^{\mathrm{T}} \tag{7.6.39}$$

and that corresponding to $\mu = -\mathrm{i}$ is

$$\boldsymbol{\psi}_{\mathrm{i}}^{(0)} = \{1, \quad -\mathrm{i}, \quad 0, \quad 0\}^{\mathrm{T}} \tag{7.6.40}$$

The solutions of the original problem (7.6.10) thus formed are

$$\boldsymbol{v}_{\mathrm{i}}^{(0)} = \mathrm{e}^{\mathrm{i}\varphi}\boldsymbol{\psi}_{\mathrm{i}}^{(0)} \quad \text{and} \quad \boldsymbol{v}_{-\mathrm{i}}^{(0)} = \mathrm{e}^{-\mathrm{i}\varphi}\boldsymbol{\psi}_{-\mathrm{i}}^{(0)} \tag{7.6.41}$$

respectively. These two solutions are mutually complex conjugate eigen-solutions. Separating the real and imaginary parts yields

$$\boldsymbol{v}_{\mathrm{iR}}^{(0)} = \{\cos\varphi,\ -\sin\varphi,\ 0,\ 0\}^{\mathrm{T}} \tag{7.6.42a}$$

$$\boldsymbol{v}_{\mathrm{iI}}^{(0)} = \{\sin\varphi,\ \cos\varphi,\ 0,\ 0\}^{\mathrm{T}} \tag{7.6.42b}$$

which correspond to rigid body translations along two perpendicular directions.

Obviously $\boldsymbol{\psi}_{\mathrm{i}}^{(0)}$ and $\boldsymbol{\psi}_{-\mathrm{i}}^{(0)}$ are not mutually symplectic adjoint but rather they are symplectic orthogonal. Therefore, there exist next order Jordan form solutions. For instance, for the first-order eigen-solution corresponding to $\mu = \mathrm{i}$, the equation is

$$\boldsymbol{H}\boldsymbol{\psi}_{\mathrm{i}}^{(1)} = \mathrm{i}\boldsymbol{\psi}_{\mathrm{i}}^{(1)} + \boldsymbol{\psi}_{\mathrm{i}}^{(0)} \tag{7.6.43}$$

This is an inhomogeneous equation. The general solution for the corresponding homogeneous equation is given by Eqs. (7.6.35)–(7.6.37) while the particular solution of the inhomogeneous equation is

$$\left\{\frac{2}{1-\nu}\mathrm{i}\xi,\quad -\frac{1+\nu+2\xi}{1-\nu},\quad 0,\quad 0\right\}^{\mathrm{T}} \tag{7.6.44}$$

Superposing Eqs. (7.6.44) and (7.6.35) yields the general solution of Eq. (7.6.43). Substituting the general solution into the side boundary conditions (7.6.8) yields

$$\boldsymbol{\psi}_{\mathrm{i}}^{(1)} = \left\{\mathrm{i}u_\rho^{(1)},\quad u_\varphi^{(1)},\quad S_{\rho\varphi}^{(1)},\quad \mathrm{i}S_\varphi^{(1)}\right\}^{\mathrm{T}} \tag{7.6.45}$$

where

$$\left.\begin{aligned}
u_\rho^{(1)} &= \frac{1}{2}(1-\nu)\xi + a(1-3\nu)\mathrm{e}^{2\xi} + b(1+\nu)\mathrm{e}^{-2\xi} \\
u_\varphi^{(1)} &= -\frac{1}{2}[1+\nu+(1-\nu)\xi] + a(5+\nu)\mathrm{e}^{2\xi} + b(1+\nu)\mathrm{e}^{-2\xi} \\
S_{\rho\varphi}^{(1)} &= E\left(\frac{1}{2} + 2a\mathrm{e}^{2\xi} - 2b\mathrm{e}^{-2\xi}\right) \\
S_\varphi^{(1)} &= E\left(\frac{1}{2} + 6a\mathrm{e}^{2\xi} + 2b\mathrm{e}^{-2\xi}\right)
\end{aligned}\right\} \tag{7.6.46}$$

and

$$a = \frac{-1}{4(R_1^2 + R_2^2)}, \quad b = -aR_1^2R_2^2 \tag{7.6.47}$$

The corresponding solution of the original problem is

$$\boldsymbol{v}_{\mathrm{i}}^{(1)} = \mathrm{e}^{\mathrm{i}\varphi}(\boldsymbol{\psi}_{\mathrm{i}}^{(1)} + \varphi\boldsymbol{\psi}_{\mathrm{i}}^{(0)}) \tag{7.6.48}$$

The solution above is the Jordan form solution of $\mu = \mathrm{i}$. For the Jordan form solution of $\mu = -\mathrm{i}$, it is only required to determine the complex conjugate of Eq. (7.6.45).

The complex expression of Eq. (7.6.48) is better be transformed into the real form. Separating the real part and imaginary parts of Eq. (7.6.48), yields

$$\boldsymbol{v}_{\mathrm{iR}}^{(1)} = \begin{Bmatrix} \varphi\cos\varphi - u_\rho^{(1)}\sin\varphi \\ -\varphi\sin\varphi + u_\varphi^{(1)}\cos\varphi \\ S_{\rho\varphi}^{(1)}\cos\varphi \\ -S_\varphi^{(1)}\sin\varphi \end{Bmatrix}; \quad \boldsymbol{v}_{\mathrm{iI}}^{(1)} = \begin{Bmatrix} \varphi\sin\varphi + u_\rho^{(1)}\cos\varphi \\ \varphi\cos\varphi + u_\varphi^{(1)}\sin\varphi \\ S_{\rho\varphi}^{(1)}\sin\varphi \\ S_\varphi^{(1)}\cos\varphi \end{Bmatrix} \tag{7.6.49}$$

Both of the real solutions are the solutions of the original problem.

Besides the eigen-solutions with zero eigenvalue (7.6.23) and (7.6.29), the eigen-solutions (7.6.43) and (7.6.49) with $\mu = \pm\mathrm{i}$ cannot be covered with Saint–Venant principle yet. The six solutions formed the basic solutions of the bending problem of curved beam.

Similar to a straight beam, therefore, the bending problem of a curved beam can be solved through the expansion of the six eigen-solutions $\boldsymbol{\psi}_0^{(0)}, \boldsymbol{\psi}_0^{(1)}, \boldsymbol{\psi}_{\mathrm{i}}^{(0)}, \boldsymbol{\psi}_{-\mathrm{i}}^{(0)}, \boldsymbol{\psi}_{\mathrm{i}}^{(1)}$ and $\boldsymbol{\psi}_{-\mathrm{i}}^{(1)}$ as the basis. Consequently, all local effects which decay with respect to distance are neglected in accordance with the Saint–Venant principle.

7.6.3. *Eigen-solutions of General Nonzero Eigenvalues*

In the previous two sections, the non-decaying eigen-solutions of $\mu = 0$ and $\mu = \pm\mathrm{i}$ are analyzed. The general solution for the eigen-solutions of

$\mu \neq 0, \pm \mathrm{i}$ with four different roots as given by Eq. (7.6.34) is

$$\left.\begin{aligned} u_\rho &= A_1 \mathrm{e}^{\lambda_1 \xi} + A_2 \mathrm{e}^{\lambda_2 \xi} + A_3 \mathrm{e}^{\lambda_3 \xi} + A_4 \mathrm{e}^{\lambda_4 \xi} \\ u_\varphi &= B_1 \mathrm{e}^{\lambda_1 \xi} + B_2 \mathrm{e}^{\lambda_2 \xi} + B_3 \mathrm{e}^{\lambda_3 \xi} + B_4 \mathrm{e}^{\lambda_4 \xi} \\ S_{\rho\varphi} &= C_1 \mathrm{e}^{\lambda_1 \xi} + C_2 \mathrm{e}^{\lambda_2 \xi} + C_3 \mathrm{e}^{\lambda_3 \xi} + C_4 \mathrm{e}^{\lambda_4 \xi} \\ S_\varphi &= D_1 \mathrm{e}^{\lambda_1 \xi} + D_2 \mathrm{e}^{\lambda_2 \xi} + D_3 \mathrm{e}^{\lambda_3 \xi} + D_4 \mathrm{e}^{\lambda_4 \xi} \end{aligned}\right\} \tag{7.6.50}$$

The constants are not independent and they are required to satisfy Eq. (7.6.13). If A_j $(j = 1, 2, 3, 4)$ are chosen as the independent constants, then the constants are related by

$$\left.\begin{aligned} B_j &= \frac{(1+\lambda_j)\lambda_j^2 - (1+\nu\mu^2)\lambda_j - \mu^2 - 1}{\mu(\mu^2 + 1 + \lambda_j - \nu\lambda_j - \nu\lambda_j^2)} A_j \\ C_j &= \frac{-E\mu\lambda_j^2}{\mu^2 + 1 + \lambda_j - \nu\lambda_j - \nu\lambda_j^2} A_j \\ D_j &= \frac{E\lambda_j^2(1+\lambda_j)}{\mu^2 + 1 + \lambda_j - \nu\lambda_j - \nu\lambda_j^2} A_j \end{aligned}\right\} \quad (j = 1, 2, 3, 4) \tag{7.6.51}$$

Substituting Eqs. (7.6.49) and (7.6.50) into the boundary conditions (7.6.8) yields a set of four homogeneous equations. Setting the determinant of coefficient to zero yields the transcendental equation of eigenvalue μ. Subsequently, substituting the eigenvalues into the homogeneous equations yields ratios between the constants A_j $(j = 1, 2, 3, 4)$. Consequently, the corresponding eigenvectors are determined.

Having obtained the eigenvectors, the expansion theorem can be applied to derive the solution. The solution procedure is similar to the discussion in the previous few chapters and thus the details are omitted here.

References

1. Longfu Wang, *Theory of Elasticity*, Beijing: Science Press, 1984 (in Chinese).
2. Jialong Wu, *Elasticity*, Shanghai: Tongji University Press, 1993 (in Chinese).
3. Hongqing Zhang and Alatancang, Completeness of symplectic orthogonal system, *Journal of Dalian University of Technology*, 1995, 35(6): 754–758 (in Chinese).
4. E. Sternberg and W.T. Koiter, The wedge under a concentrated couple: a paradox in the two-dimensional theory of elasticity, *Journal of Applied Mechanics*, 1958, 25(3), 575–581.

5. J.P. Dempsey, The wedge subjected to tractions: a paradox resolved, *Journal of Elasticity*, 1981, 11(1): 1–10.
6. Weian Yao, Jordan solutions for polar coordinate Hamiltonian system and solutions of paradoxes in elastic wedge, *Acta Mechanica Sinica*, 2001, 33(1): 79–86 (in Chinese).
7. W.X. Zhong, X.S. Xu and H.W. Zhang, On a direct method for the problem of elastic curved beams, *Engineering Mechanics*, 1996, 13(4): 1–8 (in Chinese).

Chapter 8

Hamiltonian System for Bending of Thin Plates

In this chapter we introduce in detail the theory of analogy between plane elasticity and thin plate bending problems. We then present another set of fundamental equations for the classical bending theory of thin plates. We further establish the Pro-H-R variational principle and the Pro-Hu–Washizu variational principle for bending of thin plate and derive the multi-variable variational principles for thin plate bending and plane elasticity. Based on the analogy theory, subsequently, the Hamiltonian system and its symplectic geometry theory are directly applied to the thin plate bending problem to derive a system of Hamiltonian symplectic solution. Consequently the thin plate bending problem can be analyzed using a rational Hamiltonian approach.

8.1. Small Deflection Theory for Bending of Elastic Thin Plates

Plate is one of the most important structural elements and the solution for mechanics of plate has long been an important research area in solid mechanics. A plate with a ratio of thickness to minimum characteristic dimension greater than 1/5 is called a **thick plate**. A plate with a ratio smaller than 1/80 is called a **membrane plate**. For a ratio between 1/80 and 1/5, the plate is called a **thin plate**. The middle plane dividing the plate into two equal parts is called the **neutral plane**. The neutral plane is normally assigned as the xy-plane with the positive direction of the z-axis pointing downwards.

If a thin plate is stable when subject to external loads acting on the neutral plane, it becomes a plane stress problem. If all external loads are normal to the neutral plane, we have mainly bending deformation where the z-displacement of each point on the neutral plane is called the **deflection** of plate. If the deflection is less than or equal to 1/5 of the thickness, it is a **small deflection problem**.

The basic assumption of small deflection theory of thin plate was first established by Kirchhoff and hence it is called the **Kirchhoff hypothesis**. It states that a straight line normal to the neutral plane remains straight and normal to the deflected plane after deformation. Besides, the length of line is invariant before and after deformation which is commonly known as transverse inextensibility. According to this hypothesis we have

$$\gamma_{xz} = \gamma_{yz} = \varepsilon_z = 0 \tag{8.1.1}$$

We can also deduce that there is only transverse displacement w during bending for every point on the neutral plane. Bending occurs without displacements along the x- and y-direction on the neutral plane. Hence

$$(u)_{z=0} = (v)_{z=0} = 0, \quad (w)_{z=0} = w(x, y) \tag{8.1.2}$$

Because of $\varepsilon_z = 0$, the displacement w is independent of the transverse z-coordinate and it is only a function of the in-plane coordinates x and y, or

$$w = w(x, y) \tag{8.1.3}$$

From $\gamma_{xz} = \gamma_{yz} = 0$ and the geometric relations (2.2.2), we have

$$\frac{\partial u}{\partial z} = -\frac{\partial w}{\partial x}, \qquad \frac{\partial v}{\partial z} = -\frac{\partial w}{\partial y} \tag{8.1.4}$$

Integrating the above expression with respect to z and making use of Eq. (8.1.2) yield

$$u = -z\frac{\partial w}{\partial x}, \qquad v = -z\frac{\partial w}{\partial y} \tag{8.1.5}$$

Further applying the geometric relations (2.2.2) yields

$$\varepsilon_x = -z\kappa_x, \quad \varepsilon_y = -z\kappa_y, \quad \gamma_{xy} = 2z\kappa_{xy} \tag{8.1.6}$$

where

$$\kappa_x = \frac{\partial^2 w}{\partial x^2}, \quad \kappa_y = \frac{\partial^2 w}{\partial y^2}, \quad \kappa_{xy} = -\frac{\partial^2 w}{\partial x \partial y} \tag{8.1.7}$$

are curvature and twisting curvature of the plate, respectively. Equation (8.1.7) is the curvature-deflection relation which can also be expressed in terms of operator matrix $\boldsymbol{K}(\partial)$ as

$$\boldsymbol{\kappa} = \boldsymbol{K}(\partial)w \tag{8.1.7$'$}$$

where

$$\boldsymbol{\kappa} = \{\kappa_y, \kappa_x, \kappa_{xy}\}^{\mathrm{T}} \tag{8.1.8}$$

and the operator matrix $\boldsymbol{K}(\partial)$ is

$$\boldsymbol{K}(\partial) = \left\{ \begin{array}{c} \dfrac{\partial^2}{\partial y^2} \\ \dfrac{\partial^2}{\partial x^2} \\ -\dfrac{\partial^2}{\partial x \partial y} \end{array} \right\} \tag{8.1.9}$$

From Eqs. (8.1.5) and (8.1.6), the displacements u, v and the strain components $\varepsilon_x, \varepsilon_y, \gamma_{xy}$ are linearly distributed through the thickness of plate. These quantities vanish on the neural mid-plane and they have maxima on the top and bottom surfaces.

Since the normal stress perpendicular to the neutral plane is considerably small and negligible as compared with σ_x, σ_y and τ_{xy}, the stress-strain relations (2.3.13) can be simplified as

$$\left.\begin{aligned} \sigma_x &= \frac{E}{1-\nu^2}(\varepsilon_x + \nu\varepsilon_y) \\ \sigma_y &= \frac{E}{1-\nu^2}(\varepsilon_y + \nu\varepsilon_x) \\ \tau_{xy} &= \frac{E}{2(1+\nu)}\gamma_{xy} \end{aligned}\right\} \tag{8.1.10}$$

Substituting Eq. (8.1.6) into Eq. (8.1.10) yields

$$\left.\begin{aligned} \sigma_x &= -\frac{Ez}{1-\nu^2}(\kappa_x + \nu\kappa_y) \\ \sigma_y &= -\frac{Ez}{1-\nu^2}(\kappa_y + \nu\kappa_x) \\ \tau_{xy} &= \frac{Ez}{1+\nu}\kappa_{xy} \end{aligned}\right\} \tag{8.1.11}$$

Figure 8.1 shows a rectangular differential element of the plate formed by two pairs of planes parallel to the xz- and yz-coordinate planes.

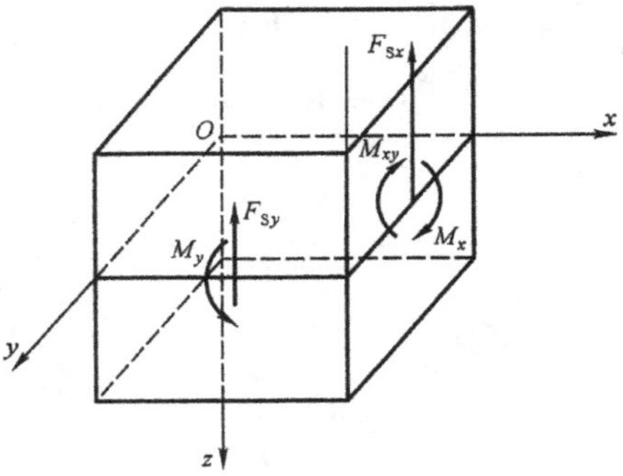

Fig. 8.1. Directions of positive internal forces on plate.

The normal stresses on the side of element result in a couple of forces (i.e. **bending moment**). The moment per unit length is

$$\left.\begin{aligned} M_x &= \int_{-h/2}^{h/2} (-\sigma_x z)\mathrm{d}z = D(\kappa_x + \nu\kappa_y) \\ M_y &= \int_{-h/2}^{h/2} (-\sigma_y z)\mathrm{d}z = D(\kappa_y + \nu\kappa_x) \end{aligned}\right\} \tag{8.1.12}$$

where D is the **flexural rigidity (bending stiffness)** of plate expressed as

$$D = \frac{Eh^3}{12(1-\nu^2)} \tag{8.1.13}$$

The shear stresses τ_{xy} also result in a couple (i.e. **torsional moment**) and the moment per unit length is

$$M_{xy} = \int_{-h/2}^{h/2} \tau_{xy} z \mathrm{d}z = D(1-\nu)\kappa_{xy} \tag{8.1.14}$$

Combining Eqs. (8.1.12) and (8.1.14) yields the moment-curvature relation

$$\boldsymbol{m} = \boldsymbol{C}\boldsymbol{\kappa} \ \text{ or } \ \boldsymbol{\kappa} = \boldsymbol{C}^{-1}\boldsymbol{m} \tag{8.1.15}$$

where

$$\boldsymbol{m} = \{M_y, M_x, 2M_{xy}\}^{\mathrm{T}} \tag{8.1.16}$$

and the elasticity coefficient matrix of material is

$$\boldsymbol{C} = D\begin{bmatrix} 1 & \nu & 0 \\ \nu & 1 & 0 \\ 0 & 0 & 2(1-\nu) \end{bmatrix} \tag{8.1.17}$$

The strain energy density in terms of curvature is

$$v_\varepsilon(\boldsymbol{\kappa}) = \frac{1}{2}\boldsymbol{\kappa}^{\mathrm{T}}\boldsymbol{C}\boldsymbol{\kappa} = \frac{1}{2}D[\kappa_x^2 + \kappa_y^2 + 2\nu\kappa_x\kappa_y + 2(1-\nu)\kappa_{xy}^2] \tag{8.1.18}$$

and the moment-curvature relation (8.1.15) can also be expressed in terms of the **strain energy density** as

$$\boldsymbol{m} = \frac{\partial v_\varepsilon(\boldsymbol{\kappa})}{\partial \boldsymbol{\kappa}} = \boldsymbol{C}\boldsymbol{\kappa} \tag{8.1.19}$$

On transforming all independent variables $\boldsymbol{\kappa}$ of strain energy density v_ε in accordance with Legendre's transformation, i.e. introducing the following function (**strain complementary energy density**)

$$v_{\mathrm{c}}(\boldsymbol{m}) = \boldsymbol{m}^{\mathrm{T}}\boldsymbol{\kappa} - v_\varepsilon(\boldsymbol{\kappa}) = \frac{1}{2}\boldsymbol{m}^{\mathrm{T}}\boldsymbol{C}^{-1}\boldsymbol{m} = \frac{6}{Eh^3}[M_x^2 + M_y^2 - 2\nu M_x M_y + 2(1+\nu)M_{xy}^2] \tag{8.1.20}$$

we can express curvature $\boldsymbol{\kappa}$ in terms of moment $\boldsymbol{m}$ as

$$\boldsymbol{\kappa} = \frac{\partial v_{\mathrm{c}}(\boldsymbol{m})}{\partial \boldsymbol{m}} = \boldsymbol{C}^{-1}\boldsymbol{m} \tag{8.1.21}$$

Referring to the element in Fig. 8.1, the following shear forces

$$F_{\mathrm{S}x} = \int_{-h/2}^{h/2}(-\tau_{xz})\mathrm{d}z, \quad F_{\mathrm{S}y} = \int_{-h/2}^{h/2}(-\tau_{yz})\mathrm{d}z \tag{8.1.22}$$

exist on the sides in addition to moment. According to Eq. (8.1.1), τ_{xz}, τ_{yz} should vanish if the stress-strain relations is applied directly. In fact these are higher-order quantities comparing with σ_x, σ_y and τ_{xy} and their effect on deformation is negligible. However, they are necessary for ensuring equilibrium and their values can be determined from the equations of equilibrium.

Consider a rectangular differential plate element with sides $\mathrm{d}x$, $\mathrm{d}y$ and thickness h for a plate with transverse load $q(x, y)$. The internal forces acting on the four sides are illustrated in Fig. 8.2.

Projecting all forces acting on the element onto the z-axis, we obtain the following equation of equilibrium

$$\frac{\partial F_{\mathrm{S}x}}{\partial x} + \frac{\partial F_{\mathrm{S}y}}{\partial y} - q = 0 \tag{8.1.23}$$

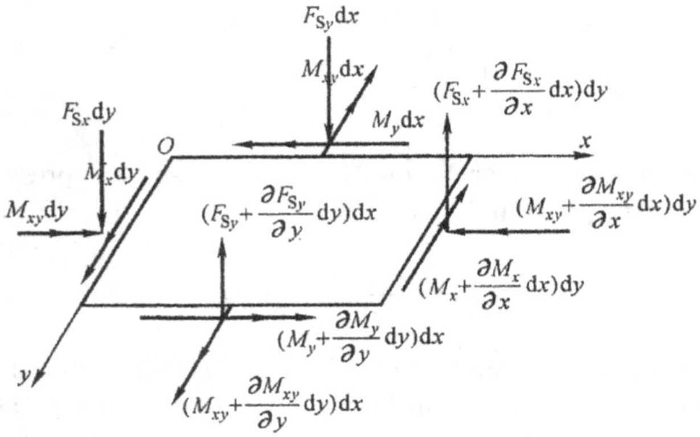

Fig. 8.2. Equilibrium of internal forces on plate element.

Taking moments of all forces acting on the element with respect to the y-axis and neglecting higher-order small quantities, we obtain the following equation of equilibrium

$$\frac{\partial M_x}{\partial x} - \frac{\partial M_{xy}}{\partial y} - F_{Sx} = 0 \tag{8.1.24}$$

In the same way, we obtain

$$\frac{\partial M_y}{\partial y} - \frac{\partial M_{xy}}{\partial x} - F_{Sy} = 0 \tag{8.1.25}$$

Since there are no forces in the x- and y-directions and no moments with respect to the z-axis on the element, Eqs. (8.1.23)–(8.1.25) completely define the state of equilibrium of the element, or they are the equilibrium equations of internal forces for plate bending. Substituting Eqs. (8.1.24) and (8.1.25) into Eq. (8.1.23), we obtain the equilibrium equation in terms of bending moment and torsional moment as

$$\frac{\partial^2 M_x}{\partial x^2} - 2\frac{\partial^2 M_{xy}}{\partial x \partial y} + \frac{\partial^2 M_y}{\partial y^2} = q \tag{8.1.26}$$

In terms of the following operator matrix

$$\hat{\boldsymbol{K}}(\partial) = \left\{ \frac{\partial^2}{\partial y^2}, \quad \frac{\partial^2}{\partial x^2}, \quad -\frac{\partial^2}{\partial x \partial y} \right\} \tag{8.1.27}$$

Equation (8.1.26) can be expressed as

$$\hat{\boldsymbol{K}}(\partial)\boldsymbol{m} = q \tag{8.1.28}$$

Finally, substituting the moment-curvature relation (8.1.15) and the curvature-deflection relation (8.1.7) into the equation above, we obtain the basic governing equation in terms of displacement for bending of thin plates as

$$\nabla^2\nabla^2 w = \frac{q}{D} \tag{8.1.29}$$

where ∇^2 is the two-dimensional Laplace operator

$$\nabla^2 = \frac{\partial^2}{\partial x^2} + \frac{\partial^2}{\partial y^2} \tag{8.1.30}$$

The various boundary conditions for thin plate are discussed here. We consider a rectangular plate as an example and assume the x- and y-axes are parallel to the sides of plate. We focus on the side AB of plate at $y = b$.

From statics viewpoint, the distributed torsional moment is equivalent to shearing force. Hence the torsional moment $M_{xy}\mathrm{d}x$ acting on one side with differential length $\mathrm{d}x$ can be replaced equivalently by two forces of magnitude M_{xy} acting on two opposite sides as shown in Fig. 8.3. The torsional moment $[M_{xy} + (\partial M_{xy}/\partial x)\mathrm{d}x]\mathrm{d}x$ acting on the adjacent side with differential length $\mathrm{d}x$ can be replaced equivalently by two forces of magnitude $M_{xy} + (\partial M_{xy}/\partial x)\mathrm{d}x$ acting on two opposite sides. On the intersecting boundary the resultant force is $(\partial M_{xy}/\partial x)\mathrm{d}x$ which can be replaced by distributed shear force $\partial M_{xy}/\partial x$ along $\mathrm{d}x$. When it is combined with the original transverse shear force $F_{\mathrm{S}y}$, we obtain the total equivalent shear force on side AB as

$$F_{\mathrm{V}y} = F_{\mathrm{S}y} - \frac{\partial M_{xy}}{\partial x} \tag{8.1.31}$$

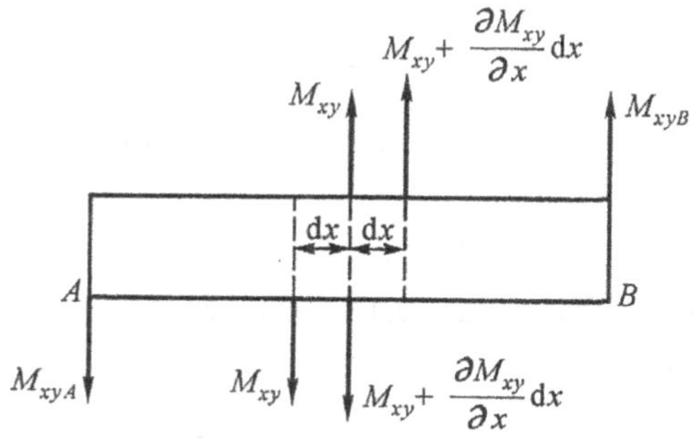

Fig. 8.3. Static equivalence for torsional moment on side AB of plate.

The positive direction of equivalent distributed shear force $F_{\mathrm{V}y}$ coincides with that of $F_{\mathrm{S}y}$. It should be noted that there are two concentrated forces $(M_{xy})_A$ and $(M_{xy})_B$ at the ends A and B of side AB. As there are also concentrated forces on the adjacent sides, there will be a resultant concentrated force at each corner, $2(M_{xy})_B$ at point B, for instance.

Thus we obtain the various boundary conditions along $y = b$ of the plate. In general, we have

(1) For a clamped edge, the deflection and rotation must be zero, i.e.

$$(w)_{y=b} = 0, \quad \left(\frac{\partial w}{\partial y}\right)_{y=b} = 0 \tag{8.1.32}$$

(2) For a simply supported edge, the deflection and bending moment must be zero, i.e.

$$(w)_{y=b} = 0, \quad (M_y)_{y=b} = 0 \tag{8.1.33}$$

(3) For a free edge, the bending moment and total equivalent shear force must be zero, i.e.

$$(M_y)_{y=b} = 0, \quad (F_{\mathrm{V}y})_{y=b} = 0 \tag{8.1.34}$$

If two adjacent sides are both free, there should be a further corner condition. Assuming point B is the corner of two adjacent free sides without support, there is

$$2(M_{xy})_B = 0 \tag{8.1.35}$$

while there is

$$(w)_B = 0 \tag{8.1.36}$$

if there is a support at point B. The boundary conditions for other edges can be obtained in a similar way.

8.2. Analogy between Plane Elasticity and Bending of Thin Plate

The fundamental equation for thin plate bending is a biharmonic equation (8.1.29) established by Lagrange and Germain in the 19th century. Since then the basic problem lies in the solution of this equation. Although very effective for isotropic plate with two opposite sides simply supported, the classical semi-inverse methods such as Naiver's method and Levy's method are difficult to be applied to cases with complicated boundary conditions, especially for bending of anisotropic plates.

The Airy stress function satisfying the biharmonic equation[1–6] has been adopted for the classical solutions of plane elasticity problems. Because the fundamental equations are the same, the plane elasticity problem and the thin plate bending problem must be similar to each other. This fact has been noticed by many researchers[7], such as Southwel[8]. As these two types of problems have their respective backgrounds and the corresponding solution methodologies, the analogy has not been fully exploited.

For the fundamental equation of thin plate bending (8.1.29), the effect of transverse load q can be solved via a particular solution and subsequently applying the principle of superposition. Hence, we consider the homogeneous equation with $q = 0$ in the first instance

$$\nabla^2\nabla^2 w = 0 \tag{8.2.1}$$

Referring to Sec. 4.1, we know that the Airy stress function φ_{f} for plane elasticity also satisfies the biharmonic equation

$$\nabla^2\nabla^2 \varphi_{\mathrm{f}} = 0 \tag{8.2.2}$$

Taking advantage of such similarity, the analogy between the two kinds of problems can be established[9]. For instance, curvatures κ_y, κ_x, κ_{xy} of plate bending correspond to stresses $\sigma_x, \sigma_y, \tau_{xy}$ of plane elasticity; and moment $M_x, M_y, 2M_{xy}$ of plate bending correspond to strains $\varepsilon_x, \varepsilon_y, \gamma_{xy}$ of plane elasticity. Except the reversal of the sign of Poisson's ratio ν, the moment-curvature relation (8.1.19) corresponds one-to-one to the strain-stress relation (4.1.3) of plane elasticity.

According to the analogy, we introduce the **bending moment functions** $\boldsymbol{\phi} = \{\phi_x, \phi_y\}^{\mathrm{T}}$ for thin plate bending corresponding to the displacement u, v of plane elasticity. Hence, there exist the following relations between moment and bending moment function for thin plate bending corresponding to the geometric relation (4.1.5) of plane elasticity,

$$M_y = \frac{\partial \phi_x}{\partial x}, \quad M_x = \frac{\partial \phi_y}{\partial y}, \quad 2M_{xy} = \frac{\partial \phi_x}{\partial y} + \frac{\partial \phi_y}{\partial x} \tag{8.2.3}$$

In terms of the operator matrix

$$\hat{\boldsymbol{E}}(\nabla) = \begin{bmatrix} \dfrac{\partial}{\partial x} & 0 \\ 0 & \dfrac{\partial}{\partial y} \\ \dfrac{\partial}{\partial y} & \dfrac{\partial}{\partial x} \end{bmatrix} \tag{8.2.4}$$

they can be expressed as

$$\boldsymbol{m} = \hat{\boldsymbol{E}}(\nabla)\boldsymbol{\phi} \tag{8.2.5}$$

The homogeneous equation (8.1.26) is satisfied by substituting Eq. (8.2.3) into the equation.

Now we introduce some properties of bending moment function.

Similar to rigid body displacements of plane elasticity, at a first instance, it should be noticed that the functions

$$\phi_x = a_0 - a_2 y, \quad \phi_y = a_1 + a_2 x \tag{8.2.6}$$

where a_0, a_1, a_2 are arbitrary constants do not result in any bending moment. These functions are called **null moment functions**.

Next, the transformation equations of bending moment functions ϕ_x, ϕ_y under rotation of coordinate system should be considered. For rotation of coordinate system at an angle α, the bending moments transform as

$$\left.\begin{aligned} M_x' &= M_x \cos^2\alpha + M_y \sin^2\alpha - 2M_{xy}\sin\alpha\cos\alpha \\ M_y' &= M_x \sin^2\alpha + M_y \cos^2\alpha + 2M_{xy}\sin\alpha\cos\alpha \\ M_{xy}' &= M_{xy}(\cos^2\alpha - \sin^2\alpha) + (M_x - M_y)\sin\alpha\cos\alpha \end{aligned}\right\} \tag{8.2.7}$$

Accordingly, ϕ_x and ϕ_y transform as

$$\left.\begin{aligned} \phi_x' &= \phi_x \cos\alpha + \phi_y \sin\alpha \\ \phi_y' &= -\phi_x \sin\alpha + \phi_y \cos\alpha \end{aligned}\right\} \tag{8.2.8}$$

It is obvious that Eq. (8.2.3) is still valid in the new coordinate system (x', y'). Hence the transformation rule for ϕ_x, ϕ_y is the same as that of vector and they can be termed **bending moment function vector**.

Next we discuss the boundary conditions. For simplicity, we consider a straight boundary with n denoting the normal and s the tangential direction and (n, s) forms a right-hand system. Denoting (ϕ_n, ϕ_s) as the bending moment function on the boundary, the shear force on the lateral edge from Eq. (8.1.24) is

$$F_{Sn} = \frac{\partial M_n}{\partial n} - \frac{\partial M_{ns}}{\partial s} = \frac{1}{2}\left(\frac{\partial^2 \phi_s}{\partial n \partial s} - \frac{\partial^2 \phi_n}{\partial s^2}\right) \tag{8.2.9}$$

From theory of thin plate bending, there are the normal bending moment $\overline{M}_n$ and equivalent shear force $\overline{F}_{Vn}$ on the boundary Γ_σ with specified forces. The boundary conditions in terms of bending moment function are

$$(M_n =)\frac{\partial \phi_s}{\partial s} = \overline{M}_n, \quad \left(F_{Vn} = F_{Sn} - \frac{\partial M_{ns}}{\partial s} =\right) - \frac{\partial^2 \phi_n}{\partial s^2} = \overline{F}_{Vn} \tag{8.2.10}$$

The differential terms in Eq. (8.2.10) are only with respect to the boundary-coordinate s and upon integration we have

$$\left.\begin{aligned} \phi_s &= \overline{\phi}_s = \int_{s_0}^{s} \overline{M}_n \mathrm{d}s' + a_1 \\ \phi_n &= \overline{\phi}_n = \int_{s_0}^{s} (s' - s)\overline{F}_{Vn} \mathrm{d}s' + a_0 + a_2 s \end{aligned}\right\} \tag{8.2.11}$$

where a_0, a_1, a_2 are undetermined constants and the rests are all known function. Since there may be separated segments of force conditions on the boundary, it is not possible to eliminate all a_0, a_1, a_2 of each segment. However, in the light of arbitrariness of null moment functions, it is always possible to eliminate the three constants of one particular segment. Referring to plane elasticity, ϕ_n, ϕ_s correspond to the normal and tangential displacements u_n, u_s. Hence the force boundary conditions of thin plate bending correspond to the displacement boundary conditions of plane elasticity. In addition, there are equilibrium conditions at corners for thin plate bending.

From the theory of thin plate bending, deflection $\overline{w}$ and rotation $\overline{\theta}_n$ should be specified on the boundary Γ_u with specified displacements. These are functions of s. Hence, the boundary conditions in terms of curvatures are

$$\kappa_{ns}\left(= -\frac{\partial^2 w}{\partial n \partial s}\right) = \overline{\kappa}_{ns}\left(= -\frac{\partial \overline{\theta}_n}{\partial s}\right), \quad \kappa_s\left(= \frac{\partial^2 w}{\partial s^2}\right) = \overline{\kappa}_s\left(= \frac{\partial^2 \overline{w}}{\partial s^2}\right) \tag{8.2.12}$$

Integrating Eq. (8.2.12) along boundary s yields displacement boundary conditions which θ_n and w are specified. The boundary conditions (8.2.12) can also be represented in terms of the components of curvature in the original coordinate system as

$$\left.\begin{aligned}\kappa_x \sin\alpha + \kappa_{xy}\cos\alpha = \overline{\kappa}_s \sin\alpha + \overline{\kappa}_{ns}\cos\alpha \\ \kappa_{xy}\sin\alpha + \kappa_y \cos\alpha = \overline{\kappa}_s \cos\alpha - \overline{\kappa}_{ns}\sin\alpha\end{aligned}\right\} \tag{8.2.13}$$

where angle α is the anticlockwise rotation from the axis x to the normal n. Apparently, the boundary conditions (8.2.12) or (8.2.13) are the same as the boundary conditions for specified forces in plane elasticity.

There exists corresponding relationship between the fundamental equation and boundary conditions of thin plate bending and those of plane elasticity. In other words, there is analogy between thin plate bending and plane elasticity. Hence, similar to the principle of minimum potential energy in plane elasticity, we can derive the principle of minimum complementary energy for thin plate bending in terms of bending moment function

$$\min_{\phi_x,\phi_y} E_{\mathrm{pc}} = \min_{\phi_x,\phi_y} (V_{\mathrm{c}} + E_{\mathrm{c}}) \tag{8.2.14}$$

where the strain complementary energy is

$$\begin{aligned} V_{\mathrm{c}} = \iint_V \frac{6}{Eh^3}\Bigg[&\left(\frac{\partial \phi_x}{\partial x}\right)^2 + \left(\frac{\partial \phi_y}{\partial y}\right)^2 - 2\nu\frac{\partial \phi_x}{\partial x}\frac{\partial \phi_y}{\partial y} \\ &+ \frac{1+\nu}{2}\left(\frac{\partial \phi_y}{\partial x} + \frac{\partial \phi_x}{\partial y}\right)^2\Bigg]\mathrm{d}x\mathrm{d}y \end{aligned} \tag{8.2.15}$$

and the complementary energy for support displacement is

$$E_{\mathrm{c}} = -\int_{\Gamma_u} (M_n\overline{\theta}_n - F_{\mathrm{V}n}\overline{w})\mathrm{d}s = -\int_{\Gamma_u} \left(\overline{\theta}_n \frac{\partial \phi_s}{\partial s} + \overline{w}\frac{\partial^2 \phi_n}{\partial s^2}\right)\mathrm{d}s \tag{8.2.16}$$

Since the specified deflection $\overline{w}$ and rotation $\overline{\theta}_n$ are both functions of s, integrating by parts yields

$$E_{\mathrm{c}} = \int_{\Gamma_u} \left(\phi_s \frac{\partial \overline{\theta}_n}{\partial s} - \phi_n \frac{\partial^2 \overline{w}}{\partial s^2}\right)\mathrm{d}s - \left[\phi_s\overline{\theta}_n + \overline{w}\frac{\partial \phi_n}{\partial s} - \phi_n \frac{\partial \overline{w}}{\partial s}\right]_{s_0}^{s_1} \tag{8.2.17}$$

where s_0, s_1 are the ends of the boundary Γ_u with specified displacements.

The variation of Eq. (8.2.14) yields the following differential equations

$$\left.\begin{aligned}\frac{\partial^2 \phi_x}{\partial x^2} + \frac{1+\nu}{2}\frac{\partial^2 \phi_x}{\partial y^2} + \frac{1-\nu}{2}\frac{\partial^2 \phi_y}{\partial x \partial y} = 0 \\ \frac{\partial^2 \phi_y}{\partial y^2} + \frac{1+\nu}{2}\frac{\partial^2 \phi_y}{\partial x^2} + \frac{1-\nu}{2}\frac{\partial^2 \phi_x}{\partial x \partial y} = 0\end{aligned}\right\} \tag{8.2.18}$$

which indicates strain compatibility. In reality, it is also possible to derive Eq. (8.2.18) from the strain compatibility equation

$$\boldsymbol{E}(\nabla)\boldsymbol{\kappa} = \mathbf{0} \tag{8.2.19}$$

where the operator matrix is

$$\boldsymbol{E}(\nabla) = \begin{bmatrix} \dfrac{\partial}{\partial x} & 0 & \dfrac{\partial}{\partial y} \\ 0 & \dfrac{\partial}{\partial y} & \dfrac{\partial}{\partial x} \end{bmatrix} \tag{8.2.20}$$

The boundary conditions (8.2.12) on the boundary Γ_u with specified displacements can also be derived directly from the variational principle because they are the natural boundary conditions of the principle of minimum complementary energy.

Similarly to the Hellinger–Reissner variational principle for plane elasticity, we can derive the **Pro-H-R variational principle** for thin plate bending as

$$\begin{aligned} \delta\Pi_2 = \delta\bigg\{ & \iint_V [\boldsymbol{\kappa}^{\mathrm{T}}\hat{\boldsymbol{E}}(\nabla)\boldsymbol{\phi} - v_\varepsilon(\boldsymbol{\kappa})]\mathrm{d}x\mathrm{d}y - \int_{\Gamma_u} (\phi_s\overline{\kappa}_{ns} + \phi_n\overline{\kappa}_s)\mathrm{d}s \\ & - \int_{\Gamma_\sigma} [\kappa_{ns}(\phi_s - \overline{\phi}_s) + \kappa_s(\phi_n - \overline{\phi}_n)]\mathrm{d}s\bigg\} = 0 \end{aligned} \tag{8.2.21}$$

where $\overline{\phi}_s, \overline{\phi}_n$ etc. are described in Eq. (8.2.11).

Considering $\phi_x, \phi_y, \kappa_y, \kappa_x, \kappa_{xy}$ as independent variables and performing $\delta\Pi_2 = 0$ yield the curvature compatibility equation (8.2.19) and the moment-curvature relation

$$\hat{\boldsymbol{E}}(\nabla)\boldsymbol{\phi} = \frac{\partial v_\varepsilon(\boldsymbol{\kappa})}{\partial \boldsymbol{\kappa}} \tag{8.2.22}$$

as well as the boundary conditions (8.2.12) on Γ_u with specified displacements and boundary conditions (8.2.11) on Γ_σ with specified forces.

Similar to the Hu–Washizu variational principle in plane elasticity, we may derive the **Pro-Hu–Washizu variational principle** for thin plate bending[10] as

$$\delta\Pi_3 = \delta\left\{\iint_V \left[\boldsymbol{\kappa}^{\mathrm{T}}\hat{\boldsymbol{E}}(\nabla)\boldsymbol{\phi} - \boldsymbol{\kappa}^{\mathrm{T}}\boldsymbol{m} + v_{\mathrm{c}}(\boldsymbol{m})\right]\mathrm{d}x\mathrm{d}y \right.$$
$$- \int_{\Gamma_\sigma} \left[\kappa_{ns}(\phi_s - \overline{\phi}_s) + \kappa_s(\phi_n - \overline{\phi}_n)\right]\mathrm{d}s$$
$$\left. - \int_{\Gamma_u} (\phi_s\overline{\kappa}_{ns} + \phi_n\overline{\kappa}_s)\,\mathrm{d}s\right\} = 0 \tag{8.2.23}$$

Again, considering $\phi_x, \phi_y, \kappa_y, \kappa_x, \kappa_{xy}$ and M_y, M_x, M_{xy} as independent variables and performing $\delta\Pi_3 = 0$ yield the curvature compatibility equation (8.2.19), the relationship of bending moment functions and bending moments (8.2.5) and the moment-curvature relationship (8.1.21), as well as the boundary conditions (8.2.12) on Γ_u with specified displacements and (8.2.11) on Γ_σ with specified forces.

It should be emphasized that Eqs. (8.2.21) and (8.2.23) are respectively named the Pro-H-R variational principle and Pro-Hu–Washizu variational principle for plate bending because they have been derived from the analogy between plate bending and plane elasticity. Different from the Hellinger–Reissner variational principle and Hu–Washizu variational principle for plate bending, deflection w does not appear in the Pro-H-R variational principle and Pro-Hu–Washizu variational principle for plate bending. It could be solved after obtaining curvature $\boldsymbol{\kappa}$. Certainly, it is also possible to derive the Pro-H-R variational principle and Pro-Hu–Washizu variational principle for plane elasticity from the Hellinger–Reissner variational principle and Hu–Washizu variational principle for plate bending. The details are omitted here.

The analogy between plane elasticity and plate bending is listed in the Table 8.1. They have isomorphism when expressed in terms of mathematical terms.

The analogy theory for plate bending and plane elasticity finds important applications in finite element analysis. It is well known that, the finite element method for plane elasticity has been well developed as compared to the finite element method for plate bending. From the established analogy principle above, the derivation of elements for plate bending may refer to the similar methods and expressions of elements for plane elasticity[10]. The analogy principle has proven that for each plane compatible element there exists a corresponding equilibrium plate element, and vice versa[11]. Most of the previous research focused on modeling of equilibrium plate element but

Table 8.1. The analogy between plane elasticity and thin plate bending.

Plane elasticity	**Thin plate bending**
Airy stress function φ_f	Transverse deflection $w(x,y)$
In-plane displacement vector $\boldsymbol{u} = \{u, v\}^T$	Bending moment function vector $\boldsymbol{\phi} = \{\phi_x, \phi_y\}^T$
Strain $\boldsymbol{\varepsilon} = \{\varepsilon_x, \varepsilon_y, \gamma_{xy}\}^T$	Bending moment $\boldsymbol{m} = \{M_y, M_x, 2M_{xy}\}^T$
Stress $\boldsymbol{\sigma} = \{\sigma_x, \sigma_y, \tau_{xy}\}^T$	Curvature $\boldsymbol{\kappa} = \{\kappa_y, \kappa_x, \kappa_{xy}\}^T$
Strain-displacement relation $\boldsymbol{\varepsilon} = \hat{\boldsymbol{E}}(\nabla)\boldsymbol{u}$	Relationship between moment and moment function $\boldsymbol{m} = \hat{\boldsymbol{E}}(\nabla)\boldsymbol{\phi}$
Relationship between stress function and stress $\boldsymbol{\sigma} = \boldsymbol{K}(\partial)\varphi_f$	Deflection-curvature relation $\boldsymbol{\kappa} = \boldsymbol{K}(\partial)w$
Strain-stress relation $\boldsymbol{\varepsilon} = \boldsymbol{C}^{-1}\boldsymbol{\sigma}$	Moment-curvature relation $\boldsymbol{m} = \boldsymbol{C}\boldsymbol{\kappa}$
Rigid body displacement	Null moment function
Boundary conditions for specified forces Γ_σ $\left.\begin{matrix}\sigma_x\cos\alpha + \tau_{xy}\sin\alpha = 0\\ \tau_{xy}\cos\alpha + \sigma_y\sin\alpha = 0\end{matrix}\right\}$	Boundary conditions for specified displacements Γ_u $\left.\begin{matrix}\kappa_y\cos\alpha + \kappa_{xy}\sin\alpha = 0\\ \kappa_{xy}\cos\alpha + \kappa_x\sin\alpha = 0\end{matrix}\right\}$
Boundary conditions for specified displacements Γ_u, $u = \overline{u}, v = \overline{v}$	Boundary conditions for specified forces Γ_σ, $\phi_s = \overline{\phi}_s, \phi_n = \overline{\phi}_n$
The principle of minimum potential energy	The principle of minimum complementary energy
H-R variational principle: $u, v; \sigma_x, \sigma_y, \tau_{xy}$; without φ_f	Pro-H-R variational principle: $\phi_x, \phi_y; \kappa_y, \kappa_x, \kappa_{xy}$; without w
Hu–Washizu variational principle: $u, v; \sigma_x, \sigma_y, \tau_{xy}; \varepsilon_x, \varepsilon_y, \gamma_{xy}$; without φ_f	Pro-Hu–Washizu variational principle: $\phi_x, \phi_y; \kappa_y, \kappa_x, \kappa_{xy}; M_y, M_x, 2M_{xy}$; without w
Pro-H-R variational principle	H-R variational principle
Pro-Hu–Washizu variational principle	Hu–Washizu variational principle

the outcome has been discouraging[12]. The recent works of the authors and others have verified that the difficulty of satisfying C^1 continuity in plate elements can be avoided at the outset if we begins from the analogy principle. Following the methods and expressions of plane elastic elements (compatible or incompatible) and via appropriate transformation, it is possible to construct a series of plate bending elements with excellent properties.

8.3. Multi-Variable Variational Principles for Thin Plate Bending and Plane Elasticity

The variational principle is a basic issue in elasticity. It is of great importance not only in theory but also in practical applications. The classical variational principles include (i) the principle of minimum potential energy and the principle of minimum complementary energy with one kind of variables; (ii) the Hellinger–Reissner variational principle with two kinds of variables; and (iii) the Hu–Washizu variational principle with three kinds of variables. These variational principles have attracted close attention and wide applications.

The Hu–Washizu variational principle has been satisfactory. The only concern is the exclusion of the stress function widely used in classical stress analysis and solution. The stress functions are, however, not physical quantities and it is not a problem if they do not appear in the variational functional. In other words, the stress function can be considered as some kind of auxiliary variables but not the regular variables. Actually, there has been much success in adopting displacement functions rather than stress functions in finite element numerical analysis. The use of stress functions has been less common now. From the analogy between plate bending and plane elasticity, stress functions appear in the variational principle in a natural manner and it results in the Pro-Hu–Washizu variational principle. The Pro-Hu–Washizu variational principle for plate bending corresponds to the Hu–Washizu variational principle for plane elasticity while, likewise, the Hu–Washizu variational principle for plate bending corresponds to the Pro-Hu–Washizu variational principle for plane elasticity. The two cases above can be synthesized to obtain the multi-variable variational principle for plate bending and plane elasticity[13] involving stress function and residual strain besides displacement, stress and strain. It also contains the five kinds of fundamental equations in elasticity including equilibrium equation, strain-displacement relation, stress-strain relation, compatibility equation, and stress–stress function relation. It is the most generalized variational principle at present.

8.3.1. *Multi-Variable Variational Principles for Plate Bending*

For plate bending problems with residual deformation, there are five kinds of fundamental equations in the domain. They can be classified into:

(1) Equation of equilibrium

$$\hat{\boldsymbol{K}}(\partial)\boldsymbol{m}_a = \hat{\boldsymbol{K}}(\partial)\boldsymbol{m}_{0a} = q \tag{8.3.1}$$

where $\boldsymbol{m}_{0a} = \{M_{0ay}, M_{0ax}, 2M_{0axy}\}^{\mathrm{T}}$ is a particular solution of the inhomogeneous bending moment vector resulting from an external load q;

while $\boldsymbol{m}_a$ is a general solution of the inhomogeneous bending moment vector resulting from the external load. There are infinite general solutions which can be regard as the variational variables.

(2) Relation between bending moment function and bending moment

$$\boldsymbol{m} = \boldsymbol{m}_a + \hat{\boldsymbol{E}}(\nabla)\boldsymbol{\phi} \tag{8.3.2}$$

(3) Curvature-deflection relation

$$\boldsymbol{\kappa} = \boldsymbol{\kappa}_0 + \boldsymbol{K}(\partial)w - \frac{\partial v_c(\boldsymbol{m}_a - \boldsymbol{m}_{0a})}{\partial \boldsymbol{m}_a} \tag{8.3.3}$$

where $\boldsymbol{\kappa}_0$ is the residual deformation caused by factors with known quantity such as temperature difference, etc.

(4) Compatibility equation

$$\boldsymbol{E}(\nabla)(\boldsymbol{\kappa} - \boldsymbol{\kappa}_0) = \boldsymbol{0} \tag{8.3.4}$$

(5) Curvature-moment relation

$$\boldsymbol{\kappa} = \frac{\partial v_c(\boldsymbol{m})}{\partial \boldsymbol{m}} \tag{8.3.5}$$

Next we discuss the boundary conditions. The boundaries are not limited to straight edges but they can be any smooth curves without corners for simplicity. The positive normal of the plate boundary curve is denoted as n and the tangential direction is denoted as s. The (n, s) system forms a right-hand system and the angle between n- and x-axis is α. The radius of curvature of the boundary curve is ρ and the outside of convex is taken as positive. For an arbitrary function g along the boundary, we have

$$\left.\begin{aligned} \frac{\partial g}{\partial n} &= \frac{\partial g}{\partial x}\cos\alpha + \frac{\partial g}{\partial y}\sin\alpha \\ \frac{\partial g}{\partial s} &= -\frac{\partial g}{\partial x}\sin\alpha + \frac{\partial g}{\partial y}\cos\alpha \end{aligned}\right\} \tag{8.3.6}$$

Additionally according to the definition of curvature, we have

$$\frac{\partial \alpha}{\partial s} = \frac{1}{\rho} \tag{8.3.7}$$

Since transformation of bending moment function is the same as transformation of a vector, the bending moment functions along the boundary are

$$\phi_n = \phi_x \cos\alpha + \phi_y \sin\alpha, \quad \phi_s = -\phi_x \sin\alpha + \phi_y \cos\alpha \tag{8.3.8}$$

and then

$$\begin{aligned}
\frac{\partial \phi_s}{\partial s} &= -\frac{\partial \phi_x}{\partial s}\sin\alpha + \frac{\partial \phi_y}{\partial s}\cos\alpha - \frac{1}{\rho}(\phi_x \cos\alpha + \phi_y \sin\alpha) \\
&= \frac{\partial \phi_x}{\partial x}\sin^2\alpha + \frac{\partial \phi_y}{\partial y}\cos^2\alpha - \left(\frac{\partial \phi_x}{\partial y} + \frac{\partial \phi_y}{\partial x}\right)\sin\alpha\cos\alpha - \frac{\phi_n}{\rho} \\
&= \hat{M}_y \sin^2\alpha + \hat{M}_x \cos^2\alpha - 2\hat{M}_{xy}\sin\alpha\cos\alpha - \frac{\phi_n}{\rho} \\
&= \hat{M}_n - \frac{\phi_n}{\rho}
\end{aligned} \tag{8.3.9}$$

where $\hat{M}_x$ etc. indicate the corresponding bending moments represented by the bending moment function vector. Hence the equation for bending moment normal to the boundary in terms of bending moment function is

$$\hat{M}_n = \frac{\partial \phi_s}{\partial s} + \frac{\phi_n}{\rho} \tag{8.3.10}$$

Similarly, we have

$$\begin{aligned}
\frac{\partial \phi_n}{\partial s} - \frac{\phi_s}{\rho} &= \frac{\partial \phi_x}{\partial s}\cos\alpha + \frac{\partial \phi_y}{\partial s}\sin\alpha \\
&= (\hat{M}_x - \hat{M}_y)\frac{1}{2}\sin(2\alpha) + \frac{\partial \phi_x}{\partial y}\cos^2\alpha - \frac{\partial \phi_y}{\partial x}\sin^2\alpha \\
&= (\hat{M}_x - \hat{M}_y)\frac{1}{2}\sin(2\alpha) + \frac{1}{2}\left(\frac{\partial \phi_x}{\partial y} + \frac{\partial \phi_y}{\partial x}\right)\cos(2\alpha) \\
&\quad + \frac{1}{2}\left(\frac{\partial \phi_x}{\partial y} - \frac{\partial \phi_y}{\partial x}\right) = \hat{M}_{ns} + \frac{1}{2}\left(\frac{\partial \phi_x}{\partial y} - \frac{\partial \phi_y}{\partial x}\right)
\end{aligned} \tag{8.3.11}$$

Taking partial derivative of both sides with respect to s yields

$$\begin{aligned}
&\frac{\partial}{\partial s}\left(\frac{\partial \phi_n}{\partial s} - \frac{\phi_s}{\rho}\right) \\
&\quad = \frac{\partial \hat{M}_{ns}}{\partial s} - \frac{1}{2}\left(\frac{\partial^2 \phi_x}{\partial x \partial y} - \frac{\partial^2 \phi_y}{\partial x^2}\right)\sin\alpha + \frac{1}{2}\left(\frac{\partial^2 \phi_x}{\partial y^2} - \frac{\partial^2 \phi_y}{\partial x \partial y}\right)\cos\alpha
\end{aligned}$$

$$= \frac{\partial \hat{M}_{ns}}{\partial s} - \left(\frac{\partial \hat{M}_y}{\partial y} - \frac{\partial \hat{M}_{xy}}{\partial x}\right)\sin\alpha - \left(\frac{\partial \hat{M}_x}{\partial x} - \frac{\partial \hat{M}_{xy}}{\partial y}\right)\cos\alpha$$

$$= \frac{\partial \hat{M}_{ns}}{\partial s} - \hat{F}_{Sn} = -\hat{F}_{Vn} \tag{8.3.12}$$

Hence, the total distributed shear forces along the boundary in terms of bending moment function is

$$\hat{F}_{Vn} = -\frac{\partial}{\partial s}\left(\frac{\partial \phi_n}{\partial s} - \frac{\phi_s}{\rho}\right) \tag{8.3.13}$$

The tangential curvature and normal twisting curvature along the boundary in terms of deflection w are derived as follows. From

$$\frac{\partial}{\partial s}\left(\frac{\partial w}{\partial n}\right) = \frac{\partial}{\partial s}\left(\frac{\partial w}{\partial x}\cos\alpha + \frac{\partial w}{\partial y}\sin\alpha\right)$$

$$= \frac{\partial}{\partial s}\left(\frac{\partial w}{\partial x}\right)\cos\alpha + \frac{\partial}{\partial s}\left(\frac{\partial w}{\partial y}\right)\sin\alpha + \frac{1}{\rho}\frac{\partial w}{\partial s}$$

$$= (\hat{\kappa}_y - \hat{\kappa}_x)\frac{1}{2}\sin(2\alpha) - \hat{\kappa}_{xy}\cos(2\alpha) + \frac{1}{\rho}\frac{\partial w}{\partial s}$$

$$= -\hat{\kappa}_{ns} + \frac{1}{\rho}\frac{\partial w}{\partial s} \tag{8.3.14}$$

the twisting curvature along the boundary can be expressed as

$$\hat{\kappa}_{ns} = -\frac{\partial \theta_n}{\partial s} + \frac{1}{\rho}\frac{\partial w}{\partial s} \tag{8.3.15}$$

Similarly, we derive

$$\frac{\partial^2 w}{\partial s^2} = \frac{\partial}{\partial s}\left(-\frac{\partial w}{\partial x}\sin\alpha + \frac{\partial w}{\partial y}\cos\alpha\right)$$

$$= -\frac{\partial}{\partial s}\left(\frac{\partial w}{\partial x}\right)\sin\alpha + \frac{\partial}{\partial s}\left(\frac{\partial w}{\partial y}\right)\cos\alpha - \frac{1}{\rho}\frac{\partial w}{\partial n}$$

$$= \hat{\kappa}_x \sin^2\alpha + 2\hat{\kappa}_{xy}\cos\alpha\sin\alpha + \hat{\kappa}_y\cos^2\alpha - \frac{1}{\rho}\frac{\partial w}{\partial n}$$

$$= \hat{\kappa}_s - \frac{1}{\rho}\frac{\partial w}{\partial n} \tag{8.3.16}$$

and the tangential curvature along the boundary is

$$\hat{\kappa}_s = \frac{\partial^2 w}{\partial s^2} + \frac{\theta_n}{\rho} \tag{8.3.17}$$

Having derived the above equation for the boundary, the boundary conditions according to the classical plate theory can be obtained. For instance,

(1) The boundary conditions for deflection and rotation on the displacement boundary Γ_u are

$$w = \overline{w} \quad \text{and} \quad \theta_n = \frac{\partial w}{\partial n} = \overline{\theta}_n \tag{8.3.18}$$

(2) The boundary conditions for bending moment and total distributed shear forces on the force boundary Γ_σ are

$$\left.\begin{aligned} M_n &= M_{an} + \frac{\partial \phi_s}{\partial s} + \frac{\phi_n}{\rho} = \overline{M}_n \\ F_{\mathrm{V}n} &= F_{\mathrm{V}an} - \frac{\partial}{\partial s}\left(\frac{\partial \phi_n}{\partial s} - \frac{\phi_s}{\rho}\right) = \overline{F}_{\mathrm{V}n} \end{aligned}\right\} \tag{8.3.19}$$

(3) The boundary conditions for deflection and bending moment on the boundary Γ_s with simply support are

$$w = \overline{w} \quad \text{and} \quad M_n = M_{an} + \frac{\partial \phi_s}{\partial s} + \frac{\phi_n}{\rho} = \overline{M}_n \tag{8.3.20}$$

where $\overline{w}, \overline{\theta}_n; \overline{M}_n, \overline{F}_{\mathrm{V}n}$ are specified functions along the boundary.

Besides, there are the boundary conditions for the complete boundary $\Gamma = \Gamma_u + \Gamma_\sigma + \Gamma_s$

$$\frac{\partial v_{\mathrm{c}}(\boldsymbol{m}_a - \boldsymbol{m}_{0a})}{\partial M_{as}} = 0, \quad \frac{\partial v_{\mathrm{c}}(\boldsymbol{m}_a - \boldsymbol{m}_{0a})}{\partial (2M_{ans})} = 0 \tag{8.3.21}$$

which, in terms of mechanics, imply the same tangential curvature and normal twisting curvature generated by the general and particular solutions of the inhomogeneous bending moment vector.

Denoting

$$\tilde{\boldsymbol{\varepsilon}} \equiv \boldsymbol{m}_a - \boldsymbol{m}_{0a}, \quad \tilde{\boldsymbol{\sigma}} \equiv \frac{\partial v_{\mathrm{c}}(\boldsymbol{m}_a - \boldsymbol{m}_{0a})}{\partial \boldsymbol{m}_a} = \frac{\partial v_{\mathrm{c}}(\tilde{\boldsymbol{\varepsilon}})}{\partial \tilde{\boldsymbol{\varepsilon}}} \tag{8.3.22}$$

and using Eqs. (8.3.1), (8.3.3) and (8.3.4) yield

$$\hat{\boldsymbol{K}}(\partial)\tilde{\boldsymbol{\varepsilon}} = 0, \quad \boldsymbol{E}(\nabla)\tilde{\boldsymbol{\sigma}} = \boldsymbol{0} \tag{8.3.23}$$

The boundary conditions (8.3.21) on boundary Γ can be rewritten as

$$\tilde{\sigma}_n = \tilde{\tau}_{ns} = 0 \tag{8.3.24}$$

Equations (8.3.22)–(8.3.24) are analogous to a plane elasticity problem without surface forces and with free boundaries. The unique solution can only be no "deformation" but with "rigid body displacement" in the domain. Hence we have

$$\tilde{\boldsymbol{\varepsilon}} = \boldsymbol{m}_a - \boldsymbol{m}_{0a} = \boldsymbol{0} \tag{8.3.25}$$

It indicates that Eqs. (8.3.1) to (8.3.5) and boundary conditions (8.3.18) to (8.3.21) constitute all definite conditions for plate bending problems.

Finally the **multi-variable variational principle** with five kinds of independent variables $w, \boldsymbol{\kappa}, \boldsymbol{m}_a, \boldsymbol{\phi}, \boldsymbol{m}$ is

$$\delta \Pi_m = 0 \tag{8.3.26}$$

where the functional is

$$\begin{aligned}
\Pi_m = & \iint_\Omega \{\boldsymbol{m}_a^{\mathrm{T}} \boldsymbol{K}(\partial) w + \boldsymbol{\kappa}^{\mathrm{T}} \boldsymbol{m} - v_{\mathrm{c}}(\boldsymbol{m}) - v_{\mathrm{c}}(\boldsymbol{m}_a - \boldsymbol{m}_{0a}) \\
& - (\boldsymbol{\kappa} - \boldsymbol{\kappa}_0)^{\mathrm{T}} [\boldsymbol{m}_a + \hat{\boldsymbol{E}}(\nabla)\boldsymbol{\phi}] - q w\} \mathrm{d}x \mathrm{d}y \\
& + \int_{\Gamma_u} \left[(w - \overline{w}) F_{\mathrm{V}an} - (\theta_n - \overline{\theta}_n) M_{an} + \phi_n \left(\frac{\partial^2 \overline{w}}{\partial s^2} + \frac{\overline{\theta}_n}{\rho} \right) \right. \\
& \left. - \phi_s \left(\frac{\partial \overline{\theta}_n}{\partial s} - \frac{1}{\rho} \frac{\partial \overline{w}}{\partial s} \right) \right] \mathrm{d}s \\
& + \int_{\Gamma_\sigma} \left[w \overline{F}_{\mathrm{V}n} - \theta_n \overline{M}_n + \phi_n \left(\frac{\partial^2 w}{\partial s^2} + \frac{\theta_n}{\rho} \right) - \phi_s \left(\frac{\partial \theta_n}{\partial s} - \frac{1}{\rho} \frac{\partial w}{\partial s} \right) \right] \mathrm{d}s \\
& + \int_{\Gamma_s} \left[(w - \overline{w}) F_{\mathrm{V}an} - \theta_n \overline{M}_n + \phi_n \left(\frac{\partial^2 \overline{w}}{\partial s^2} + \frac{\theta_n}{\rho} \right) \right. \\
& \left. - \phi_s \left(\frac{\partial \theta_n}{\partial s} - \frac{1}{\rho} \frac{\partial \overline{w}}{\partial s} \right) \right] \mathrm{d}s
\end{aligned} \tag{8.3.27}$$

On performing variation on the functional Π_m, the terms in domain Ω yield Eqs. (8.3.1) to (8.3.5). When transformed into the natural coordinate system, the terms on boundary Γ generated during integration by parts yields

$$\oint_{\Gamma} [M_{an}\delta\theta_n - F_{\mathrm{V}an}\delta w - (\kappa_s - \kappa_{0s})\delta\phi_n - (\kappa_{ns} - \kappa_{0ns})\delta\phi_s]\, \mathrm{d}s \qquad (8.3.28)$$

Associating Eq. (8.3.28) with the integral term of Γ_u and rearranging yield

$$\int_{\Gamma_u} \left[(w - \overline{w})\delta F_{\mathrm{V}an} - (\theta_n - \overline{\theta}_n)\delta M_{an} - \delta\phi_n \left(\kappa_s - \kappa_{0s} - \frac{\partial^2 \overline{w}}{\partial s^2} - \frac{\overline{\theta}_n}{\rho} \right) \right.$$
$$\left. -\delta\phi_s \left(\kappa_{ns} - \kappa_{0ns} + \frac{\partial \overline{\theta}_n}{\partial s} - \frac{1}{\rho}\frac{\partial \overline{w}}{\partial s} \right) \right] \mathrm{d}s \qquad (8.3.29)$$

and hence we obtain the boundary conditions (8.3.18) on Γ_u for specified deflection and rotation and

$$\kappa_s = \kappa_{0s} + \frac{\partial^2 \overline{w}}{\partial s^2} + \frac{\overline{\theta}_n}{\rho}, \quad \kappa_{ns} = \kappa_{0ns} - \frac{\partial \overline{\theta}_n}{\partial s} + \frac{1}{\rho}\frac{\partial \overline{w}}{\partial s} \qquad (8.3.30)$$

Subsequently, associating Eq. (8.3.28) with the integral term of Γ_σ and rearranging yield

$$\int_{\Gamma_\sigma} \left[\left(M_{an} + \frac{\partial \phi_s}{\partial s} + \frac{\phi_n}{\rho} - \overline{M}_n \right) \delta\theta_n - \left(F_{\mathrm{V}an} - \frac{\partial}{\partial s}\left(\frac{\partial \phi_n}{\partial s} - \frac{\phi_s}{\rho} \right) - \overline{F}_{\mathrm{V}n} \right) \delta w \right.$$
$$\left. - \left(\kappa_s - \kappa_{0s} - \frac{\partial^2 w}{\partial s^2} - \frac{\theta_n}{\rho} \right) \delta\phi_n - \left(\kappa_{ns} - \kappa_{0ns} + \frac{\partial \theta_n}{\partial s} - \frac{1}{\rho}\frac{\partial w}{\partial s} \right) \delta\phi_s \right] \mathrm{d}s$$
$$+ \left[\phi_n \delta \frac{\partial w}{\partial s} - \left(\frac{\partial \phi_n}{\partial s} - \frac{\phi_s}{\rho} \right) \delta w - \phi_s \delta\theta_n \right]_a^b \qquad (8.3.31)$$

where a, b are the starting and ending points of the corresponding boundary segments. From Eq. (8.3.31), we obtain the boundary conditions (8.3.19) on Γ_σ for specified normal bending moment and total distributed shear forces and

$$\kappa_s = \kappa_{0s} + \frac{\partial^2 w}{\partial s^2} + \frac{\theta_n}{\rho}, \quad \kappa_{ns} = \kappa_{0ns} - \frac{\partial \theta_n}{\partial s} + \frac{1}{\rho}\frac{\partial w}{\partial s} \qquad (8.3.32)$$

Finally, associating Eq. (8.3.28) with the integral term of Γ_s and rearranging yield

$$\int_{\Gamma_s}\left[(w-\overline{w})\delta F_{\mathrm{V}an}+\left(M_{an}+\frac{\partial\phi_s}{\partial s}+\frac{\phi_n}{\rho}-\overline{M}_n\right)\delta\theta_n\right.$$
$$-\delta\phi_n\left(\kappa_s-\kappa_{0s}-\frac{\partial^2\overline{w}}{\partial s^2}-\frac{\theta_n}{\rho}\right)$$
$$\left.-\delta\phi_s\left(\kappa_{ns}-\kappa_{0ns}+\frac{\partial\theta_n}{\partial s}-\frac{1}{\rho}\frac{\partial\overline{w}}{\partial s}\right)\right]\mathrm{d}s-[\phi_s\delta\theta_n]_a^b \quad (8.3.33)$$

From Eq. (8.3.33), we obtain the boundary conditions (8.3.20) on Γ_s for deflection and bending moment specified and

$$\kappa_s=\kappa_{0s}+\frac{\partial^2\overline{w}}{\partial s^2}+\frac{\theta_n}{\rho},\qquad \kappa_{ns}=\kappa_{0ns}-\frac{\partial\theta_n}{\partial s}+\frac{1}{\rho}\frac{\partial\overline{w}}{\partial s} \quad (8.3.34)$$

The normal twisting curvature and tangential curvature on the boundary in terms of deflection w are Eqs. (8.3.15) and (8.3.17), respectively, hence Eq. (8.3.3) on the boundary expressed in terms of natural coordinate is

$$\left.\begin{aligned}\kappa_s&=\kappa_{0s}+\frac{\partial^2 w}{\partial s^2}+\frac{\theta_n}{\rho}-\frac{\partial v_{\mathrm{c}}(\boldsymbol{m}_a-\boldsymbol{m}_{0a})}{\partial M_{as}}\\ \kappa_{ns}&=\kappa_{0ns}-\frac{\partial\theta_n}{\partial s}+\frac{1}{\rho}\frac{\partial w}{\partial s}-\frac{\partial v_{\mathrm{c}}(\boldsymbol{m}_a-\boldsymbol{m}_{0a})}{\partial(2M_{ans})}\end{aligned}\right\} \quad (8.3.35)$$

Substituting Eqs. (8.3.30), (8.3.32) and (8.3.34) into Eq. (8.3.35) yields the boundary conditions (8.3.21) for the complete boundary Γ.

The terminal term $[\phi_s\delta\theta_n]_a^b$ on Γ_σ and Γ_s generated during variation cancels out because of the presence of $\delta\theta_n=0(\theta_n=\overline{\theta}_n)$ at the intersections of any two of Γ_σ, Γ_s and Γ_u; or the presence of terms with opposite signs at the intersections. Similarly, the terminal term on Γ_σ

$$\left[\phi_n\delta\frac{\partial w}{\partial s}-\left(\frac{\partial\phi_n}{\partial s}-\frac{\phi_s}{\rho}\right)\delta w\right]_a^b \quad (8.3.36)$$

cancels out because of the presence of $\delta w=\delta(\partial w/\partial s)=0$ $(w=\overline{w})$ at the intersections of Γ_σ and either Γ_u or Γ_s; or the presence of terms with opposite signs at the intersections. Therefore the summation of the terminal terms in Eqs. (8.3.31) and (8.3.33) vanishes at all times and there is no new condition.

The derivation above shows that Eq. (8.3.26) includes and only includes all the five kinds of fundamental equations (8.3.1) to (8.3.5) and the boundary conditions (8.3.18) to (8.3.21) for plate bending. It is the most generalized multi-variable variational principle for plate bending.

The Hu–Washizu variational principle for plate bending can be derived from the multi-variable variational principle for plate bending. Taking $\boldsymbol{\kappa}_0 = 0$ (without initial incompatibility), $\boldsymbol{\phi} = \mathbf{0}$, $\boldsymbol{m}_a = \boldsymbol{m}_{0a}$, and taking extremum with respect to $\boldsymbol{m}$ in Eq. (8.3.26), the Hu–Washizu variational principle

$$\delta\Pi_3 = \delta\left\{\iint_\Omega \{\boldsymbol{m}_a^{\mathrm{T}}\boldsymbol{K}(\partial)w + v_\varepsilon(\boldsymbol{\kappa}) - \boldsymbol{\kappa}^{\mathrm{T}}\boldsymbol{m}_a - qw\}\mathrm{d}x\mathrm{d}y \right.$$
$$+ \int_{\Gamma_u} [(w-\overline{w})F_{\mathrm{V}an} - (\theta_n - \overline{\theta}_n)M_{an}]\mathrm{d}s$$
$$\left. + \int_{\Gamma_\sigma} [w\overline{F}_{\mathrm{V}n} - \theta_n\overline{M}_n]\mathrm{d}s + \int_{\Gamma_s} [(w-\overline{w})F_{\mathrm{V}an} - \theta_n\overline{M}_n]\mathrm{d}s\right\} = 0 \tag{8.3.37}$$

can be obtained. All other classical variational principles can certainly be derived accordingly and the details are omitted here.

8.3.2. *Multi-Variable Variational Principle for Plane Elasticity*

For plane elasticity problems with residual deformation, there are five kinds of fundamental equations in the domain. They can be classified into:

(1) Strain compatibility equation

$$\hat{\boldsymbol{K}}(\partial)\boldsymbol{\varepsilon}_a = \hat{\boldsymbol{K}}(\partial)\boldsymbol{\varepsilon}_{0a} \tag{8.3.38}$$

where $\boldsymbol{\varepsilon}_{0a}$ is a known quantity of the initial incompatible strain caused by factors such as temperature, etc.

(2) Strain-displacement relation

$$\boldsymbol{\varepsilon} = \boldsymbol{\varepsilon}_a + \hat{\boldsymbol{E}}(\nabla)\boldsymbol{u} \tag{8.3.39}$$

(3) Relation between stress and stress function

$$\boldsymbol{\sigma} = \boldsymbol{\sigma}_0 + \boldsymbol{K}(\partial)\varphi_{\mathrm{f}} - \frac{\partial v_\varepsilon(\boldsymbol{\varepsilon}_a - \boldsymbol{\varepsilon}_{0a})}{\partial\boldsymbol{\varepsilon}_a} \tag{8.3.40}$$

where $\boldsymbol{\sigma}_0$ is an inhomogeneous particular solution corresponding to the known body force $\boldsymbol{F}$.

(4) Equation of equilibrium

$$\boldsymbol{E}(\nabla)\boldsymbol{\sigma} = \boldsymbol{E}(\nabla)\boldsymbol{\sigma}_0 = \boldsymbol{F} \tag{8.3.41}$$

(5) Stress-strain relation

$$\boldsymbol{\sigma} = \frac{\partial v_\varepsilon(\boldsymbol{\varepsilon})}{\partial \boldsymbol{\varepsilon}} \tag{8.3.42}$$

In general, the boundary conditions are:

(1) Normal and tangential displacements on the displacement boundary Γ_u

$$u_n = \overline{u}_n \quad \text{and} \quad u_s = \overline{u}_s \tag{8.3.43}$$

(2) Normal and shear stresses on the force boundary Γ_σ

$$\sigma_n = \overline{\sigma}_n \quad \text{and} \quad \tau_{ns} = \overline{\tau}_{ns} \tag{8.3.44}$$

(3) Normal displacement and shear stress on the simply support boundary Γ_s

$$u_n = \overline{u}_n \quad \text{and} \quad \tau_{ns} = \overline{\tau}_{ns} \tag{8.3.45}$$

where $\overline{u}_n, \overline{u}_s, \overline{\sigma}_n, \overline{\tau}_{ns}$ are specified functions on the boundary.

Besides, there are the boundary conditions for the complete boundary $\Gamma = \Gamma_u + \Gamma_\sigma + \Gamma_s$

$$\varepsilon_{as} - \varepsilon_{0as} = \eta_{ans} - \eta_{0ans} = 0 \tag{8.3.46}$$

where $\varepsilon_{as}, \eta_{ans}$ etc. denote the extensional strain along the boundary curve and the change of curve of the boundary curve due to the corresponding deformation quantities[7]. We have

$$\begin{aligned}
\varepsilon_{as} &= \varepsilon_{ax}\sin^2\alpha - \gamma_{axy}\sin\alpha\cos\alpha + \varepsilon_{ay}\cos^2\alpha \\
\gamma_{ans} &= (\varepsilon_{ay} - \varepsilon_{ax})\sin(2\alpha) + \gamma_{axy}\cos(2\alpha) \\
\eta_{ans} &= -\frac{1}{2}\frac{\partial\gamma_{ans}}{\partial s} + \left(\frac{\partial\varepsilon_{ay}}{\partial x} - \frac{1}{2}\frac{\partial\gamma_{axy}}{\partial y}\right)\cos\alpha + \left(\frac{\partial\varepsilon_{ax}}{\partial y} - \frac{1}{2}\frac{\partial\gamma_{axy}}{\partial x}\right)\sin\alpha
\end{aligned} \tag{8.3.47}$$

where α is the angle between the normal of boundary n and the positive direction of x-axis.

Denoting

$$\tilde{\boldsymbol{m}} \equiv \boldsymbol{\varepsilon}_a - \boldsymbol{\varepsilon}_{0a}, \quad \tilde{\boldsymbol{\kappa}} \equiv \frac{\partial v_\varepsilon(\boldsymbol{\varepsilon}_a - \boldsymbol{\varepsilon}_{0a})}{\partial \boldsymbol{\varepsilon}_a} = \frac{\partial v_\varepsilon(\hat{\boldsymbol{m}})}{\partial \hat{\boldsymbol{m}}} \tag{8.3.48}$$

where $\tilde{\boldsymbol{m}} = \{\tilde{M}_y \; \tilde{M}_x \; 2\tilde{M}_{xy}\}^{\mathrm{T}}$, and using Eqs. (8.3.38), (8.3.40) and (8.3.41) yield

$$\hat{\boldsymbol{K}}(\partial)\tilde{\boldsymbol{m}} = 0, \quad \boldsymbol{E}(\nabla)\tilde{\boldsymbol{\kappa}} = \boldsymbol{0} \tag{8.3.49}$$

The boundary conditions (8.3.46) on Γ can be rewritten as

$$\tilde{M}_n = \tilde{F}_{\mathrm{V}n} = 0 \tag{8.3.50}$$

Equations (8.3.48)–(8.3.50) are completely analogous to a plate bending problem without transverse loads and with free boundaries. The unique solution can only be no "deformation", i.e. no "bending moment", in the domain. Hence we have

$$\tilde{\boldsymbol{m}}_a = \boldsymbol{\varepsilon}_a - \boldsymbol{\varepsilon}_{0a} = \boldsymbol{0} \tag{8.3.51}$$

It indicates that Eqs. (8.3.38) to (8.3.42) and boundary conditions (8.3.43) to (8.3.46) constitute all definite conditions for plane elasticity problems.

Finally the multi-variable variational principle with five kinds of independent variables $\varphi_{\mathrm{f}}, \boldsymbol{\sigma}, \boldsymbol{\varepsilon}_a, \boldsymbol{u}, \boldsymbol{\varepsilon}$ is

$$\delta \Pi_M = 0 \tag{8.3.52}$$

where the functional is

$$\begin{aligned} \Pi_M = & \iint_\Omega \{(\boldsymbol{\varepsilon}_a - \boldsymbol{\varepsilon}_{0a})^{\mathrm{T}} \boldsymbol{K}(\partial)\varphi_{\mathrm{f}} + \boldsymbol{\sigma}^{\mathrm{T}}\boldsymbol{\varepsilon} - v_\varepsilon(\varepsilon) - v_\varepsilon(\boldsymbol{\varepsilon}_a - \boldsymbol{\varepsilon}_{0a}) \\ & - (\boldsymbol{\sigma} - \boldsymbol{\sigma}_0)^{\mathrm{T}}[\boldsymbol{\varepsilon}_a + \hat{\boldsymbol{E}}(\nabla)\boldsymbol{u}]\}\mathrm{d}x\mathrm{d}y \\ & + \int_{\Gamma_u} [(u_n - \overline{u}_n)\sigma_n - u_n\sigma_{0n} + (u_s - \overline{u}_s)\tau_{ns} - u_s\tau_{0ns}]\mathrm{d}s \\ & + \int_{\Gamma_\sigma} [u_n(\overline{\sigma}_n - \sigma_{0n}) + u_s(\overline{\tau}_{ns} - \tau_{0ns})]\mathrm{d}s \\ & + \int_{\Gamma_s} [(u_n - \overline{u}_n)\sigma_n - u_n\sigma_{0n} + u_s(\overline{\tau}_{ns} - \tau_{0ns})]\mathrm{d}s \end{aligned} \tag{8.3.53}$$

On performing variation on the functional Π_M, the terms in domain Ω yield Eqs. (8.3.38) to (8.3.42). When transformed into the natural coordinate system, the terms on boundary Γ generated during integration by parts yields

$$\oint_{\Gamma}\left[(\varepsilon_{as}-\varepsilon_{0as})\delta\left(\frac{\partial\varphi_{\mathrm{f}}}{\partial n}\right)-(\eta_{ans}-\eta_{0ans})\delta\varphi_{\mathrm{f}}\right.$$
$$\left.-(\sigma_n-\sigma_{0n})\delta u_n-(\tau_{ns}-\tau_{0ns})\delta u_s\right]\mathrm{d}s \qquad (8.3.54)$$

The first two terms provide the boundary conditions (8.3.46) and associating the remaining two terms with the terminal terms of integration yields the boundary conditions (8.3.43) to (8.3.45).

The derivation above shows that Eq. (8.3.52) includes and only includes all the five kinds of fundamental equations (8.3.38) to (8.3.42) and the boundary conditions (8.3.43) to (8.3.46) for plane elasticity. It is the most generalized multi-variable variational principle for plane elasticity.

Similarly, taking $\boldsymbol{\varepsilon}_a = \boldsymbol{\varepsilon}_{0a} \equiv \mathbf{0}$, i.e. eliminaing initially incompatible quantities in the multi-variable variational principle for plane elasticity (8.3.52), the Hu–Washizu variational principle for plane elasticity

$$\delta\Pi_{p3}=\delta\left\{\iint_{\Omega}[\boldsymbol{\sigma}^{\mathrm{T}}\boldsymbol{\varepsilon}-v_{\varepsilon}(\boldsymbol{\varepsilon})-\boldsymbol{\sigma}^{\mathrm{T}}\hat{\boldsymbol{E}}(\nabla)\boldsymbol{u}-\boldsymbol{F}^{\mathrm{T}}\boldsymbol{u}]\mathrm{d}x\mathrm{d}y\right.$$
$$+\int_{\Gamma_u}[(u_n-\overline{u}_n)\sigma_n+(u_s-\overline{u}_s)\tau_{ns}]\mathrm{d}s$$
$$+\int_{\Gamma_\sigma}[u_n\overline{\sigma}_n+u_s\overline{\tau}_{ns}]\mathrm{d}s$$
$$\left.+\int_{\Gamma_s}[(u_n-\overline{u}_n)\sigma_n+u_s\overline{\tau}_{ns}]\mathrm{d}s\right\}=0 \qquad (8.3.55)$$

can be obtained. All other classical variational principle can certainly be derived accordingly.

The multi-variable variational principle are not only applicable to elasticity problems with residual deformation but also applicable to the derivation of multi-variable variational principle for shallow shell according to the analogy between plate bending and plane elasticity. Consequently, it offers an opportunity for solving relevant problems of shallow shell.

8.4. Symplectic Solution for Rectangular Plates

A system of symplectic solution can be derived by introducing plane elasticity problems into the Hamiltonian system. A system of symplectic solution for thin plate bending can also be established via the analogy between thin plate bending and plane elasticity. Similar to plane elasticity, the system of symplectic solution is applicable to both rectangular domain and sectorial domain. The analytical solutions for rectangular thin plate bending are presented here and the following three sections.

The relevant equations are introduced into the symplectic system in virtue of the Pro-H-R variational principle (8.2.21). Here, we let $y(b_1 \leq y \leq b_2)$ be transverse direction and its corresponding boundary conditions can be free, simply support or clamped. The x-coordinate is treated as time coordinate in the Hamiltonian system where an overdot indicates differentiation with respect to the x-coordinate, or $(\dot{}) = \partial/\partial x$. From Eqs. (8.1.12) and (8.2.3), we have

$$\kappa_x = \frac{1}{D}\frac{\partial \phi_y}{\partial y} - \nu\kappa_y \tag{8.4.1}$$

Substituting Eq. (8.4.1) into Eq. (8.2.21) to eliminate κ_x yields the Hamilton mixed energy variational principle as

$$\delta\left\{\int_{x_0}^{x_f}\int_{b_1}^{b_2}\left[\kappa_y\dot{\phi}_x + \kappa_{xy}\dot{\phi}_y - \nu\kappa_y\frac{\partial \phi_y}{\partial y} + \kappa_{xy}\frac{\partial \phi_x}{\partial y}\right.\right.$$
$$\left. + \frac{1}{2D}\left(\frac{\partial \phi_y}{\partial y}\right)^2 - D(1-\nu)\kappa_{xy}^2 - \frac{D(1-\nu^2)}{2}\kappa_y^2\right]\mathrm{d}y\mathrm{d}x$$
$$\left. - \int_{\Gamma_\sigma}[\kappa_{ns}(\phi_s - \overline{\phi}_s) + \kappa_s(\phi_n - \overline{\phi}_n)]\mathrm{d}s - \int_{\Gamma_u}(\phi_s\overline{\kappa}_{ns} + \phi_n\overline{\kappa}_s)\mathrm{d}s\right\} = 0 \tag{8.4.2}$$

where ϕ_x, ϕ_y, κ_y, κ_{xy} are independent variables for variation. The variation of Eq. (8.4.2) in the domain yields the Hamiltonian dual equations

$$\dot{\boldsymbol{v}} = \boldsymbol{H}\boldsymbol{v} \tag{8.4.3}$$

where the Hamiltonian operator matrix $\boldsymbol{H}$ is

$$\boldsymbol{H} = \begin{bmatrix} 0 & \nu\dfrac{\partial}{\partial y} & D(1-\nu^2) & 0 \\ -\dfrac{\partial}{\partial y} & 0 & 0 & 2D(1-\nu) \\ 0 & 0 & 0 & -\dfrac{\partial}{\partial y} \\ 0 & -\dfrac{1}{D}\dfrac{\partial^2}{\partial y^2} & \nu\dfrac{\partial}{\partial y} & 0 \end{bmatrix} \tag{8.4.4}$$

and $\boldsymbol{v} = \{\phi_x, \phi_y, \kappa_y, \kappa_{xy}\}^{\mathrm{T}}$ is the full state vector.

It should be noted that the full state vector does not include deflection w which can be obtained after solving $\boldsymbol{v}$. To be specific, deflection w can be obtained by first solving κ_x according to Eq. (8.4.1) and then directly integrating $\kappa_x, \kappa_y, \kappa_{xy}$.

For the purpose of discussing the property of operator matrix $\boldsymbol{H}$, we introduce the unit symplectic matrix

$$\boldsymbol{J} = \begin{bmatrix} \boldsymbol{0} & \boldsymbol{I}_2 \\ -\boldsymbol{I}_2 & \boldsymbol{0} \end{bmatrix} \tag{8.4.5}$$

and denote

$$\begin{aligned} \langle \boldsymbol{v}_1, \boldsymbol{v}_2 \rangle &\overset{\text{def}}{=} \int_{b_1}^{b_2} \boldsymbol{v}_1^{\mathrm{T}} \boldsymbol{J} \boldsymbol{v}_2 \mathrm{d}y \\ &= \int_{b_1}^{b_2} (\phi_{x1}\kappa_{y2} + \phi_{y1}\kappa_{xy2} - \kappa_{y1}\phi_{x2} - \kappa_{xy1}\phi_{y2})\,\mathrm{d}y \end{aligned} \tag{8.4.6}$$

Obviously Eq. (8.4.6) satisfies the four conditions of symplectic inner product (1.3.2). Hence the full state vectors $\boldsymbol{v}$ forms a symplectic geometric space in accordance with the definition of symplectic inner product (8.4.6).

Integration by parts yields

$$\begin{aligned} \langle \boldsymbol{v}_1, \boldsymbol{H}\boldsymbol{v}_2 \rangle = \langle \boldsymbol{v}_2, \boldsymbol{H}\boldsymbol{v}_1 \rangle &+ \left[\phi_{y2}\left(\frac{1}{D}\frac{\partial \phi_{y1}}{\partial y} - \nu\kappa_{y1} \right) + \phi_{x2}\kappa_{xy1} \right]_{b_1}^{b_2} \\ &- \left[\phi_{y1}\left(\frac{1}{D}\frac{\partial \phi_{y2}}{\partial y} - \nu\kappa_{y2} \right) + \phi_{x1}\kappa_{xy2} \right]_{b_1}^{b_2} \end{aligned} \tag{8.4.7}$$

For $\boldsymbol{v}_1, \boldsymbol{v}_2$ satisfying the corresponding homogeneous boundary conditions for free, simply supported or clamped boundary, there exists an identity

$$\langle \boldsymbol{v}_1, \boldsymbol{H}\boldsymbol{v}_2\rangle \equiv \langle \boldsymbol{v}_2, \boldsymbol{H}\boldsymbol{v}_1\rangle \tag{8.4.8}$$

Hence $\boldsymbol{H}$ is a Hamiltonian operator matrix.

Having transformed the equation into Eq. (8.4.3), the method of separation of variables can be applied directly. Assuming

$$\boldsymbol{v}(x,y) = \xi(x)\boldsymbol{\psi}(y) \tag{8.4.9}$$

and substituting the above expression into Eq. (8.4.3) yield

$$\xi(x) = \mathrm{e}^{\mu x} \tag{8.4.10}$$

and the eigenvalue equation

$$\boldsymbol{H}\boldsymbol{\psi}(y) = \mu\boldsymbol{\psi}(y) \tag{8.4.11}$$

where μ is the unknown eigenvalue and $\boldsymbol{\psi}(y)$ is the eigenvector satisfying the boundary conditions on both sides $y = b_1$ or b_2.

The Hamiltonian operator matrix $\boldsymbol{H}$ has its special characteristics which have been repeatedly presented in the previous chapters, i.e.:

(1) If μ is an eigenvalue of a Hamiltonian matrix, $-\mu$ is also an eigenvalue. The infinite eigenvalues can be divided into two sets

$$(\alpha)\ \mu_i,\quad \mathrm{Re}(\mu_i) < 0 \quad \text{or} \quad \mathrm{Re}(\mu_i) = 0 \wedge \mathrm{Im}(\mu_i) < 0 \quad (i = 1, 2, \ldots) \tag{8.4.12a}$$

$$(\beta)\ \mu_{-i} = -\mu_i \tag{8.4.12b}$$

The eigenvalues in the (α)-set are arranged in ascending order according to $|\mu_i|$.

(2) The eigenvectors of Hamiltonian operator matrix are mutually adjoint symplectic orthogonal. Let $\boldsymbol{\psi}_i$ and $\boldsymbol{\psi}_j$ be respectively the eigenvectors of the eigenvalues μ_i and μ_j, then for $\mu_i + \mu_j \neq 0$, they are symplectic orthogonal

$$\langle \boldsymbol{\psi}_i, \boldsymbol{\psi}_j\rangle = \int_{b_1}^{b_2} \boldsymbol{\psi}_i^{\mathrm{T}} \boldsymbol{J} \boldsymbol{\psi}_j \mathrm{d}x = 0 \tag{8.4.13}$$

The eigenvector which is symplectic adjoint with $\boldsymbol{\psi}_i$ is the eigenvector of eigenvalue $-\mu_i$ (or the Jordan form eigenvector).

Having derived the adjoint symplectic orthonormal relations, every full state vector $\boldsymbol{v}$ can be expanded by the eigen-solutions, i.e.

$$\boldsymbol{v} = \sum_{i=1}^{\infty} (a_i \boldsymbol{\psi}_i + b_i \boldsymbol{\psi}_{-i}) \tag{8.4.14}$$

where a_i and b_i are undetermined constants, and $\boldsymbol{\psi}_i, \boldsymbol{\psi}_{-i}$ are eigenvectors which fulfill the following normal adjoint symplectic orthonormal relations

$$\left.\begin{aligned} &\langle \boldsymbol{\psi}_i, \boldsymbol{\psi}_j \rangle = \langle \boldsymbol{\psi}_{-i}, \boldsymbol{\psi}_{-j} \rangle = 0 \\ &\langle \boldsymbol{\psi}_i, \boldsymbol{\psi}_{-j} \rangle = \delta_{ij} \end{aligned}\right\} \quad (i, j = 1, 2, \ldots) \tag{8.4.15}$$

The eigenvalues and eigenvectors for different boundary conditions are different. Similar to the plane elasticity problems, the eigen-solutions with nonzero eigenvalues can be obtained by first expanding the eigenvalue equation (8.4.11) as

$$\left.\begin{array}{lllll} 0 & +\nu \dfrac{\mathrm{d}\phi_y}{\mathrm{d}y} & +D(1-\nu^2)\kappa_y & +0 & = \mu\phi_x \\ -\dfrac{\mathrm{d}\phi_x}{\mathrm{d}y} & +0 & +0 & +2D(1-\nu)\kappa_{xy} & = \mu\phi_y \\ 0 & +0 & +0 & -\dfrac{\mathrm{d}\kappa_{xy}}{\mathrm{d}y} & = \mu\kappa_y \\ 0 & -\dfrac{1}{D}\dfrac{\mathrm{d}^2\phi_y}{\mathrm{d}y^2} & +\nu\dfrac{\mathrm{d}\kappa_y}{\mathrm{d}y} & +0 & = \mu\kappa_{xy} \end{array}\right\} \tag{8.4.16}$$

This is a set of simultaneous ordinary differential equations with respect to y. The eigenvalue λ for the y-direction is required first. The eigenvalue equation is

$$\det\begin{bmatrix} -\mu & \nu\lambda & D(1-\nu^2) & 0 \\ -\lambda & -\mu & 0 & 2D(1-\nu) \\ 0 & 0 & -\mu & -\lambda \\ 0 & -\lambda^2/D & \nu\lambda & -\mu \end{bmatrix} = 0 \tag{8.4.17}$$

Expanding the determinant yields the eigenvalue equation

$$(\lambda^2 + \mu^2)^2 = 0 \tag{8.4.18}$$

with repeated roots $\lambda = \pm\mu i$ as the eigenvalues. Hence, the general solutions of nonzero eigenvalues are

$$\left.\begin{aligned} \phi_x &= A_1\cos(\mu y) + B_1\sin(\mu y) + C_1 y\sin(\mu y) + D_1 y\cos(\mu y) \\ \phi_y &= A_2\sin(\mu y) + B_2\cos(\mu y) + C_2 y\cos(\mu y) + D_2 y\sin(\mu y) \\ \kappa_y &= A_3\cos(\mu y) + B_3\sin(\mu y) + C_3 y\sin(\mu y) + D_3 y\cos(\mu y) \\ \kappa_{xy} &= A_4\sin(\mu y) + B_4\cos(\mu y) + C_4 y\cos(\mu y) + D_4 y\sin(\mu y) \end{aligned}\right\} \tag{8.4.19}$$

The constants are not all independent. There are only four independent constants, for instance, A_2, B_2, C_2, D_2 are chosen as the independent constants. Substituting Eq. (8.4.19) into Eq. (8.4.16) yields the relations between these constants as

$$\left.\begin{aligned} A_1 &= -A_2 - \frac{3+\nu}{\mu(1-\nu)}C_2 \\ A_3 &= -\frac{\mu}{D(1-\nu)}A_2 - \frac{3-\nu}{D(1-\nu)^2}C_2 \\ A_4 &= \frac{\mu}{D(1-\nu)}A_2 + \frac{2}{D(1-\nu)^2}C_2 \end{aligned}\right\} \tag{8.4.20a}$$

$$\left.\begin{aligned} C_1 &= C_2 \\ C_3 &= \frac{\mu}{D(1-\nu)}C_2 \\ C_4 &= \frac{\mu}{D(1-\nu)}C_2 \end{aligned}\right\} \tag{8.4.20b}$$

and

$$\left.\begin{aligned} B_1 &= B_2 - \frac{3+\nu}{\mu(1-\nu)}D_2 \\ B_3 &= \frac{\mu}{D(1-\nu)}B_2 - \frac{3-\nu}{D(1-\nu)^2}D_2 \\ B_4 &= \frac{\mu}{D(1-\nu)}B_2 - \frac{2}{D(1-\nu)^2}D_2 \end{aligned}\right\} \tag{8.4.20c}$$

$$\left.\begin{aligned} D_1 &= -D_2 \\ D_3 &= -\frac{\mu}{D(1-\nu)}D_2 \\ D_4 &= \frac{\mu}{D(1-\nu)}D_2 \end{aligned}\right\} \tag{8.4.20d}$$

It should be noted that Eq. (8.4.20) is only the relation between the constants in the solution (8.4.19) of the basic eigenvectors with nonzero eigenvalues μ. If Jordan form eigen-solution exists, we should solve the following equation

$$\boldsymbol{H}\boldsymbol{\psi}^{(k)} = \mu\boldsymbol{\psi}^{(k)} + \boldsymbol{\psi}^{(k-1)} \quad (k = 1, 2, \ldots) \tag{8.4.21}$$

where superscript k denotes the kth order Jordan form eigen-solution. The Jordan form eigen-solution is related to its sub-order eigen-solutions. The general solution is formed by superposing a particular solution resulted from the inhomogeneous term $\boldsymbol{\psi}^{(k-1)}$ and the solution of (8.4.19).

Substituting the general solution (8.4.19) and Eq. (8.4.20) into the corresponding boundary conditions on both sides $y = b_1$ or b_2 yields the transcendental equation for nonzero eigenvalues and the relevant eigenvectors. Then solution can be obtained by the method of eigenvector expansion. Some typical boundary conditions are discussed in the following sections and other boundary conditions can be solved in a similar way.

8.5. Plates with Two Opposite Sides Simply Supported

Bending for plate simply supported on both opposite sides has been a well developed subject. This subject is chosen again here for solution because it is a classical case corresponding to the solution of Jordan form with nonzero eigenvalues. Besides, the methodology presented can be applied to plates with different boundary conditions for which the classical semi-inverse solution methodology fails.

For a plate with two opposite sides $y = 0$ and $y = b$ simply supported, the boundary conditions are

$$w = 0, \quad M_y = 0 \quad \text{at } y = 0 \text{ or } b \tag{8.5.1}$$

Expressed in terms of a full state vector they are

$$\phi_x = 0, \quad \frac{1}{D}\frac{\partial \phi_y}{\partial y} - \nu\kappa_y = 0 \quad \text{at } y = 0 \tag{8.5.2a}$$

$$\phi_x = a_1, \quad \frac{1}{D}\frac{\partial \phi_y}{\partial y} - \nu\kappa_y = 0 \quad \text{at } y = b \tag{8.5.2b}$$

It has been introduced in Sec. 8.2 that for boundaries with specified force conditions, the unknown constant a_1 can be determined according to Eq. (8.2.11) while the unknown constants on $y = 0$ have been eliminated through null moment functions.

Eigen-solutions are only applicable to homogeneous equations and homogeneous boundary conditions. The unknown constants in the boundary conditions should be solved first because they are inhomogeneous terms.

The a_1 term can be solved from the following equation

$$\boldsymbol{H}\boldsymbol{\psi}_0 = \boldsymbol{0} \tag{8.5.3}$$

with boundary conditions on two sides as

$$\phi_x = 0, \quad \frac{1}{D}\frac{\partial \phi_y}{\partial y} - \nu\kappa_y = 0 \quad \text{at } y = 0 \tag{8.5.4a}$$

$$\phi_x = 1, \quad \frac{1}{D}\frac{\partial \phi_y}{\partial y} - \nu\kappa_y = 0 \quad \text{at } y = b \tag{8.5.4b}$$

The solution is

$$\boldsymbol{\psi}_0 = \left\{\frac{y}{b},\ 0,\ 0,\ \frac{1}{2bD(1-\nu)}\right\}^{\mathrm{T}} \tag{8.5.5}$$

and the corresponding solution for Eq. (8.4.3) is

$$\boldsymbol{v}_0 = \boldsymbol{\psi}_0 \tag{8.5.6}$$

From Eq. (8.4.1) and the curvature-deflection relation (8.1.7), the deflection of plate after integration is

$$w = -\frac{xy}{2bD(1-\nu)} + c_1 x + c_2 y + c_3 \tag{8.5.7}$$

As it does not satisfy the boundary conditions $w = 0$ on both sides in Eq. (8.5.1), the solution of a_1 does not contain physical meaning and it should be abandoned. This solution is actually a spurious solution of the original problem due to the replacement of $w = 0$ by $\kappa_x = 0$ in the boundary conditions.

Therefore with respect to bending of plate simply supported on opposite sides we can only discuss the solution with homogeneous boundary conditions as

$$\phi_x = 0, \quad \frac{1}{D}\frac{\partial \phi_y}{\partial y} - \nu\kappa_y = 0 \quad \text{on } y = 0 \text{ or } b \tag{8.5.8}$$

It is easy to verify that the eigen-solutions of zero eigenvalue for the eigenvalue problem (8.4.11) with boundary conditions (8.5.8) does not have physical meaning. The general eigen-solutions of nonzero eigenvalues are Eqs. (8.4.19) and (8.4.20). Substituting the solutions into the homogeneous boundary conditions (8.5.8), and equating the determinant of coefficient matrix to zero yield the transcendental equation of nonzero eigenvalues for plates simply supported on opposite sides as

$$\sin^2(\mu b) = 0 \tag{8.5.9}$$

The solutions are real double roots

$$\mu_n = \frac{n\pi}{b} \quad (n = \pm 1, \pm 2, \ldots) \tag{8.5.10}$$

and the corresponding basic eigenvector is

$$\boldsymbol{\psi}_n^{(0)} = \begin{Bmatrix} \phi_x \\ \phi_y \\ \kappa_y \\ \kappa_{xy} \end{Bmatrix} = \begin{Bmatrix} \dfrac{D(1-\nu)}{\mu_n}\sin(\mu_n y) \\ \dfrac{D(1-\nu)}{\mu_n}\cos(\mu_n y) \\ \sin(\mu_n y) \\ \cos(\mu_n y) \end{Bmatrix} \tag{8.5.11}$$

Then the solution to Eq. (8.4.3) is

$$\boldsymbol{v}_n^{(0)} = \exp(\mu_n x)\boldsymbol{\psi}_n^{(0)} \tag{8.5.12}$$

From Eq. (8.4.1) and the curvature-deflection relation (8.1.7), the deflection of plate after integration is

$$w_n^{(0)} = -\frac{1}{\mu_n^2}\exp(\mu_n x)\sin(\mu_n y) + c_1 x + c_2 y + c_3 \tag{8.5.13}$$

where the integration constants can be determined by the boundary conditions $w = 0$ on both sides $y = 0$ or b. We obtain $c_1 = c_2 = c_3 = 0$, and then

$$w_n^{(0)} = -\frac{1}{\mu_n^2}\exp(\mu_n x)\sin(\mu_n y) \tag{8.5.14}$$

Because the eigenvalue μ_n is a double root, the first-order Jordan form eigen-solution exists. From Eq. (8.4.21), the first-order Jordan form eigen-solution can be obtained by solving

$$\boldsymbol{H}\boldsymbol{\psi}_n^{(1)} = \mu_n\boldsymbol{\psi}_n^{(1)} + \boldsymbol{\psi}_n^{(0)} \tag{8.5.15}$$

and imposing the boundary conditions (8.5.8). The solution is

$$\boldsymbol{\psi}_n^{(1)} = \begin{Bmatrix} -\dfrac{3+\nu}{2\mu_n^2} D\sin(\mu_n y) \\ \dfrac{3+\nu}{2\mu_n^2} D\cos(\mu_n y) \\ -\dfrac{1}{2\mu_n}\sin(\mu_n y) \\ \dfrac{1}{2\mu_n}\cos(\mu_n y) \end{Bmatrix} \tag{8.5.16}$$

and the solution to Eq. (8.4.3) is

$$\boldsymbol{v}_n^{(1)} = \exp(\mu_n x)(\boldsymbol{\psi}_n^{(1)} + x\boldsymbol{\psi}_n^{(0)}) \tag{8.5.17}$$

From Eq. (8.4.1), curvature-deflection relation (8.1.7) and boundary conditions $w = 0$ at both sides $y = 0$ or b, the deflection of plate after integration is

$$w_n^{(1)} = \frac{1 - 2\mu_n x}{2\mu_n^3}\exp(\mu_n x)\sin(\mu_n y) \tag{8.5.18}$$

These eigenvectors are adjoint symplectic orthogonal because $\boldsymbol{H}$ is a Hamiltonian operator matrix. Obviously the eigenvector symplectic adjoint with $\boldsymbol{\psi}_n^{(0)}$ must be $\boldsymbol{\psi}_{-n}^{(1)}$, i.e.

$$\langle \boldsymbol{\psi}_n^{(0)},\ \boldsymbol{\psi}_{-n}^{(1)}\rangle = -\frac{2Db}{\mu_n^2} \neq 0 \quad (n = \pm 1, \pm 2, \ldots) \tag{8.5.19}$$

while the other eigenvectors are symplectic orthogonal to each other.

From the eigenvalues and eigenvectors with adjoint symplectic orthogonality property, the general solution for plate bending simply supported on both opposite sides can be expressed as

$$\boldsymbol{v} = \sum_{n=1}^{\infty} [f_n^{(0)}\boldsymbol{v}_n^{(0)} + f_n^{(1)}\boldsymbol{v}_n^{(1)} + f_{-n}^{(0)}\boldsymbol{v}_{-n}^{(0)} + f_{-n}^{(1)}\boldsymbol{v}_{-n}^{(1)}] \tag{8.5.20}$$

according to the expansion theorem. The equation above strictly satisfies the homogeneous differential equation (8.4.3) in the domain and the homogeneous boundary conditions (8.5.8) while $f_n^{(k)}$ $(k = 0, 1; n = \pm 1, \pm 2, \ldots)$ are unknown constants which can be determined by the boundary conditions on both ends $x = x_0$ or x_f.

After determining the constants $f_n^{(k)}$, the solution of the original problem (8.1.29) is

$$w = \overline{w} + \sum_{n=1}^{\infty} \{f_n^{(0)} w_n^{(0)} + f_n^{(1)} w_n^{(1)} + f_{-n}^{(0)} w_{-n}^{(0)} + f_{-n}^{(1)} w_{-n}^{(1)}\} \tag{8.5.21}$$

where $\overline{w}$ is a particular solution with respect to the transverse load q.

It has been mentioned earlier that the classical semi-inverse Naiver's method and Levy's method are very effective for plates simply supported on both opposite sides. For instance, the solution in Fourier series by applying Levy's method is

$$\begin{aligned} w = \overline{w} + \sum_{n=1}^{\infty} \{ & [A_n \mathrm{ch}(\mu_n x) + B_n \mu_n x \mathrm{sh}(\mu_n x) \\ & + C_n \mathrm{sh}(\mu_n x) + D_n \mu_n x \mathrm{ch}(\mu_n x)] \sin(\mu_n x)\} \end{aligned} \tag{8.5.22}$$

Although the four basic functions in the outer parentheses in Eqs. (8.5.21) and (8.5.22) are different, they constitute the identical sub-space. The expanded solution of Eq. (8.5.21) is completely equivalent to the classical Levy solution. Hence the various classical analytical solutions for plate bending simply supported on both opposite sides can also be derived from Eq. (8.5.21) or Eq. (8.5.20).

For example, the particular solution for a fully simply supported plate $-a/2 \le x \le a/2, 0 \le y \le b$ with uniformly distributed load q is

$$\overline{w} = \frac{q}{24D}(y^4 - 2by^3 + b^3 y) \tag{8.5.23}$$

and the corresponding curvatures and bending moments are

$$\overline{\kappa}_y = \frac{q}{2D} y(y-b); \quad \overline{\kappa}_x = \overline{\kappa}_{xy} = 0 \tag{8.5.24}$$

$$\overline{M}_x = \frac{1}{2} q\nu y(y-b); \quad \overline{M}_y = \frac{1}{2} qy(y-b); \quad \overline{M}_{xy} = 0 \tag{8.5.25}$$

respectively. Through this particular solution, the problem can be transformed into a homogeneous equation (8.2.1), and in symplectic space the solution to Eq. (8.4.3) is required. Here the boundary conditions for simple supports at $x = \pm a/2$ should be rewritten as

$$M_x = -\overline{M}_x, \quad \kappa_y = -\overline{\kappa}_y \quad \text{at } x = \pm a/2 \tag{8.5.26}$$

after eliminating the effects resulted from the particular solution. Substituting Eq. (8.5.20) into the boundary conditions (8.5.26), the solution is

$$\left.\begin{aligned} &f_n^{(0)} = f_{-n}^{(0)} = f_n^{(1)} = f_{-n}^{(1)} = 0 && (n = 2, 4, 6, \ldots) \\ &f_n^{(0)} = -f_{-n}^{(0)} = -\frac{q[3 + 2\alpha_n \mathrm{th}(\alpha_n)]}{2Db\mu_n^3 \mathrm{ch}(\alpha_n)} && (n = 1, 3, 5, \ldots) \\ &f_n^{(1)} = f_{-n}^{(1)} = \frac{q}{Db\mu_n^2 \mathrm{ch}(\alpha_n)} && (n = 1, 3, 5, \ldots) \end{aligned}\right\} \tag{8.5.27}$$

where

$$\alpha_n = \frac{an\pi}{2b} \quad (n = 1, 3, 5, \ldots) \tag{8.5.28}$$

The result is absolutely identical to the classical Levy solution.

Although the solutions in dual system derived from expansion of eigenvectors are the same as the classical Levy solution, the respective theoretical foundations are essentially different in principle. The classical Levy solution is effective for plates simply supported on both opposite sides because the eigenvalues are all real and therefore the solution by expansion is very convenient. However, the semi-inverse solution procedure is difficult for other types of boundary conditions. As the solution methodology by expansion of eigenvectors presented here has been derived via a complete rational approach, it can be generalized accordingly to solve problems with other types of boundary conditions. The method will be shown in the next two sections.

8.6. Plates with Two Opposite Sides Free

For a plate with two opposite sides $y = \pm b$ free, the boundary conditions are

$$M_y = 0, \quad F_{Vy} = 0 \quad \text{at } y = \pm b \tag{8.6.1}$$

Expressed in terms of a full state vector they are

$$\phi_x = 0, \quad \phi_y = 0 \quad \text{at } y = -b \tag{8.6.2a}$$

$$\phi_x = \tilde{a} = a_1 - a_2 b, \quad \phi_y = a_0 + a_2 x \quad \text{at } y = b \tag{8.6.2b}$$

As these are force boundary conditions, from Eq. (8.2.11), there are unknown constants $a_0, \tilde{a}_1, a_2$ where $\tilde{a}_1$ has been divided into two parts according to the null moment functions (8.2.6). These constants are inhomogeneous terms which should be solved first[9] similar to the previous section.

First, the a_0 term should be solved from the following equation

$$\boldsymbol{H}\boldsymbol{\psi}_0^0 = \boldsymbol{0} \tag{8.6.3}$$

with boundary conditions

$$\phi_x = 0, \quad \phi_y = 0 \quad \text{at } y = -b \tag{8.6.4a}$$

$$\phi_x = 0, \quad \phi_y = 1 \quad \text{at } y = b \tag{8.6.4b}$$

The solution is

$$\boldsymbol{\psi}_0^0 = \left\{0, \quad \frac{y+b}{2b}, \quad \frac{-\nu}{2bD(1-\nu^2)}, \quad 0\right\}^{\mathrm{T}} \tag{8.6.5}$$

and the corresponding solution for Eq. (8.4.3) is

$$\boldsymbol{v}_0^0 = \boldsymbol{\psi}_0^0 \tag{8.6.6}$$

From Eq. (8.2.3), the corresponding bending moments are

$$M_{x0}^0 = \frac{1}{2b}, \quad M_{y0}^0 = 0, \quad M_{xy0}^0 = 0 \tag{8.6.7}$$

and from Eqs. (8.1.24) and (8.1.25), the shear forces are

$$F_{\mathrm{S}x0}^0 = 0, \quad F_{\mathrm{S}y0}^0 = 0 \tag{8.6.8}$$

Further from the curvature-deflection relation (8.1.7), the deflection of plate after integration is

$$w_0^0 = \frac{x^2 - \nu y^2}{4bD(1-\nu^2)} + \text{ rigid body displacement} \tag{8.6.9}$$

The physical interpretation of $\boldsymbol{v}_0^0$ is pure bending.

Next the a_1 term should be solved from the following equation

$$\boldsymbol{H}\boldsymbol{\psi}_0^1 = \boldsymbol{0} \tag{8.6.10}$$

with boundary conditions

$$\phi_x = 0, \quad \phi_y = 0 \quad \text{at } y = -b \tag{8.6.11a}$$

$$\phi_x = 1, \quad \phi_y = 0 \quad \text{at } y = b \tag{8.6.11b}$$

The solution is

$$\boldsymbol{\psi}_0^1 = \left\{ \frac{y+b}{2b}, \quad 0, \quad 0, \quad \frac{1}{4bD(1-\nu)} \right\}^{\mathrm{T}} \tag{8.6.12}$$

and the correspondingly solution for Eq. (8.4.3) is

$$\boldsymbol{v}_0^1 = \boldsymbol{\psi}_0^1 \tag{8.6.13}$$

From Eq. (8.2.3), the corresponding bending moments are

$$M_{x0}^1 = 0, \quad M_{y0}^1 = 0, \quad M_{xy0}^1 = \frac{1}{4b} \tag{8.6.14}$$

and from Eqs. (8.1.24) and (8.1.25), the shear forces are

$$F_{\mathrm{S}x0}^1 = 0, \quad F_{\mathrm{S}y0}^1 = 0 \tag{8.6.15}$$

Further from the curvature-deflection relation, the deflection of plate is

$$w_0^1 = -\frac{xy}{4bD(1-\nu)} + \text{ rigid body displacement} \tag{8.6.16}$$

The physical interpretation of $\boldsymbol{v}_0^1$ is pure torsion.

Finally, the a_2 term should be solved. As a_2 in the expression of ϕ_y in Eq. (8.6.2b) has a multiplier x, the solution corresponds to the next-order Jordan form of the inhomogeneous solution $\boldsymbol{\psi}_0^0$ with respect to $\phi_y = 1$. Hence the a_2 term should be solved from the following equation

$$\boldsymbol{H}\boldsymbol{\psi}_0^2 = \boldsymbol{\psi}_0^0 \tag{8.6.17}$$

The coefficient of a_2 in the expression of ϕ_x in Eq. (8.6.2b) is $-b$ and its influence only exists in the boundary conditions of $\boldsymbol{\psi}_0^2$. Hence boundary conditions on both opposite sides are

$$\phi_x = 0, \quad \phi_y = 0 \quad \text{at } y = -b \tag{8.6.18a}$$

$$\phi_x = -b, \quad \phi_y = 0 \quad \text{at } y = b \tag{8.6.18b}$$

The solution is

$$\boldsymbol{\psi}_0^2 = \left\{ \frac{(1-\nu)(b^2-y^2)}{4b(1+\nu)} - \frac{y+b}{2}, \ 0, \ 0, \ \frac{\nu y}{2bD(1-\nu^2)} \right\}^{\mathrm{T}} \tag{8.6.19}$$

and the correspondingly solution for Eq. (8.4.3) is

$$\boldsymbol{v}_0^2 = \boldsymbol{\psi}_0^2 + x\boldsymbol{\psi}_0^0 \tag{8.6.20}$$

From Eq. (8.2.3), the bending moments are

$$M_{x0}^2 = \frac{x}{2b}, \quad M_{y0}^2 = 0, \quad M_{xy0}^2 = \frac{\nu y}{2b(1+\nu)} \tag{8.6.21}$$

and from Eqs. (8.1.24) and (8.1.25), the shear forces are

$$F_{Sx0}^2 = \frac{1}{2b(1+\nu)}, \quad F_{Sy0}^2 = 0 \tag{8.6.22}$$

Further from the curvature-deflection relation, the deflection of plate is

$$w_0^2 = \frac{x^3 - 3\nu x y^2}{12bD(1-\nu^2)} + \text{rigid body displacement} \tag{8.6.23}$$

The physical interpretation of $\boldsymbol{v}_0^2$ is constant shear bending in the direction of x-axis.

Solving the inhomogeneous boundary terms a_0, a_1, a_2 yields three solutions with specific physical interpretation for thin plate bending. They are the pure bending solution, pure torsion solution and constant shear bending solution. It should be noted that from the fundamental equations of thin plate (8.1.29), it is also possible to derive into the Hamiltonian system and solve by the method of separable variables. Then there are six eigen-solutions of zero eigenvalue, i.e. three plate rigid body displacements $w = c_1 x + c_2 y + c_3$ and the solutions corresponding to $\boldsymbol{v}_0^0, \boldsymbol{v}_0^1, \boldsymbol{v}_0^2$. Due to application of analogy principle, the plate deflection has been replaced by curvature and direct appearance of three rigid body displacements of thin plate have also been avoided. Hence the dual solutions are represented by the particular solutions of inhomogeneous boundary terms.

After obtaining the three inhomogeneous particular solutions, we then discuss the solutions corresponding to the homogeneous boundary conditions.

The homogeneous boundary conditions for plate bending with both opposite sides free are

$$\phi_x = 0, \quad \phi_y = 0 \quad \text{at } y = \pm b \tag{8.6.24}$$

Obviously the eigen-solutions satisfying Eqs. (8.4.11) and (8.6.24) are only eigen-solutions of nonzero eigenvalue. They can be divided into two sets, i.e. the symmetric and the antisymmetric solutions with respect to x.

Substituting solutions with only A and C terms in Eq. (8.4.19) into the homogeneous boundary conditions (8.6.24) and equating the determinant of coefficient matrix to zero yield the transcendental equation of nonzero eigenvalues for symmetric plate deformation with both opposite sides free as

$$2\mu b(1-\nu) = (3+\nu)\sin(2\mu b) \tag{8.6.25}$$

Let eigenvalues μ_n be a solution to Eq. (8.6.25), the corresponding eigen-solution for symmetric deformation is

$$\boldsymbol{\psi}_n = \left\{ \begin{array}{l} -\dfrac{3+\nu}{1-\nu}\sin^2(\mu_n b)\cos(\mu_n y) + \mu_n y\sin(\mu_n y) \\ -\dfrac{3+\nu}{1-\nu}\cos^2(\mu_n b)\sin(\mu_n y) + \mu_n y\cos(\mu_n y) \\ \dfrac{\mu_n}{D(1-\nu)^2}\{[(3+\nu)\cos^2(\mu_n b) - 3 + \nu]\cos(\mu_n y) \\ \qquad +(1-\nu)\mu_n y\sin(\mu_n y)\} \\ \dfrac{\mu_n}{D(1-\nu)^2}\{[2-(3+\nu)\cos^2(\mu_n b)]\sin(\mu_n y) \\ \qquad +(1-\nu)\mu_n y\cos(\mu_n y)\} \end{array} \right\} \tag{8.6.26}$$

and the solution for the corresponding problem (8.4.3) is

$$\boldsymbol{v}_n = \exp(\mu_n x)\boldsymbol{\psi}_n \tag{8.6.27}$$

Further from Eq. (8.4.1) and the curvature-deflection relation (8.1.7), the deflection of plate after integration is

$$w_n = \exp(\mu_n x)\left\{\frac{1+\nu-(3+\nu)\cos^2(\mu_n b)}{D(1-\nu)^2\mu_n}\cos(\mu_n y) - \frac{y\sin(\mu_n y)}{D(1-\nu)}\right\} \tag{8.6.28}$$

It is also possible to apply the Newton method commonly used in plane elasticity to solve the eigenvalues. For instance, the first several eigenvalues for a plate with Poisson's ratio $\nu = 0.3$ are listed in Table 8.2.

Apparently, for each n $(n > 1)$, there are two symplectic adjoint eigenvalues μ_n and $-\mu_n$ and their respective complex conjugate eigenvalues.

Table 8.2. Nonzero eigenvalues for symmetric deformation of thin plate with both opposite sides free ($\nu = 0.3$).

$n =$	1	2	3	4	5
$\mathrm{Re}(\mu_n b) =$	1.2830	$\pi + 0.6973$	$2\pi + 0.7191$	$3\pi + 0.7313$	$4\pi + 0.7393$
$\mathrm{Im}(\mu_n b) =$	0	0.5446	0.8808	1.0730	1.2101

There are four eigenvalues in total. As the eigenvalue μ_1 corresponding to $n = 1$ is real, it has only a symplectic adjoint eigenvalue $-\mu_1$. In this case, there are only two eigenvalues. These eigenvalues are all single roots.

Substituting solutions with only B and D terms in Eq. (8.4.19) into the homogeneous boundary conditions (8.6.24) and equating the determinant of coefficient matrix to zero yield the transcendental equation of nonzero eigenvalues for antisymmetric plate deformation with both opposite sides free as

$$2\mu b(1-\nu) + (3+\nu)\sin(2\mu b) = 0 \tag{8.6.29}$$

The corresponding eigen-solution for antisymmetric deformation is

$$\overline{\boldsymbol{\psi}}_n = \begin{Bmatrix} -\dfrac{3+\nu}{1-\nu}\cos^2(\mu_n b)\sin(\mu_n y) - \mu_n y\cos(\mu_n y) \\ \dfrac{3+\nu}{1-\nu}\sin^2(\mu_n b)\cos(\mu_n y) + \mu_n y\sin(\mu_n y) \\ \dfrac{\mu_n}{D(1-\nu)^2}\{[(3+\nu)\sin^2(\mu_n b) - 3 + \nu]\sin(\mu_n y) \\ \qquad - (1-\nu)\mu_n y\cos(\mu_n y)\} \\ \dfrac{\mu_n}{D(1-\nu)^2}\{[(3+\nu)\sin^2(\mu_n b) - 2]\cos(\mu_n y) \\ \qquad + (1-\nu)\mu_n y\sin(\mu_n y)\} \end{Bmatrix} \tag{8.6.30}$$

and the solution for the corresponding problem (8.4.3) is

$$\overline{\boldsymbol{v}}_n = \exp(\mu_n x)\overline{\boldsymbol{\psi}}_n \tag{8.6.31}$$

From Eq. (8.4.1) and the curvature-deflection relation (8.1.7), the deflection of plate after integration is

$$\overline{w}_n = \exp(\mu_n x)\left\{\frac{1+\nu-(3+\nu)\sin^2(\mu_n b)}{D(1-\nu)^2\mu_n}\sin(\mu_n y) + \frac{y\cos(\mu_n y)}{D(1-\nu)}\right\} \tag{8.6.32}$$

Similarly, the first several eigenvalues for a plate with Poisson's ratio $\nu = 0.3$, for instance, can be obtained by applying the Newton method and they are listed in Table 8.3.

Again and apparently, for each $n(> 2)$, there are two symplectic adjoint eigenvalues μ_n and $-\mu_n$ and their respective complex conjugate eigenvalues. There are four eigenvalues in total. As the eigenvalues corresponding to $n = 1, 2$ are real, they have only symplectic adjoint eigenvalues. In these cases, there are only two eigenvalues in each case. These eigenvalues are all single roots.

Table 8.3. The nonzero eigenvalues for antisymmetrical deformation of thin plate with both sides free ($\nu = 0.3$).

$n =$	1	2	3	4
$\mathrm{Re}(\mu_n b) =$	$0.5\pi + 0.5690$	$0.5\pi + 0.7863$	$1.5\pi + 0.7100$	$2.5\pi + 0.7259$
$\mathrm{Im}(\mu_n b) =$	0	0	0.7439	0.9865

From the eigenvalues and eigenvectors obtained and based on the adjoint symplectic orthogonality property, the solution can be established by using the expansion theorem.

A simple example for pure bending of a semi-infinite rectangular thin plate is presented here. We take $b = 1$ and $x = 0$ clamped, while $x \to \infty$ is a free end with unit bending moment. The bending moment distribution at the clamped end is solved.

According to the definition of problem, there is only bending moment at $x \to \infty$ and the deformation is symmetric with respect to the x-axis. Hence the expanded equation is constructed from Eq. (8.6.6) and the symmetric eigen-solutions (8.6.27) of nonzero eigenvalue with $\mathrm{Re}(\mu_n) < 0$ in Eq. (8.6.27) as

$$\boldsymbol{v} = 2\boldsymbol{v}_0^0 + \sum_{n=1} f_n \exp(\mu_n x)\boldsymbol{\psi}_n \tag{8.6.33}$$

The expanded equation (8.6.33) satisfies the differential equation in the domain and the boundary conditions on sides $y = \pm b$ and at the infinite end $x \to \infty$. The clamped boundary conditions at $x = 0$ is used to determine the constants $f_n, (n = 1, 2, \ldots)$. In practical applications, it is only necessary to solve the first k terms in Eq. (8.6.33). Then the variational formula for the boundary conditions at $x = 0$ is

$$\int_{-b}^{b} [\kappa_y \delta\phi_x + \kappa_{xy}\delta\phi_y]_{x=0} \mathrm{d}y = 0 \tag{8.6.34}$$

Since there are complex eigenvalues and eigen-solutions, in practice, Eqs. (8.6.33) and (8.6.34) should be transformed into a real canonical equation before solving by expansion of eigenvectors. The relevant details can be referred to Chapter 3.

For a thin plate with Poisson's ration $\nu = 0.3$, the bending moment distribution at the clamped end by using $k = 11$ and 21 in the computation is presented in Fig. 8.4.

The figure shows that there is a stress singularity at the corner and the bending moment $M_x \to -\infty$. The fluctuation of bending moment

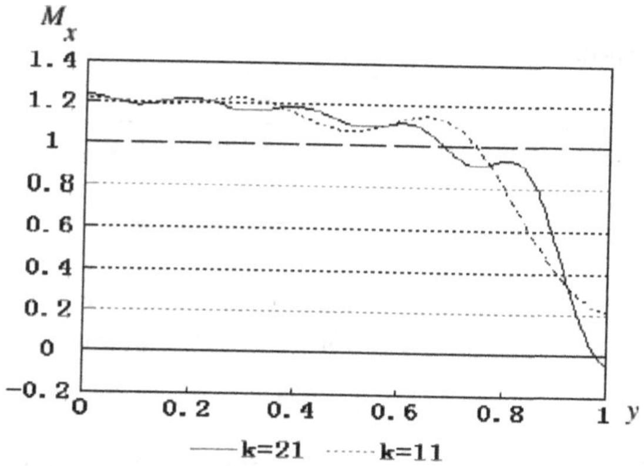

Fig. 8.4. Bending moment distribution at the clamped end for pure bending of a semi-infinite thin plate.

distribution appears and it is due to truncation of series expansion in computation.

8.7. Plate with Two Opposite Sides Clamped

For a plate with two opposite sides $y = \pm b$ clamped, the boundary conditions are

$$w = 0, \quad \theta = \frac{\partial w}{\partial y} = 0 \quad \text{at } y = \pm b \tag{8.7.1}$$

Expressed in terms of a full state vector they are

$$\frac{1}{D}\frac{\partial \phi_y}{\partial y} - \nu\kappa_y = 0, \quad \kappa_{xy} = 0 \quad \text{at } y = \pm b \tag{8.7.2}$$

There are six eigen-solutions with zero eigenvalue (including Jordan form eigen-solution) but they are all solutions without real physical interpretation[14]. There are simply the spurious solutions of the original problem due to the replacement of $w = \partial w/\partial y = 0$ by $\kappa_x = \kappa_{xy} = 0$. Hence only the eigen-solutions of nonzero eigenvalues for plate with opposite sides clamped are discussed here. Likewise the solutions can be divided into two sets: (1) solutions for symmetric deformation with respect to x-axis; and (2) solutions for antisymmetric deformation with respect to x-axis.

Substituting solutions with only A and C terms in Eq. (8.4.19) into the homogeneous boundary conditions (8.7.2) and equating the determinant of

Table 8.4. Nonzero eigenvalues of symmetric deformation for plate with both opposite sides clamped.

$n =$	1	2	3	4	5
$\mathrm{Re}(\mu_n b)$	$\frac{\pi}{2}+0.5354$	$\frac{3\pi}{2}+0.6439$	$\frac{5\pi}{2}+0.6827$	$\frac{7\pi}{2}+0.7036$	$\frac{9\pi}{2}+0.7169$
$\mathrm{Im}(\mu_n b)$	1.1254	1.5516	1.7755	1.9294	2.0469

coefficient matrix to zero yield the transcendental equation of nonzero eigenvalues for symmetric plate deformation with both opposite sides clamped as

$$2\mu b + \sin(2\mu b) = 0 \tag{8.7.3}$$

It is identical to Eq. (4.6.10). The first several eigenvalues are listed in Table 8.4.

Only roots in the first quadrant are listed in Table 8.4. For each μ_n there are a corresponding symplectic adjoint eigenvalue $-\mu_n$ and their complex conjugate eigenvalues. There is a total of four eigenvalues. It is obvious from Eq. (8.7.3) that these nonzero eigenvalues are all single roots.

Having solved eigenvalues μ_n, the corresponding eigenvectors can be expressed as

$$\boldsymbol{\psi}_n = \begin{Bmatrix} -\left[\cos^2(\mu_n b) + \dfrac{1+\nu}{1-\nu}\right]\cos(\mu_n y) + \mu_n y \sin(\mu_n y) \\ \left[\cos^2(\mu_n b) - \dfrac{2}{1-\nu}\right]\sin(\mu_n y) + \mu_n y \cos(\mu_n y) \\ \dfrac{\mu_n}{D(1-\nu)}\{-[1+\cos^2(\mu_n b)]\cos(\mu_n y) + \mu_n y \sin(\mu_n y)\} \\ \dfrac{\mu_n}{D(1-\nu)}\{\cos^2(\mu_n b)\sin(\mu_n y) + \mu_n y \cos(\mu_n y)\} \end{Bmatrix} \tag{8.7.4}$$

and the solution for the corresponding problem (8.4.3) is

$$\boldsymbol{v}_n = \exp(\mu_n x)\boldsymbol{\psi}_n \tag{8.7.5}$$

Further from the curvature-deflection relation and boundary conditions (8.7.1), the deflection of plate is

$$w_n = -\frac{\exp(\mu_n x)}{D\mu_n(1-\nu)}\{\sin^2(\mu_n b)\cos(\mu_n y) + \mu_n y \sin(\mu_n y)\} \tag{8.7.6}$$

Substituting solutions with only B and D terms in Eq. (8.4.19) into the homogeneous boundary conditions (8.7.2) yield the transcendental equation of nonzero eigenvalues for antisymmetric plate deformation with both

Table 8.5. Nonzero eigenvalues of antisymmetric deformation for plate with both opposite sides clamped.

$n =$	1	2	3	4	5
$\mathrm{Re}(\overline{\mu}_n b)$	$\pi + 0.6072$	$2\pi + 0.6668$	$3\pi + 0.6954$	$4\pi + 0.7109$	$5\pi + 0.7219$
$\mathrm{Im}(\overline{\mu}_n b)$	1.3843	1.6761	1.8584	1.9916	2.0966

opposite sides clamped as

$$2\mu b = \sin(2\mu b) \tag{8.7.7}$$

It is identical to Eq. (4.6.26). The first several eigenvalues is listed in Table 8.5.

Apparently, for each $\overline{\mu}_n$ there are a corresponding symplectic adjoint eigenvalue $-\overline{\mu}_n$ and their complex conjugate eigenvalues. There is a total of four eigenvalues. Obviously the nonzero eigenvalues of antisymmetric deformation are also single roots.

Having solved eigenvalues $\overline{\mu}_n$, the corresponding eigenvectors can be expressed as

$$\overline{\psi}_n = \begin{Bmatrix} -\left[\sin^2(\overline{\mu}_n b) + \dfrac{1+\nu}{1-\nu}\right]\sin(\overline{\mu}_n y) - \overline{\mu}_n y\cos(\overline{\mu}_n y) \\ \left[\dfrac{2}{1-\nu} - \sin^2(\overline{\mu}_n b)\right]\cos(\overline{\mu}_n y) + \overline{\mu}_n y\sin(\overline{\mu}_n y) \\ -\dfrac{\overline{\mu}_n}{D(1-\nu)}\{[1+\sin^2(\overline{\mu}_n b)]\sin(\overline{\mu}_n y) + \overline{\mu}_n y\cos(\overline{\mu}_n y)\} \\ \dfrac{\overline{\mu}_n}{D(1-\nu)}\{-\sin^2(\overline{\mu}_n b)\cos(\overline{\mu}_n y) + \overline{\mu}_n y\sin(\overline{\mu}_n y)\} \end{Bmatrix} \tag{8.7.8}$$

and the solution for the corresponding problem (8.4.3) is

$$\bar{v}_n = \exp(\mu_n x)\overline{\psi}_n \tag{8.7.9}$$

Further from the curvature-deflection relation and boundary conditions (8.7.1), the deflection of plate is

$$\overline{w}_n = \frac{\exp(\overline{\mu}_n x)}{D\overline{\mu}_n(1-\nu)}\{-\cos^2(\overline{\mu}_n b)\sin(\overline{\mu}_n y) + \overline{\mu}_n y\cos(\overline{\mu}_n y)\} \tag{8.7.10}$$

Based on the eigenvalues and eigenvectors, the general solution for plate with both opposite sides clamped can be solved from the expansion theorem. Thus analytical solution of the original problem can be obtained.

A specific example for bending of a fully clamped rectangular plate under uniformly distributed load q is presented. As the problem is symmetric with respect to x-axis, the expanded expression can only be constructed from the symmetric eigen-solutions of nonzero eigenvalues (8.7.3) to (8.7.6) as

$$\boldsymbol{v} = \sum_{n=1}^{\infty} (f_n \boldsymbol{v}_n + \tilde{f}_n \tilde{\boldsymbol{v}}_n + f_{-n} \boldsymbol{v}_{-n} + \tilde{f}_{-n} \tilde{\boldsymbol{v}}_{-n}) \tag{8.7.11}$$

and the deflection of plate is

$$w = w^* + \sum_{n=1}^{\infty} (f_n w_n + \tilde{f}_n \tilde{w}_n + f_{-n} w_{-n} + \tilde{f}_{-n} \tilde{w}_{-n}) \tag{8.7.12}$$

where $\boldsymbol{v}_n, \boldsymbol{v}_{-n}$ are eigenvectors corresponding to the eigenvalue μ_n listed in Table 8.4 and their respective symplectic adjoint eigenvalue $-\mu_n$, while $\tilde{\boldsymbol{v}}_n, \tilde{\boldsymbol{v}}_{-n}$ are eigenvectors corresponding to the complex conjugate eigenvalues $\pm\mu_n$. A particular solution caused by distributed load q in the domain is

$$w^* = \frac{q}{24D}(y^2 - b^2)^2 \tag{8.7.13}$$

The corresponding curvatures are

$$\kappa_x^* = \kappa_{xy}^* = 0, \quad \kappa_y^* = \frac{q}{6D}(3y^2 - b^2) \tag{8.7.14}$$

Equation (8.7.11) satisfies the homogeneous differential equation in the domain and the homogeneous boundary conditions on two sides (8.7.2). $f_i, \tilde{f}_i$ $(i = \pm 1, \pm 2, \ldots)$ are the unknown constants which can be determined by the boundary conditions $\kappa_y = \kappa_{xy} = 0$ at two ends $x = \pm a$. In practical applications, it is only necessary to solve the first k terms in Eq. (8.7.11). Then the variational formula for the boundary conditions at two ends $x = \pm a$ is

$$\int_{-b}^{b} [(\kappa_y - \kappa_y^*)\delta\phi_x + (\kappa_{xy} - \kappa_{xy}^*)\delta\phi_y]_{x=-a}^{x=a} \mathrm{d}y = 0 \tag{8.7.15}$$

Since there are complex eigenvalues and eigen-solutions, in practice Eq. (8.7.11) should be transformed into a real canonical equation before solving.

For a thin plate with Poisson's ratio $\nu = 0.3$, the result by using $k = 4$ is listed in Table. 8.6. The values in parentheses in the table are the solutions obtained by expansion of different series with many more terms[1]. The solutions obtained are in excellent agreement with those in Ref. 1. However, the solutions of the present method converge more quickly.

Table 8.6. Analytical solution of a fully clamped plate under uniform load.

$\frac{a}{b}$	$\frac{Dw(0,0)}{16qb^4}$	$\frac{M_x(a,0)}{4qb^2}$	$\frac{M_y(0,b)}{4qb^2}$	$\frac{M_x(0,0)}{4qb^2}$	$\frac{M_y(0,0)}{4qb^2}$
1.0	0.00127(0.00126)	0.0514(0.0513)	0.0513(0.0513)	−0.0229(−0.0231)	−0.0229(−0.0231)
1.1	0.00151(0.00150)	0.0539(0.0538)	0.0581(0.0581)	−0.0231(−0.0231)	−0.0267(−0.0264)
1.2	0.00173(0.00172)	0.0554(0.0554)	0.0639(0.0639)	−0.0228(−0.0228)	−0.0300(−0.0299)
1.3	0.00191(0.00191)	0.0563(0.0563)	0.0687(0.0687)	−0.0222(−0.0222)	−0.0327(−0.0327)
1.4	0.00207(0.00207)	0.0568(0.0568)	0.0726(0.0726)	−0.0213(−0.0212)	−0.0350(−0.0349)
1.5	0.00220(0.00220)	0.0570(0.0570)	0.0757(0.0757)	−0.0203(−0.0203)	−0.0368(−0.0368)
1.6	0.00230(0.00230)	0.0571(0.0571)	0.0780(0.0780)	−0.0193(−0.0193)	−0.0382(−0.0381)
1.7	0.00238(0.00238)	0.0571(0.0571)	0.0798(0.0799)	−0.0183(−0.0182)	−0.0393(−0.0392)
1.8	0.00245(0.00245)	0.0571(0.0571)	0.0812(0.0812)	−0.0174(−0.0174)	−0.0401(−0.0401)
1.9	0.00250(0.00249)	0.0570(0.0571)	0.0822(0.0822)	−0.0165(−0.0165)	−0.0407(−0.0407)
2.0	0.00253(0.00254)	0.0570(0.0571)	0.0829(0.0829)	−0.0158(−0.0158)	−0.0412(−0.0412)

8.8. Bending of Sectorial Plates

The bending of rectangular plates in symplectic system has been discussed in the previous several sections. The symplectic system is certainly also applicable to the bending of sectorial plates[15]. In this section the bending of sectorial plates in polar coordinates as shown in Fig. 8.5 is discussed.

The basic equations for plate bending in a polar coordinate system include:

(1) The curvature-deflection relations

$$\left.\begin{aligned} \kappa_\varphi &= \frac{1}{\rho}\frac{\partial w}{\partial \rho} + \frac{1}{\rho^2}\frac{\partial^2 w}{\partial \varphi^2} \\ \kappa_\rho &= \frac{\partial^2 w}{\partial \rho^2} \\ \kappa_{\rho\varphi} &= -\frac{\partial}{\partial \rho}\left(\frac{1}{\rho}\frac{\partial w}{\partial \varphi}\right) \end{aligned}\right\} \tag{8.8.1}$$

(2) The relations between bending moment and curvature

$$\left\{\begin{matrix} M_\varphi \\ M_\rho \\ 2M_{\rho\varphi} \end{matrix}\right\} = D \begin{bmatrix} 1 & \nu & 0 \\ \nu & 1 & 0 \\ 0 & 0 & 2(1-\nu) \end{bmatrix} \left\{\begin{matrix} \kappa_\varphi \\ \kappa_\rho \\ \kappa_{\rho\varphi} \end{matrix}\right\} \tag{8.8.2}$$

where D is flexural rigidity of plate. The strain energy density of plate is

$$v_\varepsilon(\kappa_\varphi, \kappa_\rho, \kappa_{\rho\varphi}) = \frac{1}{2}D[\kappa_\varphi^2 + \kappa_\rho^2 + 2\nu\kappa_\varphi\kappa_\rho + 2(1-\nu)\kappa_{\rho\varphi}^2] \tag{8.8.3}$$

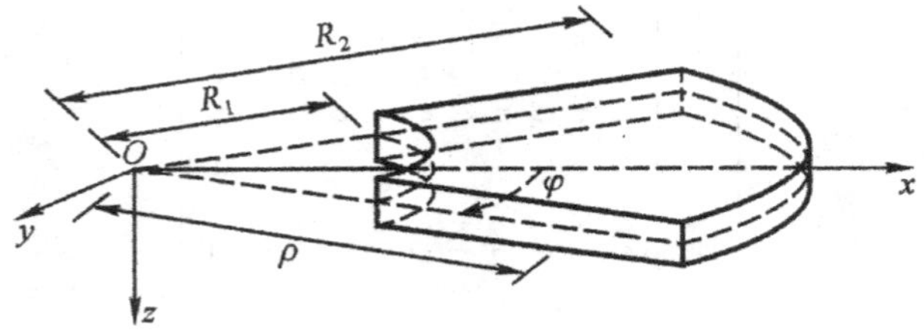

Fig. 8.5. Bending of a sectorial plate.

(3) The equations of equilibrium

$$\left.\begin{aligned} \frac{\partial M_\rho}{\partial \rho} - \frac{1}{\rho}\frac{\partial M_{\rho\varphi}}{\partial \varphi} + \frac{M_\rho - M_\varphi}{\rho} = F_{S\rho} \\ \frac{1}{\rho}\frac{\partial M_\varphi}{\partial \varphi} - \frac{2}{\rho}M_{\rho\varphi} - \frac{\partial M_{\rho\varphi}}{\partial \rho} = F_{S\varphi} \end{aligned}\right\} \tag{8.8.4a}$$

and

$$\frac{\partial F_{S\rho}}{\partial \rho} + \frac{F_{S\rho}}{\rho} + \frac{1}{\rho}\frac{\partial F_{S\varphi}}{\partial \varphi} = q \tag{8.8.4b}$$

The total distributed shear forces as a result of continuous distributed torsional moment and transverse shear forces along the boundary are

$$F_{V\rho} = F_{S\rho} - \frac{1}{\rho}\frac{\partial M_{\rho\varphi}}{\partial \varphi}, \quad F_{V\varphi} = F_{S\varphi} - \frac{\partial M_{\rho\varphi}}{\partial \rho} \tag{8.8.5}$$

The positive directions of internal forces are shown in Fig. 8.1. As load q can be treated by a particular solution according to the superposition principle, it is possible to consider the homogeneous equation $q = 0$ first.

According to the analogy between plate bending and plane elasticity, deflection of plate bending w corresponds to the Airy stress function of plane elasticity. Hence the displacements u_ρ, u_θ of plane elasticity in polar coordinates correspond to bending moment functions ϕ_ρ, ϕ_φ of plate bending in polar coordinates. The relations between bending moment and bending moment function are

$$M_\varphi = \frac{\partial \phi_\rho}{\partial \rho}, \quad M_\rho = \frac{1}{\rho}\left(\phi_\rho + \frac{\partial \phi_\varphi}{\partial \varphi}\right), \quad M_{\rho\varphi} = \frac{1}{2\rho}\left(\frac{\partial \phi_\rho}{\partial \varphi} + \rho\frac{\partial \phi_\varphi}{\partial \rho} - \phi_\varphi\right) \tag{8.8.6}$$

It can be verified through substitution that the homogeneous equations (8.8.4) is satisfied.

Similar to rigid body displacements in plane elasticity, the functions

$$\phi_\rho = a_0 \sin\varphi + a_1 \cos\varphi, \quad \phi_\varphi = a_0 \cos\varphi - a_1 \sin\varphi + a_2\rho \tag{8.8.7}$$

do not result in any bending moment. Here a_0, a_1, a_2 are arbitrary constants. And equivalently Eq. (8.8.7) are the **null moment functions** in polar coordinates.

Similar to the Hellinger–Reissner variational principle for plane elasticity in polar coordinates, the plate bending Pro-H-R variational principle in polar coordinates is

$$\delta \int_{-\alpha}^{\alpha} \int_{R_1}^{R_2} \left\{ \kappa_\varphi \frac{\partial \phi_\rho}{\partial \rho} + \frac{\kappa_\rho}{\rho} \left(\phi_\rho + \frac{\partial \phi_\varphi}{\partial \varphi} \right) + \kappa_{\rho\varphi} \left(\frac{\partial \phi_\varphi}{\partial \rho} - \frac{\phi_\varphi}{\rho} + \frac{1}{\rho} \frac{\partial \phi_\rho}{\partial \varphi} \right) - v_\varepsilon \right\} \rho \mathrm{d}\rho \mathrm{d}\varphi = 0 \tag{8.8.8}$$

Deflection w is not present in this variational principle. It can be obtained by integration after solving the curvature. Considering $\kappa_\varphi, \kappa_\rho, \kappa_{\rho\varphi}, \phi_\rho, \phi_\varphi$ as independent variables, variation of Eq. (8.8.8) yields the relations between bending moment and curvature

$$\frac{\partial \phi_\rho}{\partial \rho} = \frac{\partial v_\varepsilon}{\partial \kappa_\varphi}, \quad \frac{1}{\rho} \left(\phi_\rho + \frac{\partial \phi_\varphi}{\partial \varphi} \right) = \frac{\partial v_\varepsilon}{\partial \kappa_\rho}, \quad \frac{1}{\rho} \left(\frac{\partial \phi_\rho}{\partial \varphi} + \rho \frac{\partial \phi_\varphi}{\partial \rho} - \phi_\varphi \right) = \frac{\partial v_\varepsilon}{\partial \kappa_{\rho\varphi}} \tag{8.8.9}$$

and the curvature compatibility equation

$$\left. \begin{aligned} \frac{\partial \kappa_\varphi}{\partial \rho} + \frac{\kappa_\varphi - \kappa_\rho}{\rho} + \frac{1}{\rho} \frac{\partial \kappa_{\rho\varphi}}{\partial \varphi} &= 0 \\ \frac{1}{\rho} \frac{\partial \kappa_\rho}{\partial \varphi} + \frac{\partial \kappa_{\rho\varphi}}{\partial \rho} + \frac{2\kappa_{\rho\varphi}}{\rho} &= 0 \end{aligned} \right\} \tag{8.8.10}$$

The boundary conditions for sides $\varphi = \pm\alpha$ are the same as those of rectangular plates because the sides are straight edges. For example, the deflection $\overline{w}$ and normal rotation $\overline{\theta}_n$ are specified along the displacement boundary Γ_u. The boundary conditions expressed in terms of tangential curvature κ_s and normal curvature κ_{ns} are

$$\kappa_s = \overline{\kappa}_s = \frac{\partial^2 \overline{w}}{\partial s^2}, \quad \kappa_{ns} = \overline{\kappa}_{ns} = -\frac{\partial \overline{\theta}_n}{\partial s} \tag{8.8.11}$$

Similarly, the normal bending moment $\overline{M}_n$ and equivalent shear force $\overline{F}_{\mathrm{V}n}$ are specified along the force boundary Γ_σ. The boundary conditions expressed in terms of bending moment functions are

$$\left. \begin{aligned} \phi_s = \overline{\phi}_s &= \int_{s_0}^{s} \overline{M}_n \mathrm{d}s' + a_1 \\ \phi_n = \overline{\phi}_n &= \int_{s_0}^{s} (s' - s) \overline{F}_{\mathrm{V}n} \mathrm{d}s' + a_0 + a_2 s \end{aligned} \right\} \tag{8.8.12}$$

Equations (8.8.1) to (8.8.10) constitute another set of basic equations for sectorial plates. As these basic equations are in analogy to those for plane elasticity in polar coordinates, the various plane elasticity solution methodologies for a sectorial domain are completely applicable.

8.8.1. *Derivation of Hamiltonian System*

Similar to plane elasticity in a sectorial domain, we introduce the transformation

$$\xi = \ln \rho, \qquad \text{i.e. } \rho = \mathrm{e}^{\xi} \tag{8.8.13}$$

and denote

$$\xi_1 = \ln R_1, \quad \xi_2 = \ln R_2 \tag{8.8.14}$$

$$s_\rho = \rho \kappa_\rho, \quad s_\varphi = \rho \kappa_\varphi, \quad s_{\rho\varphi} = \rho \kappa_{\rho\varphi} \tag{8.8.15}$$

Hence the Pro-H-R variational principle (8.8.8) can be expressed as

$$\begin{aligned}\delta \int_{-\alpha}^{\alpha} \int_{\xi_1}^{\xi_2} \Big\{ & s_\varphi \frac{\partial \phi_\rho}{\partial \xi} + s_\rho \left(\phi_\rho + \frac{\partial \phi_\varphi}{\partial \varphi} \right) + s_{\rho\varphi} \left(\frac{\partial \phi_\varphi}{\partial \xi} - \phi_\varphi + \frac{\partial \phi_\rho}{\partial \varphi} \right) \\ & - \frac{1}{2} D [s_\varphi^2 + s_\rho^2 + 2\nu s_\varphi s_\rho + 2(1-\nu) s_{\rho\varphi}^2] \Big\} \mathrm{d}\xi \mathrm{d}\varphi = 0\end{aligned} \tag{8.8.16}$$

We further treat ξ as the time coordinate in Hamiltonian system, and denote $\dot{(\,)} = \partial/\partial\xi$. The variation of Eq. (8.8.16) with respect to s_ρ yields

$$s_\rho = \frac{1}{D} \left(\phi_\rho + \frac{\partial \phi_\varphi}{\partial \varphi} \right) - \nu s_\varphi \tag{8.8.17}$$

Substituting into Eq. (8.8.16) and eliminating s_ρ yield

$$\delta \int_{-\alpha}^{\alpha} \int_{\xi_1}^{\xi_2} \{ \boldsymbol{p}^{\mathrm{T}} \dot{\boldsymbol{q}} - \mathscr{H}(\boldsymbol{q}, \boldsymbol{p}) \} \mathrm{d}\xi \mathrm{d}\varphi = 0 \tag{8.8.18}$$

where

$$\boldsymbol{q} = \{\phi_\rho, \phi_\varphi\}^{\mathrm{T}}, \quad \boldsymbol{p} = \{s_\varphi, s_{\rho\varphi}\}^{\mathrm{T}} \tag{8.8.19}$$

and the Hamiltonian density function is

$$\begin{aligned}\mathscr{H}(\boldsymbol{q}, \boldsymbol{p}) = {} & \nu s_\varphi \left(\phi_\rho + \frac{\partial \phi_\varphi}{\partial \varphi} \right) + s_{\rho\varphi} \left(\phi_\varphi - \frac{\partial \phi_\rho}{\partial \varphi} \right) - \frac{1}{2D} \left(\phi_\rho + \frac{\partial \phi_\varphi}{\partial \varphi} \right)^2 \\ & + \frac{1-\nu}{2} D \left[(1+\nu) s_\varphi^2 + 2 s_{\rho\varphi}^2 \right]\end{aligned} \tag{8.8.20}$$

The variation of Eq. (8.8.18) with respect to $\boldsymbol{q}, \boldsymbol{p}$ yields the system of dual equations

$$\dot{\boldsymbol{v}} = \boldsymbol{H}\boldsymbol{v} \tag{8.8.21}$$

where $\boldsymbol{v} = \{\phi_\rho \ \phi_\varphi \ s_\varphi \ s_{\rho\varphi}\}^{\mathrm{T}}$ is the full state vector. The operator matrix is

$$\boldsymbol{H} = \begin{bmatrix} \nu & \nu\dfrac{\partial}{\partial\varphi} & D\left(1-\nu^2\right) & 0 \\ -\dfrac{\partial}{\partial\varphi} & 1 & 0 & 2D\left(1-\nu\right) \\ \dfrac{1}{D} & \dfrac{1}{D}\dfrac{\partial}{\partial\varphi} & -\nu & -\dfrac{\partial}{\partial\varphi} \\ -\dfrac{1}{D}\dfrac{\partial}{\partial\varphi} & -\dfrac{1}{D}\dfrac{\partial^2}{\partial\varphi^2} & \nu\dfrac{\partial}{\partial\varphi} & -1 \end{bmatrix} \tag{8.8.22}$$

Similar to the foregoing sections, Eq. (8.8.21) can be solved by the method of separation of variables. Let

$$\boldsymbol{v}(\xi, \varphi) = \mathrm{e}^{\mu\xi}\boldsymbol{\psi}(\varphi) \tag{8.8.23}$$

where μ is the unknown eigenvalue and $\boldsymbol{\psi}(x)$ is the eigenvector which is only a function of φ. The eigenvalue equation is

$$\boldsymbol{H}\boldsymbol{\psi}(\varphi) = \mu\boldsymbol{\psi}(\varphi) \tag{8.8.24}$$

The eigenvector $\boldsymbol{\psi}(\varphi)$ is also required to satisfy the corresponding homogeneous boundary conditions on two sides $\varphi = \pm\alpha$.

For $\boldsymbol{v}_1, \boldsymbol{v}_2$ satisfying the homogeneous boundary conditions for free, clamped or hinged support boundary conditions on sides $\varphi = \pm\alpha$, respectively, it is easy to verify that there exists an identity

$$\int_{-\alpha}^{\alpha} \boldsymbol{v}_1^{\mathrm{T}}\boldsymbol{J}\boldsymbol{H}\boldsymbol{v}_2 \mathrm{d}\varphi \equiv \int_{-\alpha}^{\alpha} \boldsymbol{v}_2^{\mathrm{T}}\boldsymbol{J}\boldsymbol{H}\boldsymbol{v}_1 \mathrm{d}\varphi \tag{8.8.25}$$

where

$$\boldsymbol{J} = \begin{bmatrix} \boldsymbol{0} & \boldsymbol{I} \\ -\boldsymbol{I} & \boldsymbol{0} \end{bmatrix} \tag{8.8.26}$$

Hence $\boldsymbol{JH}$ is a symmetric operator, or equivalently $\boldsymbol{H}$ is a Hamiltonian operator matrix.

For some specified problems with special boundary conditions there may exist zero and $\mu = \pm 1$ eigenvalues. Such solutions can be obtained via rational derivation and the solutions are relevant to problems to be discussed.

Similar to plane elasticity, the general solution for eigen-solutions of general nonzero eigenvalues is

$$\left.\begin{aligned}\phi_\rho &= A_1\cos[(1+\mu)\varphi] + B_1\sin[(1+\mu)\varphi]\\ &\quad + C_1\cos[(1-\mu)\varphi] + D_1\sin[(1-\mu)\varphi]\\ \phi_\varphi &= A_2\sin[(1+\mu)\varphi] + B_2\cos[(1+\mu)\varphi]\\ &\quad + C_2\sin[(1-\mu)\varphi] + D_2\cos[(1-\mu)\varphi]\\ s_\varphi &= A_3\cos[(1+\mu)\varphi] + B_3\sin[(1+\mu)\varphi]\\ &\quad + C_3\cos[(1-\mu)\varphi] + D_3\sin[(1-\mu)\varphi]\\ s_{\rho\varphi} &= A_4\sin[(1+\mu)\varphi] + B_4\cos[(1+\mu)\varphi]\\ &\quad + C_4\sin[(1-\mu)\varphi] + D_4\cos[(1-\mu)\varphi]\end{aligned}\right\} \tag{8.8.27}$$

The constants are not all independent. There are only four independent constants. Substituting Eq. (8.8.27) into Eq. (8.8.24) yields

$$\left.\begin{aligned}A_2 &= -A_1\\ A_3 &= \frac{\mu}{D(1-\nu)}A_1\\ A_4 &= -\frac{\mu}{D(1-\nu)}A_1\end{aligned}\right\};\qquad \left.\begin{aligned}C_2 &= \frac{-3-\nu-\mu+\nu\mu}{3+\nu-\mu+\nu\mu}C_1\\ C_3 &= \frac{\mu(3-\mu)}{D(3+\nu-\mu+\nu\mu)}C_1\\ C_4 &= \frac{\mu(1-\mu)}{D(3+\nu-\mu+\nu\mu)}C_1\end{aligned}\right\} \tag{8.8.28}$$

and

$$\left.\begin{aligned}B_1 &= B_2\\ B_3 &= \frac{\mu}{D(1-\nu)}B_2\\ B_4 &= \frac{\mu}{D(1-\nu)}B_2\end{aligned}\right\};\qquad \left.\begin{aligned}D_1 &= \frac{3+\nu-\mu+\nu\mu}{3+\nu+\mu-\nu\mu}D_2\\ D_3 &= \frac{\mu(3-\mu)}{D(3+\nu+\mu-\nu\mu)}D_2\\ D_4 &= \frac{-\mu(1-\mu)}{D(3+\nu+\mu-\nu\mu)}D_2\end{aligned}\right\} \tag{8.8.29}$$

Further substituting Eqs. (8.8.28) and (8.8.29) into the corresponding homogeneous boundary conditions for two sides $\varphi = \pm\alpha$ yields the general transcendental equation for nonzero eigenvalues and the corresponding eigenvectors. The eigenvalues and eigenvectors for different boundary conditions are different. The typical sectorial plate with two opposite sides free is discussed in the following section.

8.8.2. *Sectorial Plate with Two Opposite Sides Free*

Consider a sectorial plate with two opposite sides free. From Eq. (8.8.12) and transformation (8.8.13), the boundary conditions can be expressed in terms of a full state vector as

$$\phi_\rho = \phi_\varphi = 0 \quad \text{at } \varphi = -\alpha \tag{8.8.30a}$$

and

$$\phi_\rho = \tilde{a}_1 \quad \phi_\varphi = \tilde{a}_0 + a_2 e^\xi \quad \text{at } \varphi = \alpha \tag{8.8.30b}$$

For force boundary conditions, there are three unkown constants according to Eq. (8.8.12). Three constants for $\varphi = -\alpha$ have been eliminated by virtue of arbitrariness of null moment functions while for convenience of discussion two constants $\tilde{a}_0, \tilde{a}_1$ for $\varphi = \alpha$ have been transformed into

$$\phi_\rho = a_0 \sin\alpha + a_1 \cos\alpha, \quad \phi_\varphi = a_0 \cos\alpha - a_1 \sin\alpha + a_2 e^\xi \quad \text{at } \varphi = \alpha \tag{8.8.30b$'$}$$

according to the null moment functions (8.8.7). As eigen-solutions deal with homogeneous equations and homogeneous boundary conditions, the undetermined inhomogeneous constants in boundary conditions should be solved first similar to in the solution procedure for rectangular plate.

First the a_0 term should be solve from the following equation

$$\boldsymbol{H}\boldsymbol{\psi} = \boldsymbol{0} \tag{8.8.31}$$

with the boundary conditions

$$\phi_\rho(-\alpha) = \phi_\varphi(-\alpha) = 0, \quad \phi_\rho(\alpha) = \sin\alpha, \quad \phi_\varphi(\alpha) = \cos\alpha \tag{8.8.32}$$

The solution is

$$\tilde{\boldsymbol{\psi}}_0 = \begin{Bmatrix} \dfrac{1}{k_0}[(3+\nu)\varphi\sin\varphi - (1-\nu)\sin^2\alpha\cos\varphi] + \dfrac{1}{2}\sin\varphi \\ \dfrac{1}{k_0}[(3+\nu)\varphi\cos\varphi - (1-\nu)\cos^2\alpha\sin\varphi] + \dfrac{1}{2}\cos\varphi \\ -\dfrac{2\nu\cos\varphi}{D(1-\nu)k_0} \\ \dfrac{2\sin\varphi}{D(1-\nu)k_0} \end{Bmatrix} \tag{8.8.33}$$

where

$$k_0 = 2(3+\nu)\alpha - (1-\nu)\sin(2\alpha) \tag{8.8.34}$$

The solution for the original problem (8.8.21) is

$$\tilde{\boldsymbol{v}}_0 = \tilde{\boldsymbol{\psi}}_0 \tag{8.8.35}$$

and the corresponding deflection and bending moment are

$$\tilde{w}_0 = \frac{\rho[2(\ln\rho - 1)\cos\varphi - (1+\nu)\varphi\sin\varphi]}{D(1-\nu)k_0} + \text{rigid body displacement} \tag{8.8.36}$$

$$\tilde{M}_{\rho 0} = \frac{2(1+\nu)\cos\varphi}{\rho k_0}, \quad \tilde{M}_{\varphi 0} = 0, \quad \tilde{M}_{\rho\varphi 0} = \frac{2\sin\varphi}{\rho k_0} \tag{8.8.37}$$

Further from Eqs. (8.8.4a) and (8.8.5), the shear forces and total distributed shear forces along the boundary are

$$\tilde{F}_{\mathrm{S}\rho 0} = -\frac{2\cos\varphi}{\rho^2 k_0}, \quad \tilde{F}_{\mathrm{S}\varphi 0} = -\frac{2\sin\varphi}{\rho^2 k_0} \tag{8.8.38}$$

$$\tilde{F}_{\mathrm{V}\rho 0} = -\frac{4\cos\varphi}{\rho^2 k_0}, \quad \tilde{F}_{\mathrm{V}\varphi 0} = 0 \tag{8.8.39}$$

Obviously, the internal forces result in a unit bending moment on the edge surface $\xi_1 = \ln R_1 (\rho = R_1)$

$$\left.\begin{aligned} F &= \int_{-\alpha}^{\alpha} \rho\tilde{F}_{\mathrm{V}\rho 0}\mathrm{d}\varphi + [2\tilde{M}_{\rho\varphi 0}]_{-\alpha}^{\alpha} = 0 \\ M_x &= \int_{-\alpha}^{\alpha} [\rho\tilde{F}_{\mathrm{V}\rho 0} - \tilde{M}_{\rho 0}]\rho\sin\varphi\mathrm{d}\varphi + [2\rho\sin\varphi\tilde{M}_{\rho\varphi 0}]_{-\alpha}^{\alpha} = 0 \\ M_y &= \int_{-\alpha}^{\alpha} [\tilde{M}_{\rho 0} - \rho\tilde{F}_{\mathrm{V}\rho 0}]\rho\cos\varphi\mathrm{d}\varphi - [2\rho\cos\varphi\tilde{M}_{\rho\varphi 0}]_{-\alpha}^{\alpha} = 1 \end{aligned}\right\} \tag{8.8.40}$$

Taking $\xi_1 \to -\infty$, $\tilde{\boldsymbol{v}}_0$ becomes the solution of a wedge with a unit bending moment acting at the apex (Fig. 8.6).

Next the a_1 term should be solved from the following equation

$$\boldsymbol{H\psi} = \boldsymbol{0} \tag{8.8.41}$$

with boundary conditions

$$\phi_\rho(-\alpha) = \phi_\varphi(-\alpha) = 0, \quad \phi_\rho(\alpha) = \cos\alpha, \quad \phi_\varphi(\alpha) = -\sin\alpha \tag{8.8.42}$$

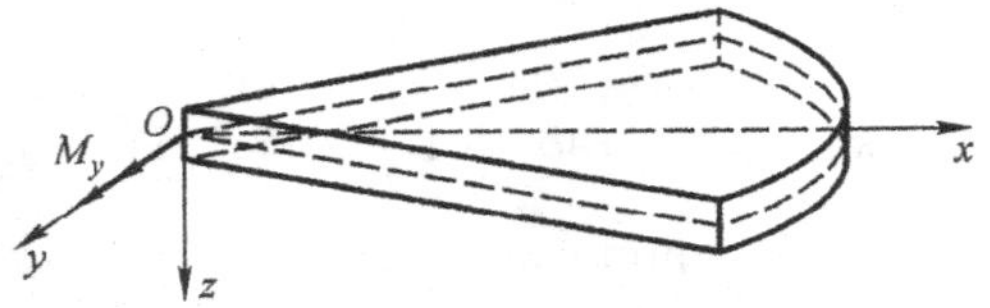

Fig. 8.6. A wedge with bending moment at the apex.

The solution is

$$\tilde{\boldsymbol{\psi}}_1 = \begin{Bmatrix} \dfrac{1}{k_1}[(3+\nu)\varphi\cos\varphi + (1-\nu)\cos^2\alpha\sin\varphi] + \dfrac{1}{2}\cos\varphi \\ -\dfrac{1}{k_1}[(3+\nu)\varphi\sin\varphi + (1-\nu)\sin^2\alpha\cos\varphi] - \dfrac{1}{2}\sin\varphi \\ \dfrac{2\nu\sin\varphi}{D(1-\nu)k_1} \\ \dfrac{2\cos\varphi}{D(1-\nu)k_1} \end{Bmatrix} \tag{8.8.43}$$

where

$$k_1 = 2(3+\nu)\alpha + (1-\nu)\sin 2\alpha \tag{8.8.44}$$

The solution for the original problem (8.8.21) is

$$\tilde{\boldsymbol{v}}_1 = \tilde{\boldsymbol{\psi}}_1 \tag{8.8.45}$$

and the corresponding deflection and bending moment are

$$\tilde{w}_1 = -\frac{\rho[2(\ln\rho - 1)\sin\varphi + (1+\nu)\varphi\cos\varphi]}{D(1-\nu)k_1} + \text{rigid body displacement} \tag{8.8.46}$$

$$\tilde{M}_{\rho 1} = -\frac{2(1+\nu)\sin\varphi}{\rho k_1}, \quad \tilde{M}_{\varphi 1} = 0, \quad \tilde{M}_{\rho\varphi 1} = \frac{2\cos\varphi}{\rho k_1} \tag{8.8.47}$$

Further from Eqs. (8.8.4a) and (8.8.5), the shear forces and total distributed shear forces along the boundary are

$$\tilde{F}_{\mathrm{S}\rho 1} = \frac{2\sin\varphi}{\rho^2 k_1}, \quad \tilde{F}_{\mathrm{S}\varphi 1} = -\frac{2\cos\varphi}{\rho^2 k_1} \tag{8.8.48}$$

$$\tilde{F}_{\mathrm{V}\rho 1} = \frac{4\sin\varphi}{\rho^2 k_1}, \quad \tilde{F}_{\mathrm{V}\varphi 1} = 0 \tag{8.8.49}$$

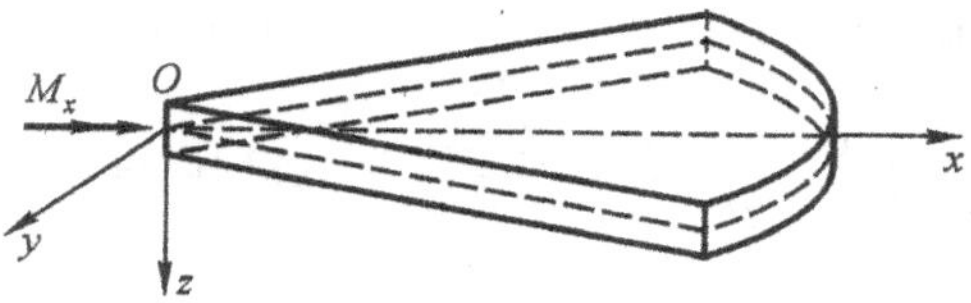

Fig. 8.7. A wedge with torsion moment at the apex.

Obviously, the internal forces result in a unit torsion moment on the edge surface $\xi_1 = \ln R_1 (\rho = R_1)$

$$\left.\begin{aligned} F &= \int_{-\alpha}^{\alpha} \rho \tilde{F}_{V\rho 1} d\varphi + [2\tilde{M}_{\rho\varphi 1}]_{-\alpha}^{\alpha} = 0 \\ M_x &= \int_{-\alpha}^{\alpha} [\rho \tilde{F}_{V\rho 1} - \tilde{M}_{\rho 1}]\rho \sin\varphi d\varphi + [2\rho\sin\varphi \tilde{M}_{\rho\varphi 1}]_{-\alpha}^{\alpha} = 1 \\ M_y &= \int_{-\alpha}^{\alpha} [\tilde{M}_{\rho 1} - \rho \tilde{F}_{V\rho 1}]\rho \cos\varphi d\varphi - [2\rho\cos\varphi \tilde{M}_{\rho\varphi 1}]_{-\alpha}^{\alpha} = 0 \end{aligned}\right\} \tag{8.8.50}$$

Taking $\xi_1 \to -\infty$, $\tilde{\boldsymbol{v}}_1$ becomes the solution of a wedge with a unit torsion moment acting at the apex (Fig. 8.7).

Finally, the a_2 term should be solved. As a_2 in the expression of ϕ_φ in Eq. (8.8.30b′) has a multiplier e^ξ, the solution corresponds to eigenvalues of 1. Hence the a_2 term should be solved from the following equation

$$\boldsymbol{H}\boldsymbol{\psi} = \boldsymbol{\psi} \tag{8.8.51}$$

with boundary conditions

$$\phi_\rho(-\alpha) = \phi_\varphi(-\alpha) = 0, \quad \phi_\rho(\alpha) = 0, \quad \phi_\varphi(\alpha) = 1 \tag{8.8.52}$$

The solution is

$$\tilde{\boldsymbol{\psi}}_2 = \left\{\begin{matrix} \dfrac{\cos(2\alpha) - \cos(2\varphi)}{2\sin(2\alpha)} \\ \dfrac{\sin(2\varphi)}{2\sin(2\alpha)} + \dfrac{1}{2} \\ -\dfrac{\cos(2\varphi)}{2D(1-\nu)\sin(2\alpha)} + \dfrac{\cos(2\alpha)}{2D(1+\nu)\sin(2\alpha)} \\ \dfrac{\sin(2\varphi)}{2D(1-\nu)\sin(2\alpha)} \end{matrix}\right\} \tag{8.8.53}$$

The solution for the original problem (8.8.21) is

$$\tilde{\boldsymbol{v}}_2 = e^\xi \tilde{\boldsymbol{\psi}}_2 = \rho \tilde{\boldsymbol{\psi}}_2 \tag{8.8.54}$$

and the corresponding deflection and bending moment are

$$\tilde{w}_2 = \frac{\rho^2}{4D\sin(2\alpha)}\left[\frac{\cos(2\varphi)}{1-\nu} + \frac{\cos(2\alpha)}{1+\nu}\right] + \text{rigid body displacement} \tag{8.8.55}$$

$$\tilde{M}_{\rho 2} = \frac{\cos(2\alpha)+\cos(2\varphi)}{2\sin(2\alpha)}, \quad \tilde{M}_{\varphi 2} = \frac{\cos(2\alpha)-\cos(2\varphi)}{2\sin(2\alpha)}, \quad \tilde{M}_{\rho\varphi 2} = \frac{\sin(2\varphi)}{2\sin(2\alpha)} \tag{8.8.56}$$

Further from Eqs. (8.8.4a) and (8.8.5), the shear forces and total distributed shear forces along the boundary are

$$\tilde{F}_{\mathrm{S}\rho 2} = 0, \quad \tilde{F}_{\mathrm{S}\varphi 2} = 0 \tag{8.8.57}$$

$$\tilde{F}_{\mathrm{V}\rho 2} = -\frac{\cos(2\varphi)}{\rho\sin(2\alpha)}, \quad \tilde{F}_{\mathrm{V}\varphi 2} = 0 \tag{8.8.58}$$

Obviously, the internal forces result in a unit vertical concentrated force at the apex.

$$\left.\begin{aligned} F &= \int_{-\alpha}^{\alpha} \rho\tilde{F}_{\mathrm{V}\rho 2}\mathrm{d}\varphi + [2\tilde{M}_{\rho\varphi 2}]_{-\alpha}^{\alpha} = 1 \\ M_x &= \int_{-\alpha}^{\alpha} [\rho\tilde{F}_{\mathrm{V}\rho 2} - \tilde{M}_{\rho 2}]\rho\sin\varphi\mathrm{d}\varphi + [2\rho\sin\varphi\tilde{M}_{\rho\varphi 2}]_{-\alpha}^{\alpha} = 0 \\ M_y &= \int_{-\alpha}^{\alpha} [\tilde{M}_{\rho 2} - \rho\tilde{F}_{\mathrm{V}\rho 2}]\rho\cos\varphi\mathrm{d}\varphi - [2\rho\cos\varphi\tilde{M}_{\rho\varphi 2}]_{-\alpha}^{\alpha} = 0 \end{aligned}\right\} \tag{8.8.59}$$

Taking $\xi_1 \to -\infty$, $\tilde{\boldsymbol{v}}_2$ becomes the solution of a wedge with a unit vertical concentrated force acting at the apex (Fig. 8.8).

For $\alpha = \pi/2, \pi$, it should be noted that the solutions (8.8.55) and (8.8.56) are infinite and they are similar to the paradox for a plane elastic wedge. The solution should be in Jordan form. Solving the Jordan form yields solutions of special semi-plate and cracked plate subject to a vertical concentrated force. The details are omitted here.

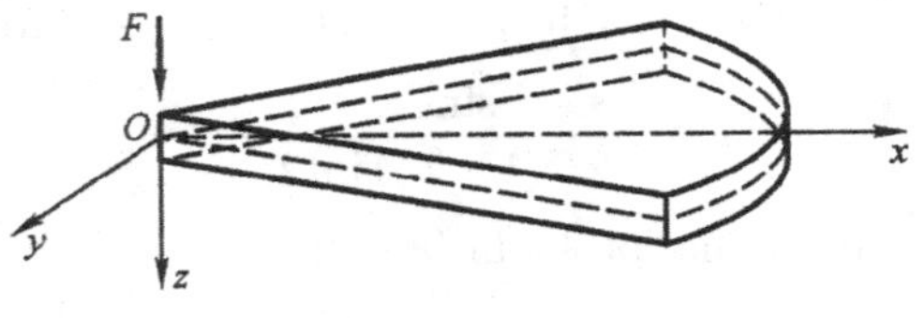

Fig. 8.8. A wedge with a vertical concentrated force at the apex.

In fact, it is certainly possible to derive the Hamiltonian system and establish the solution by the method separable variables from the biharmonic equation of deflection in polar coordinates. There are then six special eigen-solutions of zero and ± 1 eigenvalues, i.e. three plate rigid body displacements and the solutions corresponding to $\tilde{\boldsymbol{v}}_0, \tilde{\boldsymbol{v}}_1, \tilde{\boldsymbol{v}}_2$. As the method presented here replaces deflection with curvature in order to avoid three plate rigid body displacements, the corresponding dual solutions are expressed in the form of particular solutions of inhomogeneous boundary conditions.

Having established the three inhomogeneous solutions, the solution of homogeneous boundary conditions is considered below.

For a plate with both opposite sides free, the homogeneous boundary conditions are

$$\phi_\rho = \phi_\varphi = 0 \quad \text{at } \varphi = \pm\alpha \tag{8.8.60}$$

Obviously only eigen-solutions of nonzero eigenvalues satisfy Eqs. (8.8.24) and (8.8.60). The solutions can be divided into two sets: symmetric deformation and antisymmetric deformation with respect to $\varphi = 0$. It is clear from Eq. (8.8.27) that A and C correspond to eigen-solutions of symmetric deformation while B and D corresponds to eigen-solutions of antisymmetric deformation.

Substituting solutions with only A and C terms in Eq. (8.8.27) into the homogeneous boundary conditions (8.8.60) and equating the determinant of coefficient matrix to zero yield the transcendental equation of nozero eigenvalues for symmetric plate deformation with both opposite sides free as

$$(3+\nu)\sin(2\mu\alpha) - \mu(1-\nu)\sin(2\alpha) = 0 \tag{8.8.61}$$

The corresponding eigen-solution is

$$\boldsymbol{\psi}_s = \left\{\begin{array}{l} D(1-\nu)\{-(3+\nu+\mu-\nu\mu)\sin[(1-\mu)\alpha]\cos[(1+\mu)\varphi] \\ \quad +(3+\nu-\mu+\nu\mu)\sin[(1+\mu)\alpha]\cos[(1-\mu)\varphi]\} \\ D(1-\nu)(3+\nu+\mu-\nu\mu) \\ \quad \{\sin[(1-\mu)\alpha]\sin[(1+\mu)\varphi] - \sin[(1+\mu)\alpha]\sin[(1-\mu)\varphi]\} \\ \mu\{-(3+\nu+\mu-\nu\mu)\sin[(1-\mu)\alpha]\cos[(1+\mu)\varphi] \\ \quad +(1-\nu)(3-\mu)\sin[(1+\mu)\alpha]\cos[(1-\mu)\varphi]\} \\ \mu\{(3+\nu+\mu-\nu\mu)\sin[(1-\mu)\alpha]\sin[(1+\mu)\varphi] \\ \quad +(1-\nu)(1-\mu)\sin[(1+\mu)\alpha]\sin[(1-\mu)\varphi]\} \end{array}\right\} \tag{8.8.62}$$

The solution for the original problem (8.8.21) is

$$\boldsymbol{v}_s = \exp(\mu\xi)\boldsymbol{\psi}_s = \rho^{\mu}\boldsymbol{\psi}_s \tag{8.8.63}$$

and the corresponding deflection is

$$\begin{aligned} w_s = \rho^{\mu+1}\bigg\{ & \frac{3+\nu+\mu-\nu\mu}{1+\mu}\sin[(1-\mu)\alpha]\cos[(1+\mu)\varphi] \\ & + (1-\nu)\sin[(1+\mu)\alpha]\cos[(1-\mu)\varphi]\bigg\} \\ & + \text{rigid body displacement} \end{aligned} \tag{8.8.64}$$

Substituting solutions with only B and D terms in Eq. (8.8.27) into the homogeneous boundary conditions (8.8.60) and equating the determinant of coefficient matrix to zero yield the transcendental equation of nonzero eigenvalues for antisymmetric plate deformation with both opposite sides free as

$$(3+\nu)\sin(2\mu\alpha) + \mu(1-\nu)\sin(2\alpha) = 0 \tag{8.8.65}$$

The corresponding eigen-solution is

$$\boldsymbol{\psi}_a = \left\{\begin{array}{l} D(1-\nu)\{(3+\nu+\mu-\nu\mu)\cos[(1-\mu)\alpha]\sin[(1+\mu)\varphi] \\ \quad -(3+\nu-\mu+\nu\mu)\cos[(1+\mu)\alpha]\sin[(1-\mu)\varphi]\} \\ D(1-\nu)(3+\nu+\mu-\nu\mu) \\ \quad \{\cos[(1-\mu)\alpha]\cos[(1+\mu)\varphi] - \cos[(1+\mu)\alpha]\cos[(1-\mu)\varphi]\} \\ \mu\{(3+\nu+\mu-\nu\mu)\cos[(1-\mu)\alpha]\sin[(1+\mu)\varphi] \\ \quad -(1-\nu)(3-\mu)\cos[(1+\mu)\alpha]\sin[(1-\mu)\varphi]\} \\ \mu\{(3+\nu+\mu-\nu\mu)\cos[(1-\mu)\alpha]\cos[(1+\mu)\varphi] \\ \quad +(1-\nu)(1-\mu)\cos[(1+\mu)\alpha]\cos[(1-\mu)\varphi]\} \end{array}\right\} \tag{8.8.66}$$

The solution for the original problem (8.8.21) is

$$\boldsymbol{v}_a = \exp(\mu\xi)\boldsymbol{\psi}_a = \rho^{\mu}\boldsymbol{\psi}_a \tag{8.8.67}$$

and the corresponding deflection is

$$w_a = -\rho^{\mu+1}\left\{\frac{3+\nu+\mu-\nu\mu}{1+\mu}\cos[(1-\mu)\alpha]\sin[(1+\mu)\varphi] + (1-\nu)\cos[(1+\mu)\alpha]\sin[(1-\mu)\varphi]\right\} + \text{rigid body displacement} \tag{8.8.68}$$

Having established the eigenvalues and eigenvectors and knowing the adjoint symplectic orthogonality property, the general solution for sectorial plate with both opposite sides free can be solved according to the expansion theorem as

$$\boldsymbol{v} = a_0\tilde{\boldsymbol{v}}_0 + a_1\tilde{\boldsymbol{v}}_1 + a_2\tilde{\boldsymbol{v}}_2 + \sum_{n=1}^{\infty}(b_n\boldsymbol{v}_n + c_n\boldsymbol{v}_{-n}) \tag{8.8.69}$$

where $\boldsymbol{v}_n, \boldsymbol{v}_{-n}$ are the symplectic adjoint eigen-solutions corresponding to eigenvalues μ_n and $-\mu_n$, and a_0, a_1, a_2 and b_n, c_n $(n = 1, 2, \ldots)$ are unknwon constants. Equation (8.8.69) satisfies all governing equations in the domain and free boundary conditions on both opposite sides $\varphi = \pm\alpha$. Substituting the equation into the boundary conditions at two ends ($\rho = R_1$ and R_2) and subsequently determining the relevant constants yield the analytical solution of the problem.

This chapter presents another different set of basic governing equations and solution methodology for the classical theory of plate bending based on the analogy principle. It presents a contrasting approach with respect to the widely adopted classical methodology in Ref. 1. The classical methodology adopts displacement and biharmonic equation while the new methodology adopts force and bending moment function vector. The classical methodology adopts a trial and error approach which consequently limits the possibility of obtaining an analytical solution. For instance, analytical solutions can only be established for a rectangular plate simply supported on both opposite sides whereas for other boundary constraints the classical methodology has been impractical. Based on the Hamiltonian system, the new methodology adopts a rational approach via some effective methods such as separation of variables, eigenfunction, symplectic orthogonal system, expansion theorem, etc. to establish analytical solutions. The new methodology presents a breakthrough which has thus far restricted the applicability of the classical trial and error methodology. The solutions for rectangular plates with opposite sides free or clamped presented in this chapter, as well as solutions for sectorial plates, are examples which the classical methodology has been either impractical or inapplicable.

In fact, based on the analogy principle, it is not only possible to solve the bending problem of isotropic plates but also the bending problem of anisotropic plate[16]. The details are left for the readers.

It should also be emphasized here that the Hamiltonian system for bending of plate introduced in this chapter has been based on the analogy principle using the bending moment function vector. It is equally possible to directly derive the Hamiltonian system from the biharmonic equation of the deflection of plate bending. The latter is the displacement Hamiltonian system which is required in the research of vibration and stability problems. The relevant contents are not introduced in this chapter.

References

1. S.P. Timoshenko and S. Woinowsky-Krieger, *Theory of Plates and Shells*, New York: McGraw-Hill, 1959.
2. Zhilun Xu, *Elasticity*, Beijing: Higher Education Press, 1990 (in Chinese).
3. Jialong Wu, *Elasticity*, Shanghai: Tongji University Press, 1993 (in Chinese).
4. Longfu Wang, *Theory of Elasticity*, Beijing: Science Press, 1984 (in Chinese).
5. S.P. Timoshenko and J.N. Goodier, *Theory of Elasticity*, 3rd edn., New York: McGraw-Hill, 1970.
6. N.I. Muschelishvili, *Some Basic Problems of the Mathematical Theory of Elasticity*, Groningen, Netherlands: Noordholf, 1953.
7. Haichang Hu, *Variational Principle of Elasticity and Its Application*, Beijing: Science Press, 1981 (in Chinese).
8. R.V. Southwell, On the analogues relating flexure and extension of flat plate, *Quarterly Journal of Mechanics and Mathematics*, 1950, 3: 257.
9. Wanxie Zhong and Weian Yao, New solution system for plate bending and its application, *Acta Mechanics Sinica* 1999, 31(2): 173–184 (in Chinese).
10. Wanxie Zhong and Weian Yao, Similarity between finite elements in plane elasticity and plate bending, *Chinese Journal of Computational Mechanics* 1998, 15(1): 1–13 (in Chinese).
11. B. Fraeijs De Veubeke and O.C. Zienkiewicz, Strain-energy bounds of finite-element analysis by slab analogy, *J Strai Analysis*, 1967, 2: 265–271.
12. B. Fraeijs De Veubeke and G. Sander, An equilibrium model for plate bending, *I J Solids and Structures*, 1968, 4(4): 447–468.
13. Wanxie Zhong and Weian Yao, Multi-variable variational principles for plate bending and plane elasticity, *Acta Mechanica Sinica* 1999, 31(6): 717–723 (in Chinese).

14. Wanxie Zhong and Weian Yao, New solution system for plate bending — force method Hamiltonian system and its application, in: C.J. Cheng, S.Q. Dai and Y.L. Liu, eds., *Modern Maths and Mechanics (MMM-VII)*, Shanghai: Shanghai University Press, 1997, 121–129.
15. Weian Yao, Wanxie Zhong and Bin Su, New solution system for circular sector plate bending and its application, *Acta Mechanica Solida Sinica*, 1999, 12(4): 307–315.
16. Bin Su, *Application of Hamiltonian System in Anisotropy Plate Bending*, Master Dissertation of Dalian University of Technology, 2000 (in Chinese).

About the Authors

Weian Yao was born in Liaoning, China in 1963. He graduated with a B.Sc. degree in Computational Mathematics from Liaoning University in 1985. He was conferred a M.Eng. degree in Computational Mechanics from Dalian University of Technology in 1988 for research in shape optimization of annular fins. From 1985 to 1996, he made researches on optimization and boundary element method in Dalian University as a lecturer or assistant professor. In 1996, he shifted to Dalian University of Technology till today. During this period, his main research interests are in symplectic elasticity and computational mechanics, and he was promoted to professor in 2002 and was awarded a Ph.D. degree in Solid Mechanics from Dalian University of Technology in 2004 for research in further application of symplectic dual solution system in elasticity. He has more than 60 research publications.

Wanxie Zhong was born in Shanghai, China in 1934. Majoring in bridge and tunnel engineering, he graduated in 1956 in Tong-ji University, Shanghai. As a famous scientist of engineering mechanics, he is an academician of Chinese Academy of Sciences, and Honorary Professor of The University of Wales, UK. From 1956 to 1962, he conducted researches on solid mechanics, fluid mechanics and variational principle in The Institute of Mechanics, Chinese Academy of Sciences. In 1962, he shifted to Dalian University of Technology till today. Professor Zhong's academic interests span a very wide range. However he has paid special attention to the development of plate and shell theories and methodologies to engineering applications. In early 1970s, he led some young scholars to promote the application of computational mechanics in China. The program systems JIGFEX, DDJ, GCAD, etc. developed under his guidance resolved a great deal of key engineering analyses. The work headed by Zhong enjoys very high reputation in China. From 1986 to 2006, he was a Member of the Executive Committee of IACM. In 1980s, he founded the parametric variable variational theory which has been used in many fields such as elasto-plastic contact analysis, soil mechanics, etc. In 1990s, he founded the analogy relationship between classical structural mechanics and modern control theory, and so enables the established achievements in whichever field be readily shared by the other, and so powerfully pushed forward the development of both fields. Meanwhile, he established the precise integration scheme and first applied symplectic approach into elasticity. Professor Zhong has published more than 300 papers and 14 books.

Chee Wah Lim was born in Batu Pahat, Johor, Malaysia in 1965. He graduated with a B.Eng. degree in Mechanical Engineering (Aeronautics) from University of Technology of Malaysia in 1989 with a best academic performance award. He was conferred a M.Eng. degree in Mechanical Engineering from National University of Singapore in 1992 for research in hydrodynamic stability of potential and boundary layer flows over periodically supported compliant surfaces. Subsequently, he pursued research in vibration of isotropic and laminated plates and shells and was awarded a Ph.D. degree in Mechanical Engineering from Nanyang Technological University, Singapore in 1995. Dr. Lim continued research as a postdoctoral fellow at Department of Civil Engineering, The University of Queensland, Australia from 1995 to 1997, and as a research fellow at Department of Mechanical Engineering, The University of Hong Kong from 1998 to early 2000. He then joined Department of Building and Construction, City University of Hong Kong as an assistant professor in 2000 and was promoted to associate professor in 2003. His main research interests are in developing new models and applications of plate and shell structures, composite structures, smart piezoelectric materials and structures, MEMS and NEMS, nonlinear oscillation, nanomechanics and symplectic elasticity. Dr. Lim is the associate editor for Asia-Pacific region for *Advances in Vibration Engineering* and on the editorial boards of a few other international journals. He has published more than one hundred and fifty research papers in international refereed journals with more than one thousand self-excluded independent citations in more than 500 publications. In addition, he has presented more than sixty research papers at various international conferences including three keynote papers, nine invited papers and chairs in numerous sessions of international conferences. Dr. Lim is married to Moi P. Choo, and they have a daughter Qin Y. and a son Ying H.